C*-ALGEBRAS
AND
OPERATOR
THEORY

C*-ALGEBRAS
AND
OPERATOR
THEORY

Gerard J. Murphy

Mathematics Department
University College
Cork, Ireland

ACADEMIC PRESS, INC.

Harcourt Brace Jovanovich, Publishers

Boston   San Diego   New York
London   Sydney   Tokyo   Toronto

Transferred to digital printing 2004
Copyright © 1990 by Academic Press, Inc.

ACADEMIC PRESS, INC.
1250 Sixth Avenue, San Diego, CA 92101

United Kingdom Edition published by
ACADEMIC PRESS LIMITED
24–28 Oval Road, London NW1 7DX

Library of Congress Cataloging-in-Publication Data

Murphy, Gerard J.
    C*-algebras and operator theory / Gerard J. Murphy.
      p. cm.
    Includes bibliographical references.
    ISBN 0-12-511360-9 (alk. paper)
    1. C*-algebras. 2. Operator theory. I. Title.
QA326.M87   1990
512'.55—dc20   90-524
                                                    CIP

90 91 92 93   9 8 7 6 5 4 3 2 1

For my family

Mary, Alison, Adele, Neil

Contents

Chapter 4. Von Neumann Algebras

Chapter 5. Representations of C*-Algebras

Chapter 6. Direct Limits and Tensor Products

Chapter 7. K-Theory of C*-Algebras

Preface

This is an introductory textbook to a vast subject, which although more than fifty years old is still extremely active and rapidly expanding, and coming to have an increasingly greater impact on other areas of mathematics, as well as having applications to theoretical physics. I have attempted to give a leisurely and accessible exposition of the core material of the subject, and to cover a number of topics (the theory of C*-tensor products and K-theory) having a high contemporary profile. There was no intention to be encyclopedic, and many important topics had to be omitted in order to keep to a moderate size.

This book is aimed at the beginning graduate student and the specialist in another area who wishes to know the basics of this subject. The reader is assumed to have a good background in real and complex analysis, point set topology, measure theory, and elementary general functional analysis. Thus, such results as the Hahn–Banach extension theorem, the uniform boundedness principle, the Stone–Weierstrass theorem, and the Riesz–Kakutani theorem are assumed known. However, the theory of locally convex spaces is not presupposed, and the relevant material including the Krein–Milman theorem and the separation theorem are developed in a brief appendix. The book is arranged so that the appendix is not used until Chapter 4, and the first three chapters can, if desired, form the basis of a short course. The background material for the book is covered by the following textbooks: [Coh], [Kel], [Rud 1], and [Rud 2].

Each chapter concludes with a list of exercises arranged roughly according to the order in which the relevant item appeared in the chapter, and statements of additional results related to, and extending, the material in the text.

The symbols $\mathbf{N}, \mathbf{Z}, \mathbf{R}, \mathbf{R}^+$, and $\mathbf{C}$ refer, respectively, to the sets of non-negative integers, integers, real numbers, non-negative real numbers, and complex numbers. Other notation is explained as needed.

The reader who has finished this book and wants direction for further study may refer to the Notes section where some books are recommended.

I am indebted to many authors of books on operator theory and operator algebras. Section 7.5 of this book is based on the approach of J. Cuntz to K-theory. I should like to thank my colleagues Trevor West and Martin Mathieu for reading preliminary drafts of some of the earlier chapters.

Gerard J. Murphy

CHAPTER 1

Elementary Spectral Theory

In this chapter we cover the basic results of spectral theory. The most important of these are the non-emptiness of the spectrum, Beurling's spectral radius formula, and the Gelfand representation theory for commutative Banach algebras. We also introduce compact and Fredholm operators and analyse their elementary theory. Important concepts here are the essential spectrum and the Fredholm index.

Throughout this book the ground field for all vector spaces and algebras is the complex field **C**, unless the contrary is explicitly indicated in a particular context.

1.1. Banach Algebras

We begin by setting up the basic vocabulary needed to discuss Banach algebras and by giving some examples.

An *algebra* is a vector space A together with a bilinear map

$$A^2 \to A, \quad (a, b) \mapsto ab,$$

such that

$$a(bc) = (ab)c \quad (a, b, c \in A).$$

A *subalgebra* of A is a vector subspace B such that $b, b' \in B \Rightarrow bb' \in B$. Endowed with the multiplication got by restriction, B is itself an algebra.

A norm $\|.\|$ on A is said to be *submultiplicative* if

$$\|ab\| \leq \|a\|\|b\| \quad (a, b \in A).$$

In this case the pair $(A, \|.\|)$ is called a *normed algebra*. If A admits a unit 1 ($a1 = 1a = a$, for all $a \in A$) and $\|1\| = 1$, we say that A is a *unital normed algebra*.

1

If A is a normed algebra, then it is evident from the inequality

$$\|ab - a'b'\| \leq \|a\|\|b - b'\| + \|a - a'\|\|b'\|$$

that the multiplication operation $(a, b) \mapsto ab$ is jointly continuous.

A complete normed algebra is called a *Banach algebra*. A complete unital normed algebra is called a *unital Banach algebra*.

A subalgebra of a normed algebra is obviously itself a normed algebra with the norm got by restriction. The closure of a subalgebra is a subalgebra. A closed subalgebra of a Banach algebra is a Banach algebra.

1.1.1. Example. If S is a set, $\ell^\infty(S)$, the set of all bounded complex-valued functions on S, is a unital Banach algebra where the operations are defined pointwise:

$$(f + g)(x) = f(x) + g(x)$$
$$(fg)(x) = f(x)g(x)$$
$$(\lambda f)(x) = \lambda f(x),$$

and the norm is the sup-norm

$$\|f\|_\infty = \sup_{x \in S} |f(x)|.$$

1.1.2. Example. If Ω is a topological space, the set $C_b(\Omega)$ of all bounded continuous complex-valued functions on Ω is a closed subalgebra of $\ell^\infty(\Omega)$. Thus, $C_b(\Omega)$ is a unital Banach algebra.

If Ω is compact, $C(\Omega)$, the set of continuous functions from Ω to $\mathbf{C}$, is of course equal to $C_b(\Omega)$.

1.1.3. Example. If Ω is a locally compact Hausdorff space, we say that a continuous function f from Ω to $\mathbf{C}$ *vanishes at infinity*, if for each positive number ε the set $\{\omega \in \Omega \mid |f(\omega)| \geq \varepsilon\}$ is compact. We denote the set of such functions by $C_0(\Omega)$. It is a closed subalgebra of $C_b(\Omega)$, and therefore, a Banach algebra. It is unital if and only if Ω is compact, and in this case $C_0(\Omega) = C(\Omega)$. The algebra $C_0(\Omega)$ is one of the most important examples of a Banach algebra, and we shall see it used constantly in C*-algebra theory (the functional calculus).

1.1.4. Example. If (Ω, μ) is a measure space, the set $L^\infty(\Omega, \mu)$ of (classes of) essentially bounded complex-valued measurable functions on Ω is a unital Banach algebra with the usual (pointwise-defined) operations and the essential supremum norm $f \mapsto \|f\|_\infty$.

1.1.5. _Example._ If Ω is a measurable space, let $B_\infty(\Omega)$ denote the set of all bounded complex-valued measurable functions on Ω. Then $B_\infty(\Omega)$ is a closed subalgebra of $\ell^\infty(\Omega)$, so it is a unital Banach algebra. This example will be used in connection with the spectral theorem in Chapter 2.

1.1.6. _Example._ The set A of all continuous functions on the closed unit disc $\mathbf{D}$ in the plane which are analytic on the interior of $\mathbf{D}$ is a closed subalgebra of $C(\mathbf{D})$, so A is a unital Banach algebra, called the _disc algebra_. This is the motivating example in the theory of function algebras, where many aspects of the theory of analytic functions are extended to a Banach algebraic setting.

All of the above examples are of course _abelian_—that is, $ab = ba$ for all elements a and b—but the following examples are not, in general.

1.1.7. _Example._ If X is a normed vector space, denote by $B(X)$ the set of all bounded linear maps from X to itself (the _operators_ on X). It is routine to show that $B(X)$ is a normed algebra with the pointwise-defined operations for addition and scalar multiplication, multiplication given by $(u, v) \mapsto u \circ v$, and norm the _operator norm_:

$$\|u\| = \sup_{x \neq 0} \frac{\|u(x)\|}{\|x\|} = \sup_{\|x\| \leq 1} \|u(x)\|.$$

If X is a Banach space, $B(X)$ is complete and is therefore a Banach algebra.

1.1.8. _Example._ The algebra $M_n(\mathbf{C})$ of $n \times n$-matrices with entries in $\mathbf{C}$ is identified with $B(\mathbf{C}^n)$. It is therefore a unital Banach algebra. Recall that an _upper triangular_ matrix is one of the form

$$\begin{pmatrix} \lambda_{11} & \lambda_{12} & \cdots & \cdots & \lambda_{1n} \\ 0 & \lambda_{22} & \cdots & \cdots & \lambda_{2n} \\ 0 & 0 & \lambda_{33} & \cdots & \lambda_{3n} \\ \vdots & \vdots & & \ddots & \vdots \\ 0 & 0 & \cdots & 0 & \lambda_{nn} \end{pmatrix}$$

(all entries below the main diagonal are zero). These matrices form a subalgebra of $M_n(\mathbf{C})$.

We shall be seeing many more examples of Banach algebras as we proceed. Most often these will be non-abelian, but in the first three sections of this chapter we shall be principally concerned with the abelian case.

If $(B_\lambda)_{\lambda \in \Lambda}$ is a family of subalgebras of an algebra A, then $\cap_{\lambda \in \Lambda} B_\lambda$ is a subalgebra, also. Hence, for any subset S of A, there is a smallest subalgebra B of A containing S (namely, the intersection of all the subalgebras containing S). This algebra is called the subalgebra of A *generated* by S. If S is the singleton set $\{a\}$, then B is the linear span of all powers a^n ($n = 1, 2, \ldots$) of a. If A is a normed algebra, the closed algebra C *generated* by a set S is the smallest closed subalgebra containing S. It is plain that $C = \bar{B}$, where B is the subalgebra generated by S.

If $A = C(\mathbf{T})$, where $\mathbf{T}$ is the unit circle, and if $z \colon \mathbf{T} \to \mathbf{C}$ is the inclusion function, then the closed algebra generated by z and its conjugate $\bar{z}$ is $C(\mathbf{T})$ itself (immediate from the Stone–Weierstrass theorem).

A *left* (respectively, *right) ideal* in an algebra A is a vector subspace I of A such that

$$a \in A \text{ and } b \in I \Rightarrow ab \in I \quad (\text{respectively, } ba \in I).$$

An *ideal* in A is a vector subspace that is simultaneously a left and a right ideal in A. Obviously, 0 and A are ideals in A, called the *trivial* ideals. A *maximal* ideal in A is a proper ideal (that is, it is not A) that is not contained in any other proper ideal in A. Maximal left ideals are defined similarly.

An ideal I is *modular* if there is an element u in A such that $a - au$ and $a - ua$ are in I for all $a \in A$. It follows easily from Zorn's lemma that every proper modular ideal is contained in a maximal ideal.

If ω is an element of a locally compact Hausdorff space Ω, and $M_\omega = \{f \in C_0(\Omega) \mid f(\omega) = 0\}$, then M_ω is a modular ideal in the algebra $C_0(\Omega)$. This is so because there is an element $u \in C_0(\Omega)$ such that $u(\omega) = 1$, and hence, $f - uf \in M_\omega$ for all $f \in C_0(\Omega)$. Since M_ω is of codimension one in $C_0(\Omega)$ (as $M \oplus \mathbf{C}u = C_0(\Omega)$), it is a maximal ideal.

If I is an ideal of A, then A/I is an algebra with the multiplication given by

$$(a + I)(b + I) = ab + I.$$

If I is modular, then A/I is unital (if $a - au, a - ua \in I$ for all $a \in A$, then $u + I$ is the unit). Conversely, if A/I is unital then I is modular.

If A is unital, then obviously all its ideals are modular, and therefore, A posesses maximal ideals.

If $(I_\lambda)_{\lambda \in \Lambda}$ is a family of ideals of an algebra A, then $\cap_{\lambda \in \Lambda} I_\lambda$ is an ideal of A. Hence, if $S \subseteq A$, there is a smallest ideal I of A containing S. We call I the ideal *generated* by S. If A is a normed algebra, then the closure of an ideal is an ideal. The closed ideal J *generated* by a set S is the smallest closed ideal containing S. It is clear that J is the closure of the ideal generated by S.

1.1.1. Theorem. *If I is a closed ideal in a normed algebra A, then A/I is a normed algebra when endowed with the quotient norm*

$$\|a + I\| = \inf_{b \in I} \|a + b\|.$$

Proof. Let $\varepsilon > 0$ and suppose that a, b belong to A. Then $\varepsilon + \|a + I\| > \|a + a'\|$ and $\varepsilon + \|b + I\| > \|b + b'\|$ for some $a', b' \in I$. Hence,

$$(\varepsilon + \|a + I\|)(\varepsilon + \|b + I\|) > \|a + a'\|\|b + b'\| \geq \|ab + c\|,$$

where $c = a'b + ab' + a'b' \in I$. Thus, $(\varepsilon + \|a + I\|)(\varepsilon + \|b + I\|) \geq \|ab + I\|$. Letting $\varepsilon \to 0$, we get $\|a + I\|\|b + I\| \geq \|ab + I\|$; that is, the quotient norm is submultiplicative. $\qquad\square$

A *homomorphism* from an algebra A to an algebra B is a linear map $\varphi \colon A \to B$ such that $\varphi(ab) = \varphi(a)\varphi(b)$ for all $a, b \in A$. Its kernel $\ker(\varphi)$ is an ideal in A and its image $\varphi(A)$ is a subalgebra of B. We say φ is *unital* if A and B are unital and $\varphi(1) = 1$.

If I is an ideal in A, the quotient map $\pi \colon A \to A/I$ is a homomorphism.

If φ, ψ are continuous homomorphisms from a normed algebra A to a normed algebra B, then $\varphi = \psi$ if φ and ψ are equal on a set S that generates A as a normed algebra (that is, A is the closed algebra generated by S). This follows from the observation that the set $\{a \in A \mid \varphi(a) = \psi(a)\}$ is a closed subalgebra of A.

If A is the disc algebra and $\lambda \in \mathbf{D}$, the function

$$A \to \mathbf{C}, \quad f \mapsto f(\lambda),$$

is a continuous homomorphism. Moreover, every non-zero continuous homomorphism from A to $\mathbf{C}$ is of this form. This follows from the fact that the closed subalgebra generated by the unit and the inclusion function $z \colon \mathbf{D} \to \mathbf{C}$ is A. We show this: If $f \in A$ and $0 < r < 1$, define $f_r \in C(\mathbf{D})$ by $f_r(\lambda) = f(r\lambda)$. By uniform continuity of f on $\mathbf{D}$, we have $\lim_{r \to 1^-} \|f - f_r\|_\infty = 0$. Since f_r is extendable to an analytic function on the open disc of center 0 and radius $1/r$, it is the uniform limit on $\mathbf{D}$ of its Taylor series. Thus, f_r is the uniform limit of polynomial functions on $\mathbf{D}$, and therefore, so is f.

1.2. The Spectrum and the Spectral Radius

Let $\mathbf{C}[z]$ denote the algebra of all polynomials in an indeterminate z with complex coefficients. If a is an element of a unital algebra A and $p \in \mathbf{C}[z]$ is the polynomial

$$p = \lambda_0 + \lambda_1 z^1 + \cdots + \lambda_n z^n,$$

we set

$$p(a) = \lambda_0 1 + \lambda_1 a^1 + \cdots + \lambda_n a^n.$$

The map

$$\mathbf{C}[z] \to A, \quad p \mapsto p(a),$$

is a unital homomorphism.

We say that $a \in A$ is *invertible* if there is an element b in A such that $ab = ba = 1$. In this case b is unique and written a^{-1}. The set

$$\mathrm{Inv}(A) = \{a \in A \mid a \text{ is invertible}\}$$

is a group under multiplication.

We define the *spectrum* of an element a to be the set

$$\sigma(a) = \sigma_A(a) = \{\lambda \in \mathbf{C} \mid \lambda 1 - a \notin \mathrm{Inv}(A)\}.$$

We shall henceforth find it convenient to write $\lambda 1$ simply as λ.

1.2.1. Example. Let $A = C(\Omega)$, where Ω is a compact Hausdorff space. Then $\sigma(f) = f(\Omega)$ for all $f \in A$.

1.2.2. Example. Let $A = \ell^\infty(S)$, where S is a non-empty set. Then $\sigma(f) = (f(S))^-$ (the closure in $\mathbf{C}$) for all $f \in A$.

1.2.3. Example. Let A be the algebra of upper triangular $n \times n$-matrices. If $a \in A$, say

$$a = \begin{pmatrix} \lambda_{11} & \lambda_{12} & \cdots & \lambda_{1n} \\ 0 & \lambda_{22} & \cdots & \lambda_{2n} \\ \vdots & \vdots & \ddots & \vdots \\ 0 & \cdots & 0 & \lambda_{nn} \end{pmatrix}$$

it is elementary that

$$\sigma(a) = \{\lambda_{11}, \lambda_{22}, \ldots, \lambda_{nn}\}.$$

Similarly, if $A = M_n(\mathbf{C})$ and $a \in A$, then $\sigma(a)$ is the set of eigenvalues of a.

Thus, one thinks of the spectrum as simultaneously a generalisation of the range of a function and the set of eigenvalues of a finite square matrix.

1.2.1. Remark. If a, b are elements of a unital algebra A, then $1 - ab$ is invertible if and only if $1 - ba$ is invertible. This follows from the observation that if $1 - ab$ has inverse c, then $1 - ba$ has inverse $1 + bca$.

A consequence of this equivalence is that $\sigma(ab) \setminus \{0\} = \sigma(ba) \setminus \{0\}$ for all $a, b \in A$.

1.2.1. Theorem. *Let a be an element of a unital algebra A. If $\sigma(a)$ is non-empty and $p \in \mathbf{C}[z]$, then*

$$\sigma(p(a)) = p(\sigma(a)).$$

Proof. We may suppose that p is not constant. If $\mu \in \mathbf{C}$, there are elements $\lambda_0, \ldots, \lambda_n$ in $\mathbf{C}$, where $\lambda_0 \neq 0$, such that

$$p - \mu = \lambda_0(z - \lambda_1) \ldots (z - \lambda_n),$$

and therefore,

$$p(a) - \mu = \lambda_0(a - \lambda_1) \ldots (a - \lambda_n).$$

It is clear that $p(a) - \mu$ is invertible if and only if $a - \lambda_1, \ldots, a - \lambda_n$ are. It follows that $\mu \in \sigma(p(a))$ if and only if $\mu = p(\lambda)$ for some $\lambda \in \sigma(a)$, and therefore, $\sigma(p(a)) = p(\sigma(a))$. $\quad\square$

The spectral mapping property for polynomials is generalised to continuous functions in Chapter 2, but only for certain elements in certain algebras. There is a version of Theorem 1.2.1 for analytic functions and Banach algebras (see [Tak, Proposition 2.8], for example). We shall not need this, however.

1.2.2. Theorem. *Let A be a unital Banach algebra and a an element of A such that $\|a\| < 1$. Then $1 - a \in \mathrm{Inv}(A)$ and*

$$(1 - a)^{-1} = \sum_{n=0}^{\infty} a^n.$$

Proof. Since $\sum_{n=0}^{\infty} \|a^n\| \leq \sum_{n=0}^{\infty} \|a\|^n = (1 - \|a\|)^{-1} < +\infty$, the series $\sum_{n=0}^{\infty} a^n$ is convergent, to b say, in A, and since $(1 - a)(1 + \cdots + a^n) = 1 - a^{n+1}$ converges to $(1 - a)b = b(1 - a)$ and to 1 as $n \to \infty$, the element b is the inverse of $1 - a$. $\quad\square$

The series in Theorem 1.2.2 is called the *Neumann* series for $(1 - a)^{-1}$.

1.2.3. Theorem. *If A is a unital Banach algebra, then $\mathrm{Inv}(A)$ is open in A, and the map*

$$\mathrm{Inv}(A) \to A, \quad a \mapsto a^{-1},$$

is differentiable.

Proof. Suppose that $a \in \mathrm{Inv}(A)$ and $\|b - a\| < \|a^{-1}\|^{-1}$. Then $\|ba^{-1} - 1\| \leq \|b - a\|\|a^{-1}\| < 1$, so $ba^{-1} \in \mathrm{Inv}(A)$, and therefore, $b \in \mathrm{Inv}(A)$. Thus, $\mathrm{Inv}(A)$ is open in A.

If $b \in A$ and $\|b\| < 1$, then $1 + b \in \text{Inv}(A)$ and

$$\|(1+b)^{-1} - 1 + b\| = \|\sum_{n=0}^{\infty}(-1)^n b^n - 1 + b\| = \|\sum_{n=2}^{\infty}(-1)^n b^n\|$$

$$\leq \sum_{n=2}^{\infty} \|b\|^n = \|b\|^2/(1 - \|b\|)^{-1}.$$

Let $a \in \text{Inv}(A)$ and suppose that $\|c\| < \frac{1}{2}\|a^{-1}\|^{-1}$. Then $\|a^{-1}c\| < 1/2 < 1$, so (with $b = a^{-1}c$),

$$\|(1 + a^{-1}c)^{-1} - 1 + a^{-1}c\| \leq \|a^{-1}c\|^2/(1 - \|a^{-1}c\|)^{-1} \leq 2\|a^{-1}c\|^2,$$

since $1 - \|a^{-1}c\| > 1/2$. Now define u to be the linear operator on A given by $u(b) = -a^{-1}ba^{-1}$. Then,

$$\|(a+c)^{-1} - a^{-1} - u(c)\| = \|(1 + a^{-1}c)^{-1}a^{-1} - a^{-1} + a^{-1}ca^{-1}\|$$

$$\leq \|(1 + a^{-1}c)^{-1} - 1 + a^{-1}c\|\|a^{-1}\| \leq 2(\|a^{-1}\|^3\|c\|^2).$$

Consequently,

$$\lim_{c \to 0} \frac{\|(a+c)^{-1} - a^{-1} - u(c)\|}{\|c\|} = 0,$$

and therefore, the map $\sigma: b \mapsto b^{-1}$ is differentiable at $b = a$ with derivative $\sigma'(a) = u$. □

The algebra $\mathbf{C}[z]$ is a normed algebra where the norm is defined by setting

$$\|p\| = \sup_{|\lambda| \leq 1} |p(\lambda)|.$$

Observe that $\text{Inv}(\mathbf{C}[z]) = \mathbf{C} \setminus \{0\}$, so the polynomials $p_n = 1 + z/n$ are not invertible. But $\lim_{n \to \infty} p_n = 1$, which shows that $\text{Inv}(\mathbf{C}[z])$ is not open in $\mathbf{C}[z]$. Thus, the norm on $\mathbf{C}[z]$ is not complete.

1.2.4. Lemma. *Let A be a unital Banach algebra and let $a \in A$. The spectrum $\sigma(a)$ of a is a closed subset of the disc in the plane of centre the origin and radius $\|a\|$, and the map*

$$\mathbf{C} \setminus \sigma(a) \to A, \quad \lambda \mapsto (a - \lambda)^{-1},$$

is differentiable.

Proof. If $|\lambda| > \|a\|$, then $\|\lambda^{-1}a\| < 1$, so $1 - \lambda^{-1}a$ is invertible, and therefore, so is $\lambda - a$. Hence, $\lambda \notin \sigma(a)$. Thus, $\lambda \in \sigma(a) \Rightarrow |\lambda| \leq \|a\|$. The set $\sigma(a)$ is closed, that is, $\mathbf{C} \setminus \sigma(a)$ is open, because $\text{Inv}(A)$ is open in A. Differentiability of the map $\lambda \mapsto (a - \lambda)^{-1}$ follows from Theorem 1.2.3. □

The following result can be thought of as the fundamental theorem of Banach algebras.

1.2.5. Theorem (Gelfand). *If a is an element of a unital Banach algebra A, then the spectrum $\sigma(a)$ of a is non-empty.*

Proof. Suppose that $\sigma(a) = \emptyset$ and we shall obtain a contradiction. If $|\lambda| > 2\|a\|$, then $\|\lambda^{-1}a\| < \frac{1}{2}$, and therefore, $1 - \|\lambda^{-1}a\| > \frac{1}{2}$. Hence,

$$\|(1 - \lambda^{-1}a)^{-1} - 1\| = \|\sum_{n=1}^{\infty}(\lambda^{-1}a)^n\|$$

$$\leq \frac{\|\lambda^{-1}a\|}{1 - \|\lambda^{-1}a\|} \leq 2\|\lambda^{-1}a\| < 1.$$

Consequently, $\|(1 - \lambda^{-1}a)^{-1}\| < 2$, and therefore,

$$\|(a - \lambda)^{-1}\| = \|\lambda^{-1}(1 - \lambda^{-1}a)^{-1}\| < 2/|\lambda| < \|a\|^{-1}$$

($a \neq 0$ since $\sigma(a) = \emptyset$). Moreover, since the map $\lambda \mapsto (a - \lambda)^{-1}$ is continuous, it is bounded on the (compact) disc $2\|a\|\mathbf{D}$. Thus, we have shown that this map is bounded on all of $\mathbf{C}$; that is, there is a positive number M such that $\|(a - \lambda)^{-1}\| \leq M$ ($\lambda \in \mathbf{C}$).

If $\tau \in A^*$, the function $\lambda \mapsto \tau((a - \lambda)^{-1})$ is entire, and bounded by $M\|\tau\|$, so by Liouville's theorem in complex analysis, it is constant. In particular, $\tau(a^{-1}) = \tau((a - 1)^{-1})$. Because this is true for all $\tau \in A^*$, we have $a^{-1} = (a - 1)^{-1}$, so $a = a - 1$, which is a contradiction. $\square$

It is easy to see that there are algebras in which not all elements have non-empty spectrum. For example, if $\mathbf{C}(z)$ denotes the field of quotients of $\mathbf{C}[z]$, then $\mathbf{C}(z)$ is an algebra, and the spectrum of z in this algebra is empty.

1.2.6. Theorem (Gelfand–Mazur). *If A is a unital Banach algebra in which every non-zero element is invertible, then $A = \mathbf{C}1$.*

Proof. This is immediate from Theorem 1.2.5. $\square$

If a is an element of a unital Banach algebra A, its *spectral radius* is defined to be

$$r(a) = \sup_{\lambda \in \sigma(a)} |\lambda|.$$

By Remark 1.2.1, $r(ab) = r(ba)$ for all $a, b \in A$.

1.2.4. *Example.* If $A = C(\Omega)$, where Ω is a compact Hausdorff space, then $r(f) = \|f\|_\infty$ ($f \in A$).

1.2.5. *Example.* Let $A = M_2(\mathbf{C})$ and

$$a = \begin{pmatrix} 0 & 1 \\ 0 & 0 \end{pmatrix}.$$

Then $\|a\| = 1$, but $r(a) = 0$, since $a^2 = 0$.

1.2.7. Theorem (Beurling). *If a is an element of a unital Banach algebra A, then*

$$r(a) = \inf_{n \geq 1} \|a^n\|^{1/n} = \lim_{n \to \infty} \|a^n\|^{1/n}.$$

Proof. If $\lambda \in \sigma(a)$, then $\lambda^n \in \sigma(a^n)$, so $|\lambda^n| \leq \|a^n\|$, and therefore, $r(a) \leq \inf_{n \geq 1} \|a^n\|^{1/n} \leq \liminf_{n \to \infty} \|a^n\|^{1/n}$.

Let Δ be the open disc in $\mathbf{C}$ centered at 0 and of radius $1/r(a)$ (we use the usual convention that $1/0 = +\infty$). If $\lambda \in \Delta$, then $1 - \lambda a \in \mathrm{Inv}(A)$. If $\tau \in A^*$, then the map

$$f \colon \Delta \to \mathbf{C}, \ \lambda \mapsto \tau((1 - \lambda a)^{-1}),$$

is analytic, so there are unique complex numbers λ_n such that

$$f(\lambda) = \sum_{n=0}^{\infty} \lambda_n \lambda^n \quad (\lambda \in \Delta).$$

However, if $|\lambda| < 1/\|a\| (\leq 1/r(a))$, then $\|\lambda a\| < 1$, so

$$(1 - \lambda a)^{-1} = \sum_{n=0}^{\infty} \lambda^n a^n,$$

and therefore,

$$f(\lambda) = \sum_{n=0}^{\infty} \lambda^n \tau(a^n).$$

It follows that $\lambda_n = \tau(a^n)$ for all $n \geq 0$. Hence, the sequence $(\tau(a^n)\lambda^n)$ converges to 0 for each $\lambda \in \Delta$, and therefore *a fortiori*, it is bounded. Since this is true for each $\tau \in A^*$, it follows from the principle of uniform boundedness that $(\lambda^n a^n)$ is a bounded sequence. Hence, there is a positive number M (depending on λ, of course) such that $\|\lambda^n a^n\| \leq M$ for all $n \geq 0$, and therefore, $\|a^n\|^{1/n} \leq M^{1/n}/|\lambda|$ (if $\lambda \neq 0$). Consequently, $\limsup_{n \to \infty} \|a^n\|^{1/n} \leq 1/|\lambda|$. We have thus shown that if $r(a) < |\lambda^{-1}|$, then $\limsup_{n \to \infty} \|a^n\|^{1/n} \leq |\lambda^{-1}|$. It follows that $\limsup_{n \to \infty} \|a^n\|^{1/n} \leq r(a)$, and since $r(a) \leq \liminf_{n \to \infty} \|a^n\|^{1/n}$, therefore $r(a) = \lim_{n \to \infty} \|a^n\|^{1/n}$. $\square$

1.2.6. Example. Let A be the set of C^1-functions on the interval $[0,1]$. This is an algebra when endowed with the pointwise-defined operations, and a submultiplicative norm on A is given by

$$\|f\| = \|f\|_\infty + \|f'\|_\infty \quad (f \in A).$$

It is elementary that A is complete under this norm, and therefore, A is a Banach algebra. Let $x \colon [0,1] \to \mathbf{C}$ be the inclusion, so $x \in A$. Clearly, $\|x^n\| = 1 + n$ for all n, so $r(x) = \lim(1 + n)^{1/n} = 1 < 2 = \|x\|$.

Recall that if K is a non-empty compact set in $\mathbf{C}$, its complement $\mathbf{C} \setminus K$ admits exactly one unbounded component, and that the bounded components of $\mathbf{C} \setminus K$ are called the *holes* of K.

1.2.8. Theorem. *Let B be a closed subalgebra of a unital Banach algebra A, containing the unit of A.*

(1) *The set $\mathrm{Inv}(B)$ is a clopen subset of $B \cap \mathrm{Inv}(A)$.*

(2) *For each $b \in B$,*

$$\sigma_A(b) \subseteq \sigma_B(b) \quad \text{and} \quad \partial \sigma_B(b) \subseteq \partial \sigma_A(b).$$

(3) *If $b \in B$ and $\sigma_A(b)$ has no holes, then $\sigma_A(b) = \sigma_B(b)$.*

Proof. Clearly $\mathrm{Inv}(B)$ is an open set in $B \cap \mathrm{Inv}(A)$. To see that it is also closed, let (b_n) be a sequence in $\mathrm{Inv}(B)$ converging to a point $b \in B \cap \mathrm{Inv}(A)$. Then (b_n^{-1}) converges to b^{-1} in A, so $b^{-1} \in B$, which implies that $b \in \mathrm{Inv}(B)$. Hence, $\mathrm{Inv}(B)$ is clopen in $B \cap \mathrm{Inv}(A)$.

If $b \in B$, the inclusion $\sigma_A(b) \subseteq \sigma_B(b)$ is immediate from the inclusion $\mathrm{Inv}(B) \subseteq \mathrm{Inv}(A)$.

If $\lambda \in \partial \sigma_B(b)$, then there is a sequence (λ_n) in $\mathbf{C} \setminus \sigma_B(b)$ converging to λ. Hence, $b - \lambda_n \in \mathrm{Inv}(B)$, and $b - \lambda \notin \mathrm{Inv}(B)$, so $b - \lambda \notin \mathrm{Inv}(A)$, by Condition (1). Also, $b - \lambda_n \in \mathrm{Inv}(A)$, so $\lambda_n \in \mathbf{C} \setminus \sigma_A(b)$. Therefore, $\lambda \in \partial \sigma_A(b)$. This proves Condition (2).

If $b \in B$ and $\sigma_A(b)$ has no holes, then $\mathbf{C} \setminus \sigma_A(b)$ is connected. Since $\mathbf{C} \setminus \sigma_B(b)$ is a clopen subset of $\mathbf{C} \setminus \sigma_A(b)$ by Conditions (1) and (2), it follows that $\mathbf{C} \setminus \sigma_A(b) = \mathbf{C} \setminus \sigma_B(b)$, and therefore, $\sigma_A(b) = \sigma_B(b)$. $\square$

1.2.7. Example. Let $C = C(\mathbf{T})$ and let A be the disc algebra. If $f \in A$, let $\varphi(f)$ be its restriction to $\mathbf{T}$. One easily checks that the map

$$\varphi \colon A \to C, \quad f \mapsto \varphi(f),$$

is an isometric homomorphism onto the closed subalgebra B of C generated by the unit and the inclusion $z \colon \mathbf{T} \to \mathbf{C}$ (the equation $\|\varphi(f)\|_\infty = \|f\|_\infty$ is given by the maximum modulus principle). Clearly, $\sigma_B(z) = \sigma_A(z) = \mathbf{D}$, and $\sigma_C(z) = \mathbf{T}$.

Let a be an element of a unital Banach algebra A. Since

$$\sum_{n=0}^{\infty} \|a^n/n!\| \leq \sum_{n=0}^{\infty} \|a\|^n/n! < \infty,$$

the series $\sum_{n=0}^{\infty} a^n/n!$ is convergent in A. We denote its sum by e^a.

In proving the next theorem, we shall use some elementary results concerning differentiation. Suppose that f, g are differentiable maps from $\mathbf{R}$ to A with derivatives f', g', respectively. Then fg is differentiable and $(fg)' = fg' + f'g$. (To prove this, just mimic the proof of the scalar-valued case.) If $f' = 0$, then f is constant. We prove this: If $\tau \in A^*$, then the function $\mathbf{R} \to \mathbf{C}, t \mapsto \tau(f(t))$, is differentiable with zero derivative, and therefore, $\tau(f(t)) = \tau(f(0))$ for all t. Since τ was arbitrary, this implies that $f(t) = f(0)$.

1.2.9. Theorem. *Let A be a unital Banach algebra.*

(1) *If $a \in A$ and $f: \mathbf{R} \to A$ is differentiable, $f(0) = 1$, and $f'(t) = af(t)$ for all $t \in \mathbf{R}$, then $f(t) = e^{ta}$ for all $t \in \mathbf{R}$.*

(2) *If $a \in A$, then e^a is invertible with inverse e^{-a}, and if a, b are commuting elements of A, then $e^{a+b} = e^a e^b$.*

Proof. First we observe that if $f: \mathbf{R} \to A$ is defined by $f(t) = e^{ta}$, then $f(t) = \sum_{n=0}^{\infty} t^n a^n / n!$, so differentiating term by term we get $f'(t) = af(t)$. Now suppose f, g are any pair of differentiable maps from $\mathbf{R}$ to A such that $f'(t) = af(t)$ and $g'(t) = ag(t)$ and $f(0) = g(0) = 1$. Then the map $h: \mathbf{R} \to A$, $t \mapsto f(t)g(-t)$, is differentiable with zero derivative (apply the product rule for differentiation). Hence, $h(t) = 1$ for all $t \in \mathbf{R}$. Applying this to the map $t \mapsto e^{ta}$, we get $e^{ta}e^{-ta} = 1$; in particular, $e^a e^{-a} = 1$.

It follows that if $f: \mathbf{R} \to A$ is differentiable, $f(0) = 1$, and $f'(t) = af(t)$ for all t, then $f(t) = e^{ta}$ (set $g(t) = e^{ta}$ and get $f(t)e^{-ta} = 1$, so $f(t) = e^{ta}$).

Now suppose that a and b are commuting elements of A and set $f(t) = e^{ta}e^{tb}$. Then $f(0) = 1$ and $f'(t) = e^{ta}be^{tb} + ae^{ta}e^{tb}$ (by the product rule) $= (a + b)f(t)$. Hence, $f(t) = e^{t(a+b)}$ for all $t \in \mathbf{R}$, so, in particular, $e^{a+b} = f(1) = e^a e^b$. $\square$

We shall see later that not every invertible element is of the form e^a.

If an algebra is non-unital we can adjoin a unit to it. This is very helpful in many cases, and we shall frequently make use of it, but it does not reduce the theory to the unital case. There are situations where adjoining a unit is unnatural, such as when one is studying the group algebra $L^1(G)$ of a locally compact group G (see the addenda section of this chapter for the definition of this algebra).

If A is an algebra, we set $\tilde{A} = A \oplus \mathbf{C}$ as a vector space. We define a multiplication on $\tilde{A}$ making it a unital algebra by setting

$$(a, \lambda)(b, \mu) = (ab + \lambda b + \mu a, \lambda \mu).$$

The unit is $(0,1)$. The algebra $\tilde{A}$ is called the *unitization* of A. The map

$$A \to \tilde{A}, \quad a \mapsto (a, 0),$$

is an injective homomorphism, which we use to identify A as an ideal of $\tilde{A}$. We then write $a + \lambda$ for (a, λ). The map

$$\tilde{A} \to \mathbf{C}, \quad a + \lambda \mapsto \lambda,$$

is a unital homomorphism with kernel A, called the *canonical* homomorphism.

If A is abelian, so is $\tilde{A}$.

If A is a normed algebra, we make $\tilde{A}$ into a normed algebra by setting

$$\|a + \lambda\| = \|a\| + |\lambda|.$$

Observe that A is a closed subalgebra of $\tilde{A}$, and that $\tilde{A}$ is a Banach algebra if A is one.

If A is a non-unital Banach algebra, then for $a \in A$ we set $\sigma_A(a) = \sigma_{\tilde{A}}(a)$, and $r(a) = \sup_{\lambda \in \sigma_A(a)} |\lambda|$. Note that 0 is an element of $\sigma_A(a)$ in this case.

1.3. The Gelfand Representation

The idea of this section is to represent an abelian Banach algebra as an algebra of continuous functions on a locally compact Hausdorff space. This is an extremely useful way of looking at these algebras, but in the case of the more "complicated" algebras, the picture it presents may be of limited accuracy.

We begin by proving some results on ideals and multiplicative linear functionals.

1.3.1. Theorem. *Let I be a modular ideal of a Banach algebra A. If I is proper, so is its closure $\bar{I}$. If I is maximal, then it is closed.*

Proof. Let u be an element of A such that $a - au$ and $a - ua$ are in I for all $a \in A$. If $b \in I$ and $\|u - b\| < 1$, then the element $v = 1 - u + b$ is invertible in $\tilde{A}$. If $a \in A$, then $av = a - au + ab \in I$, so $A = Av \subseteq I$. This contradicts the assumption that I is proper, and shows that $\|u - b\| \geq 1$ for all $b \in I$. It follows that $u \notin \bar{I}$, so $\bar{I}$ is proper.

If I is maximal, then $I = \bar{I}$, as $\bar{I}$ is a proper ideal containing I. $\qquad\square$

1.3.1. Remark. If L is a left ideal of a Banach algebra A, it is *modular* if there is an element u in A such that $a - au \in L$ for all $a \in A$, and in this case its closure is a proper left ideal. Moreover, if L is a modular maximal left ideal, it is closed. The proofs are the same as for Theorem 1.3.1.

1.3.2. Lemma. *If I is a modular maximal ideal of a unital abelian algebra A, then A/I is a field.*

Proof. The algebra A/I is unital and abelian, with unit $u + I$ say. If J is an ideal of A/I and π is the quotient map from A to A/I, then $\pi^{-1}(J)$ is an ideal of A containing I. Hence, $\pi^{-1}(J) = A$ or I, by maximality of I. Therefore, $J = A/I$ or 0. Thus, A/I and 0 are the only ideals of A/I. Now suppose that $\pi(a)$ is a non-zero element of A/I. Then $J = \pi(a)(A/I)$ is a non-zero ideal of A/I, and therefore, $J = A/I$. Hence, there is an element b of A such that $(a + I)(b + I) = u + I$, so $a + I$ is invertible. This shows that A/I is a field. $\qquad\square$

Note that if $\varphi\colon A \to B$ is a homomorphism between algebras A and B and B is unital, then $\tilde{\varphi}\colon \tilde{A} \to B$, $a + \lambda \mapsto \varphi(a) + \lambda$, $(a \in A,\ \lambda \in \mathbf{C})$ is the unique unital homomorphism extending φ.

If $\varphi\colon A \to B$ is a unital homomorphism between unital algebras, then $\varphi(\mathrm{Inv}(A)) \subseteq \mathrm{Inv}(B)$, so $\sigma(\varphi(a)) \subseteq \sigma(a)$ $(a \in A)$.

A *character* on an abelian algebra A is a non-zero homomorphism $\tau\colon A \to \mathbf{C}$. We denote by $\Omega(A)$ the set of characters on A.

1.3.3. Theorem. *Let A be a unital abelian Banach algebra.*

(1) *If $\tau \in \Omega(A)$, then $\|\tau\| = 1$.*

(2) *The set $\Omega(A)$ is non-empty, and the map*

$$\tau \mapsto \ker(\tau)$$

defines a bijection from $\Omega(A)$ onto the set of all maximal ideals of A.

Proof. If $\tau \in \Omega(A)$ and $a \in A$, then $\tau(a) \in \sigma(a)$, so $|\tau(a)| \leq r(a) \leq \|a\|$. Hence, $\|\tau\| \leq 1$. Also, $\tau(1) = 1$, since $\tau(1) = \tau(1)^2$ and $\tau(1) \neq 0$. Hence, $\|\tau\| = 1$.

Let I denote the closed ideal $\ker(\tau)$. This is proper, since $\tau \neq 0$, and $I + \mathbf{C}1 = A$, since $a - \tau(a) \in I$ for all $a \in A$. It follows that I is a maximal ideal of A.

If $\tau_1, \tau_2 \in \Omega(A)$ and $\ker(\tau_1) = \ker(\tau_2)$, then for each $a \in A$ we have $\tau_1(a - \tau_2(a)) = 0$, so $\tau_1(a) = \tau_2(a)$. Thus, $\tau_1 = \tau_2$.

If I is an arbitrary maximal ideal of A, then I is closed by Theorem 1.3.1 and A/I is a unital Banach algebra in which every non-zero element is invertible, by Lemma 1.3.2. Hence, by Theorem 1.2.6 $A/I = \mathbf{C}(1 + I)$. It follows that $A = I \oplus \mathbf{C}1$. Define $\tau\colon A \to \mathbf{C}$ by $\tau(a + \lambda) = \lambda$, $(a \in I,\ \lambda \in \mathbf{C})$. Then τ is a character and $\ker(\tau) = I$.

Thus, we have shown that the map $\tau \mapsto \ker(\tau)$ is a bijection from the characters onto the maximal ideals of A.

We have seen already that A admits maximal ideals (since it is unital). Therefore, $\Omega(A) \neq \emptyset$. $\square$

1.3.4. Theorem. *Let A be an abelian Banach algebra.*

(1) *If A is unital, then*

$$\sigma(a) = \{\tau(a) \mid \tau \in \Omega(A)\} \quad (a \in A).$$

(2) *If A is non-unital, then*

$$\sigma(a) = \{\tau(a) \mid \tau \in \Omega(A)\} \cup \{0\} \quad (a \in A).$$

Proof. If A is unital and a is an element of A whose spectrum contains λ, then the ideal $I = (a - \lambda)A$ is proper, so I is contained in a maximal ideal

$\ker(\tau)$, where $\tau \in \Omega(A)$. Hence, $\tau(a) = \lambda$. This shows that the inclusion $\sigma(a) \subseteq \{\tau(a) \mid \tau \in \Omega(A)\}$ holds, and the reverse inclusion is clear.

Now suppose that A is non-unital, and let $\tau_\infty \colon \tilde{A} \to \mathbf{C}$ be the canonical homomorphism. Then $\Omega(\tilde{A}) = \{\tilde{\tau} \mid \tau \in \Omega(A)\} \cup \{\tau_\infty\}$, where $\tilde{\tau}$ is the unique character on $\tilde{A}$ extending the character τ on A. Hence, by Condition (1), $\sigma(a) = \sigma_{\tilde{A}}(a) = \{\tau(a) \mid \tau \in \Omega(\tilde{A})\} = \{\tau(a) \mid \tau \in \Omega(A)\} \cup \{0\}$ for each $a \in A$. $\qquad\square$

If A is an abelian Banach algebra, it follows from Theorem 1.3.4 that $\Omega(A)$ is contained in the closed unit ball of A^*. We endow $\Omega(A)$ with the relative weak* topology, and call the topological space $\Omega(A)$ the *character space*, or *spectrum*, of A.

1.3.5. Theorem. *If A is an abelian Banach algebra, then $\Omega(A)$ is a locally compact Hausdorff space. If A is unital, then $\Omega(A)$ is compact.*

Proof. It is easily checked that $\Omega(A) \cup \{0\}$ is weak* closed in the closed unit ball S of A^*. Since S is weak* compact (Banach–Alaoglu theorem), $\Omega(A) \cup \{0\}$ is weak* compact, and therefore, $\Omega(A)$ is locally compact.

If A is unital, then $\Omega(A)$ is weak* closed in S and thus compact. $\qquad\square$

Note that $\Omega(A)$ may be empty. This is the case for $A = 0$, for example.

Suppose that A is an abelian Banach algebra for which the space $\Omega(A)$ is non-empty. If $a \in A$, we define the function $\hat{a}$ by

$$\hat{a} \colon \Omega(A) \to \mathbf{C}, \quad \tau \mapsto \tau(a).$$

Clearly the topology on $\Omega(A)$ is the smallest one making all of the functions $\hat{a}$ continuous. The set $\{\tau \in \Omega(A) \mid |\tau(a)| \geq \varepsilon\}$ is weak* closed in the closed unit ball of A^* for each $\varepsilon > 0$, and weak* compact by the Banach–Alaoglu theorem. Hence, $\hat{a} \in C_0(\Omega(A))$.

We call $\hat{a}$ the *Gelfand transform* of a.

Although the following result is very important, its proof is easy, because we have already done most of the work needed to demonstrate it.

1.3.6. Theorem (Gelfand Representation). *Suppose that A is an abelian Banach algebra and that $\Omega(A)$ is non-empty. Then the map*

$$A \to C_0(\Omega(A)), \quad a \mapsto \hat{a},$$

is a norm-decreasing homomorphism, and

$$r(a) = \|\hat{a}\|_\infty \qquad (a \in A).$$

If A is unital, $\sigma(a) = \hat{a}(\Omega(A))$, and if A is non-unital, $\sigma(a) = \hat{a}(\Omega(A)) \cup \{0\}$, for each $a \in A$.

Proof. By Theorem 1.3.4 the spectrum $\sigma(a)$ is the range of $\hat{a}$, together with $\{0\}$ if A is non-unital. Hence, $r(a) = \|\hat{a}\|_\infty$, which implies that the map $a \mapsto \hat{a}$ is norm-decreasing. That this map is a homomorphism is easily checked.   □

The kernel of the Gelfand representation is called the *radical* of the algebra A. It consists of the elements a such that $r(a) = 0$. It therefore contains the nilpotent elements. If the radical is zero, A is said to be *semisimple*.

In a general algebra an element whose spectrum consists of the set $\{0\}$ is said to be *quasinilpotent*.

Let a, b be commuting elements of an arbitrary Banach algebra A. Then $r(a + b) \le r(a) + r(b)$, and $r(ab) \le r(a)r(b)$. To see this, we may suppose that A is unital and abelian (if necessary, adjoin a unit and restrict to the closed subalgebra generated by $1, a$, and b). Then $r(a + b) = \|(a + b)\hat{}\|_\infty \le \|\hat{a}\|_\infty + \|\hat{b}\|_\infty = r(a) + r(b)$ by Theorem 1.3.6. Similarly, $r(ab) = \|(ab)\hat{}\|_\infty \le \|\hat{a}\|_\infty \|\hat{b}\|_\infty = r(a)r(b)$. Direct proofs of the first of these inequalities (that is, where the Gelfand representation is not invoked) tend to be messy.

The spectral radius is neither subadditive nor submultiplicative in general: Let $A = M_2(\mathbf{C})$ and suppose

$$a = \begin{pmatrix} 0 & 1 \\ 0 & 0 \end{pmatrix} \quad \text{and} \quad b = \begin{pmatrix} 0 & 0 \\ 1 & 0 \end{pmatrix}.$$

Then $r(a) = r(b) = 0$, since a and b have square zero, but $r(a + b) = r(ab) = 1$.

The interpretation of the character space as a sort of generalised spectrum is motivated by the following result.

1.3.7. Theorem. *Let A be a unital Banach algebra generated by 1 and an element a. Then A is abelian and the map*

$$\hat{a} \colon \Omega(A) \to \sigma(a), \quad \tau \mapsto \tau(a),$$

is a homeomorphism.

Proof. It is clear that A is abelian and that $\hat{a}$ is a continuous bijection, and because $\Omega(A)$ and $\sigma(a)$ are compact Hausdorff spaces, $\hat{a}$ is therefore a homeomorphism.   □

To illustrate this, consider the disc algebra A. If z is its canonical generator, then since $\sigma(z) = \mathbf{D}$, we have $\Omega(A) = \mathbf{D}$ by Theorem 1.3.7. In this case if $f \in A$, then $\hat{f}(\lambda) = f(\lambda)$, so the Gelfand transform is the identity map.

We now present an interesting application of the preceding results to a problem in classical analysis.

1.3.1. *Example*. We denote by $\ell^1(\mathbf{Z})$ the set of all complex-valued functions f on $\mathbf{Z}$ such that $\sum_{n=-\infty}^{\infty} |f(n)|$ is finite. This is a Banach space when endowed with the pointwise-defined operations and the norm

$$\|f\|_1 = \sum_{n=-\infty}^{\infty} |f(n)|.$$

If $f, g \in \ell^1(\mathbf{Z})$ we define their *convolution* $f * g \colon \mathbf{Z} \to \mathbf{C}$ by the formula

$$(f * g)(m) = \sum_{n=-\infty}^{\infty} f(m-n)g(n). \tag{1}$$

If $f \in \ell^1(\mathbf{Z})$, it is bounded, so the sum in Eq. 1 exists. To see that $f * g \in \ell^1(\mathbf{Z})$ observe that

$$
\begin{aligned}
\sum_{n=-\infty}^{\infty} |(f*g)(n)| &= \sum_{n=-\infty}^{\infty} \Big| \sum_{m=-\infty}^{\infty} f(n-m)g(m) \Big| \\
&\leq \sum_{n=-\infty}^{\infty} \sum_{m=-\infty}^{\infty} |f(n-m)||g(m)| \\
&= \sum_{m=-\infty}^{\infty} \sum_{n=-\infty}^{\infty} |f(n-m)||g(m)| \\
&= \sum_{m=-\infty}^{\infty} \left(|g(m)| \sum_{n=-\infty}^{\infty} |f(n-m)| \right) \\
&= \sum_{m=-\infty}^{\infty} |g(m)| \|f\|_1 = \|f\|_1 \|g\|_1.
\end{aligned}
$$

Thus, $f * g \in \ell^1(\mathbf{Z})$ and $\|f * g\|_1 \leq \|f\|_1 \|g\|_1$. It is now a straightforward exercise to show that $\ell^1(\mathbf{Z})$ is an abelian unital Banach algebra with multiplication given by $(f, g) \mapsto f * g$. The characteristic function of the set $\{0\}$ is the unit, and if w is the characteristic function of the set $\{1\}$, then $f = \sum_{n=-\infty}^{\infty} f(n)w^n$ for all $f \in \ell^1(\mathbf{Z})$.

For $z \in \mathbf{T}$, define a character τ_z on $\ell^1(\mathbf{Z})$ by setting

$$\tau_z(f) = \sum_{n=-\infty}^{\infty} f(n)z^n.$$

We then have a map

$$\mathbf{T} \to \Omega(\ell^1(\mathbf{Z})), \ z \mapsto \tau_z,$$

which, it is easy to check, is a bijection. In fact, this is a homeomorphism, and to see this we need only show continuity, since the domain and range are compact and Hausdorff. Continuity is shown if we show the function $\mathbf{T} \to \mathbf{C}$, $z \mapsto \tau_z(f)$, is continuous when $f \in \ell^1(\mathbf{Z})$, and this follows from the observation that $\tau_z(f)$ is the uniform limit in z of the continuous functions $\sum_{|n| \leq N} f(n)z^n$ ($N = 1, 2, \ldots$), since $\sum_{n=-\infty}^{\infty} |f(n)z^n| = \|f\|_1 < \infty$.

We identify $\Omega(\ell^1(\mathbf{Z}))$ with $\mathbf{T}$ using the above homeomorphism. Thus, the Gelfand transform $\hat{f}$ of $f \in \ell^1(\mathbf{Z})$ is a continuous function on $\mathbf{T}$ such that

$$\hat{f}(z) = \sum_{n=-\infty}^{\infty} f(n)z^n.$$

It is readily verified that the numbers $f(n)$ are the Fourier coefficients of $\hat{f}$,

$$f(n) = \frac{1}{2\pi} \int_0^{2\pi} \hat{f}(e^{it}) e^{-int} \, dt.$$

Thus $\ell^1(\mathbf{Z})\hat{\ }$, the set of all Gelfand transforms, is the set of all functions $h \in C(\mathbf{T})$ whose Fourier series is absolutely convergent. One can show that not every function in $C(\mathbf{T})$ has such a Fourier series. A well-known theorem of Wiener states that if a continuous function on $\mathbf{T}$ has an absoluely convergent Fourier series and never vanishes, then its reciprocal has such a Fourier series. The proof of this is easy, using what we know about the algebra $\ell^1(\mathbf{Z})$:

Let h be a continuous function that never vanishes and that has absolutely convergent Fourier series, so $h = \hat{f}$ for some $f \in \ell^1(\mathbf{Z})$. Because $\hat{f}(z) \neq 0$ for all $z \in \mathbf{T}$, it follows from Theorem 1.3.4 that $0 \notin \sigma(f)$. Thus, f is invertible in $\ell^1(\mathbf{Z})$, with inverse g say. Then $\hat{g} = 1/h$, so $1/h$ has absolutely convergent Fourier series.

This proof is due to Gelfand.

We shall resume our study of Banach algebras in Chapter 2, but now we turn to single operator theory.

1.4. Compact and Fredholm Operators

This section is concerned with the elementary spectral theory of operators. We begin with the simplest non-trivial class of operators, the compact ones, a class that plays an important and fundamental role in operator theory. These operators behave much like operators on finite-dimensional vector spaces, and for this reason they are relatively easy to analyse.

A linear map $u: X \to Y$ between Banach spaces X and Y is *compact* if $u(S)$ is relatively compact in Y, where S is the closed unit ball of X. Equivalently, $u(S)$ is totally bounded. In this case $u(S)$ is bounded, and therefore, u is bounded.

1.4.1. Remark. Note that the range of a compact operator is separable. This is immediate from the fact that a compact metric space is separable, and that the closure of the image of the ball under a compact operator is compact.

The theory of compact operators arose out of the analysis of linear integral equations. The following example illustrates the connection.

1.4.1. Example. Let $I = [0,1]$ and let X be the Banach space $C(I)$, where the norm is the supremum norm. If $k \in C(I^2)$, define $u \in B(X)$ by setting

$$u(f)(s) = \int_0^1 k(s,t)f(t)\,dt \qquad (f \in X, s \in I).$$

We show that $u(f) \in X$. Observe first that

$$|u(f)(s) - u(f)(s')| = \left| \int_0^1 (k(s,t) - k(s',t))f(t)\,dt \right|$$

$$\leq \int_0^1 |k(s,t) - k(s',t)||f(t)|\,dt$$

$$\leq \sup_{t \in I} |k(s,t) - k(s',t)| \|f\|_\infty.$$

Now k is uniformly continuous because I^2 is compact, so if $\varepsilon > 0$, there exists $\delta > 0$ such that $\max\{|s - s'|, |t - t'|\} < \delta \Rightarrow |k(s,t) - k(s',t')| < \varepsilon$. Hence,

$$|s - s'| < \delta \Rightarrow |u(f)(s) - u(f)(s')| \leq \varepsilon \|f\|_\infty. \qquad (1)$$

Thus, $u(f)$ is continuous, that is, $u(f) \in X$, but more is true, for it is immediate from Inequality (1) that $u(S)$ is equicontinuous, where S is the closed unit ball of X. Also, $u(S)$ is pointwise-bounded, that is, $\sup_{f \in S} |u(f)(s)| < \infty$, since

$$|u(f)(s)| \leq \int_0^1 |k(s,t)f(t)|\,dt \leq \|k\|_\infty \|f\|_\infty.$$

By the Arzelà–Ascoli theorem [Rud 2, Theorem A5] the set $u(S)$ is totally bounded. Therefore, u is a compact operator on X. The function k is called the *kernel* of the operator u, and u is called an *integral* operator.

A similar example is obtained if we define $v \in B(X)$ by $v(f)(s) = \int_0^s f(t)\,dt$. If $s, s' \in I$ and $f \in X$, then $|v(f)(s) - v(f)(s')| = |\int_s^{s'} f(t)\,dt| \leq |s - s'|\|f\|_\infty$. Hence, $v(S)$ is equicontinuous and pointwise-bounded, so by the Arzelà–Ascoli theorem again, $v(S)$ is totally bounded; that is, v is compact.

Observe that v has no eigenvalues (it will follow from Theorem 1.4.11 that v is quasinilpotent). That $v(f) = 0 \Rightarrow f = 0$ is elementary. Suppose then that $\lambda \in \mathbf{C} \setminus \{0\}$ and $f \in X$ and $v(f) = \lambda f$. Then $f(0) = 0$ and (by differentiation) $f'(t) = \mu f(t)$, where $\mu = 1/\lambda$. Consequently, $f(t) = f(0)e^{\mu t} = 0$ for all t, so $f = 0$.

The operator v is called the *Volterra integral* operator on X.

If X, Y are Banach spaces, we denote by $B(X,Y)$ the vector space of all bounded linear maps from X to Y. This is a Banach space when endowed with the operator norm. The set of all compact operators from X to Y is denoted by $K(X,Y)$.

The proof of the following is a routine exercise.

1.4.1. Theorem. *Let X and Y be Banach spaces and $u \in B(X,Y)$. Then the following conditions are equivalent:*

(1) *u is compact;*
(2) *For each bounded set S in X, the set $u(S)$ is relatively compact in Y;*
(3) *For each bounded sequence (x_n) in X, the sequence $(u(x_n))$ admits a subsequence that converges in Y.*

It follows easily from Theorem 1.4.1 that $K(X,Y)$ is a vector subspace of $B(X,Y)$. Also, if $X' \xrightarrow{v} X \xrightarrow{u} Y \xrightarrow{w} Y'$ are bounded linear maps between Banach spaces and u is compact, then wu and uv are compact. Hence $K(X) = K(X,X)$ is an ideal in $B(X)$.

1.4.2. Theorem. *If X is a Banach space, then $K(X) = B(X)$ if and only if X is finite-dimensional.*

Proof. If S denotes the closed unit ball of X, then $K(X) = B(X) \Leftrightarrow \mathrm{id}_X$ is compact $\Leftrightarrow S$ is compact $\Leftrightarrow X$ is finite-dimensional.   □

1.4.3. Theorem. *If X, Y are Banach spaces, then $K(X,Y)$ is a closed vector space of $B(X,Y)$.*

Proof. We show that if a sequence (u_n) in $K(X,Y)$ converges to an operator u in $B(X,Y)$, then u is compact. Let S denote the closed unit ball of X and let $\varepsilon > 0$. Choose an integer N such that $\|u_N - u\| < \varepsilon/3$. Since $u_N(S)$ is totally bounded, there are elements $x_1, \ldots, x_n \in S$, such that for each x in S, the inequality $\|u_N(x) - u_N(x_j)\| < \varepsilon/3$ holds for some index j. Hence,

$$\|u(x) - u(x_j)\| \leq \|u(x) - u_N(x)\| + \|u_N(x) - u_N(x_j)\| + \|u_N(x_j) - u(x_j)\|$$
$$< \varepsilon/3 + \varepsilon/3 + \varepsilon/3 = \varepsilon.$$

Thus, $u(S)$ is totally bounded, and therefore, $u \in K(X,Y)$.   □

Recall that a linear map $u: X \to Y$ is of *finite rank* if $u(X)$ is finite-dimensional and that $\mathrm{rank}(u) = \dim(u(X))$.

If X and Y are Banach spaces and $u \in B(X, Y)$ is of finite rank, then $u \in K(X, Y)$. This is immediate from the fact that the closed unit ball of the finite-dimensional space $u(X)$ is compact.

It follows from this remark and Theorem 1.4.3 that norm-limits of finite-rank operators are compact, and it is natural to ask whether the converse is true. This is the case for Hilbert spaces, as we shall see in the next chapter, but it is not true for arbitrary Banach spaces. P. Enflo [Enf] has given an example of a Banach space for which there are compact operators that are not norm-limits of finite-rank operators.

If $u \colon X \to Y$ is a bounded linear map between Banach spaces, we define its *transpose* $u^* \in B(Y^*, X^*)$ by $u^*(\tau) = \tau \circ u$.

1.4.4. Theorem. *Let X, Y be Banach spaces and let $u \in K(X, Y)$. Then $u^* \in K(Y^*, X^*)$.*

Proof. Let S be the closed unit ball of X and let $\varepsilon > 0$. Since $u(S)$ is totally bounded, there exist elements $x_1, \ldots, x_n$ in S, such that if $x \in S$, then $\|u(x) - u(x_i)\| < \varepsilon/3$ for some index i. Define $v \in B(Y^*, \mathbf{C}^n)$ by setting $v(\tau) = (\tau u(x_1), \ldots, \tau u(x_n))$. Since the rank of v is finite, v is compact, and therefore $v(T)$ is totally bounded, where T is the closed unit ball of Y^*. Hence, there exist functionals $\tau_1, \ldots, \tau_m$ in T, such that if $\tau \in T$, then $\|v(\tau) - v(\tau_j)\| < \varepsilon/3$ for some index j. Observe that

$$\|v(\tau) - v(\tau_j)\| = \max_{1 \le i \le n} |u^*(\tau)(x_i) - u^*(\tau_j)(x_i)|.$$

Now suppose that $x \in S$. Then $\|u(x) - u(x_i)\| < \varepsilon/3$ for some index i, and $|u^*(\tau)(x_i) - u^*(\tau_j)(x_i)| < \varepsilon/3$. Hence,

$$|u^*(\tau)(x) - u^*(\tau_j)(x)| \le |u^*(\tau)(x) - u^*(\tau)(x_i)| + |u^*(\tau)(x_i) - u^*(\tau_j)(x_i)|$$
$$+ |u^*(\tau_j)(x_i) - u^*(\tau_j)(x)|$$
$$\le \varepsilon/3 + \varepsilon/3 + \varepsilon/3 = \varepsilon.$$

It follows that $\|u^*(\tau) - u^*(\tau_j)\| \le \varepsilon$, so $u^*(T)$ is totally bounded and therefore u^* is compact. $\qquad\square$

A linear map $u \colon X \to Y$ between Banach spaces is *bounded below* if there is a positive number δ such that $\|u(x)\| \ge \delta \|x\|$ ($x \in X$). Note that in this case $u(X)$ is necessarily closed, for if $(u(x_n))$ is a Cauchy sequence in $u(X)$, then (x_n) is a Cauchy sequence in X and therefore converges to some element $x \in X$, because X is complete. Hence, the sequence $(u(x_n))$ converges to $u(x)$ by continuity of u. Thus, $u(X)$ is complete and therefore closed in Y.

Observe that every invertible linear map is bounded below, as is every isometric linear map.

It is easily checked that $u \colon X \to Y$ is not bounded below if and only if there is a sequence of unit vectors (x_n) in X such that $\lim_{n \to \infty} u(x_n) = 0$. These remarks will be used in the following theorem.

1.4.5. Theorem. *Let u be a compact operator on a Banach space X and suppose that $\lambda \in \mathbf{C} \setminus \{0\}$.*

(1) *The space $\ker(u - \lambda)$ is finite-dimensional.*
(2) *The space $(u - \lambda)(X)$ is closed and finite-codimensional in X. (In fact, the codimension of $(u - \lambda)(X)$ in X is the dimension of $\ker(u^* - \lambda)$.)*

Proof. Let $Z = \ker(u - \lambda)$. Then $u(Z) \subseteq Z$, and the restriction u_Z of u to Z is in $K(Z)$. Since $u_Z = \lambda \operatorname{id}_Z$ and $\lambda \neq 0$, the map id_Z is compact. Hence, Z is finite-dimensional by Theorem 1.4.2.

Because Z is finite-dimensional, there is a closed vector space Y in X such that $Z \oplus Y = X$. Observe that $(u - \lambda)X = (u - \lambda)Y$, so to show that $(u - \lambda)X$ is closed in X it suffices to show that the restriction $(u - \lambda)_Y : Y \to X$ is bounded below. Suppose otherwise, and we shall obtain a contradiction. There is a sequence (x_n) of unit vectors in Y such that $\lim_{n \to \infty} \|u(x_n) - \lambda x_n\| = 0$. Using the compactness of u and going to a subsequence if necessary, we may suppose that $(u(x_n))$ is convergent. It follows from the equation $x_n = \lambda^{-1}(u(x_n) - (u - \lambda)(x_n))$ that the sequence (x_n) is convergent, to x say, and, since Y is closed in X, it contains x. Obviously, $u(x) = \lambda x$, so $x \in Y \cap \ker(u - \lambda)$ and therefore $x = 0$. However, x is the limit of unit vectors and is therefore itself a unit vector, a contradiction. This shows that $(u - \lambda)_Y$ is bounded below.

Now let $W = X/(u - \lambda)(X)$. To show that $(u - \lambda)(X)$ is finite-codimensional in X, we have to show W is finite-dimensional, and we do this by showing W^* is finite-dimensional. Let $\pi : X \to W$ be the quotient map. It is clear that the image of π^* is contained in the kernel of $u^* - \lambda$. In fact these spaces are equal. For suppose that $\sigma \in \ker(u^* - \lambda)$. Then σ annihilates $(u - \lambda)(X)$ and therefore induces a bounded linear functional $\tau : W \to \mathbf{C}$ such that $\sigma = \tau \circ \pi = \pi^*(\tau)$. Since u^* is compact by Theorem 1.4.4, $\ker(u^* - \lambda)$ is finite-dimensional by the first part of this proof. Thus, π^* has finite-dimensional range, and clearly π^* is injective, so W^* is finite-dimensional, and therefore $\dim(W) = \dim(W^*) = \dim(\pi^*(W^*)) = \dim(\ker(u^* - \lambda))$. $\qquad\square$

If $u : X \to X$ is a linear map on a vector space X, then the sequence of spaces $(\ker(u^n))$ is clearly increasing. If $\ker(u^n) \neq \ker(u^{n+1})$ for all $n \in \mathbf{N}$, we say that u has *infinite ascent* and set $\operatorname{ascent}(u) = +\infty$. Otherwise we say u has *finite ascent* and we define $\operatorname{ascent}(u)$ to be the least p such that $\ker(u^p) = \ker(u^{p+1})$. In this case, $\ker(u^p) = \ker(u^n)$ for all $n \geq p$.

The sequence of spaces $(u^n(X))$ is decreasing. We say that u has *infinite descent*, and we set $\operatorname{descent}(u) = +\infty$, if $u^n(X) \neq u^{n+1}(X)$ for all $n \in \mathbf{N}$. Otherwise, we say that u has *finite descent* and we define $\operatorname{descent}(u)$ to be the least $p \in \mathbf{N}$ such that $u^p(X) = u^{p+1}(X)$. In this case $u^p(X) = u^n(X)$ for all $n \geq p$.

We recall now a theorem of F. Riesz from elementary functional analysis [Rud 2, Lemma 4.22]: If Y is a proper closed vector subspace of a normed vector space X and $\varepsilon > 0$, then there exists a unit vector $x \in X$ such that $\|x + Y\| > 1 - \varepsilon$. This simple result plays a key role in the theory of compact operators. (The result as stated here is a slight reformulation of Lemma 4.22 of [Rud 2].)

1.4.6. Theorem. *Let u be a compact operator on a Banach space X and suppose that $\lambda \in \mathbf{C} \setminus \{0\}$. Then $u - \lambda$ has finite ascent and descent.*

Proof. Suppose the ascent is infinite, and we deduce a contradiction. If $N_n = \ker(u - \lambda)^n$, then N_{n-1} is a proper subspace of N_n, and therefore, by the theorem of Riesz discussed earlier, there is a unit vector $x_n \in N_n$ such that $\|x_n + N_{n-1}\| \geq 1/2$. If $m < n$, then

$$u(x_n) - u(x_m) = \lambda x_n + (u - \lambda)(x_n) - (u - \lambda)(x_m) - \lambda x_m = \lambda x_n - z,$$

where $z \in N_{n-1}$. Hence, $\|u(x_n) - u(x_m)\| = \|\lambda x_n - z\| = |\lambda| \|x_n - \lambda^{-1} z\| \geq |\lambda|/2 > 0$. It follows that $(u(x_n))$ has no convergent subsequence, contradicting the compactness of u. Consequently, $\mathrm{ascent}(u) < +\infty$.

The proof that $u - \lambda$ has finite descent is completely analogous and is left as an exercise. $\qquad\square$

We shall have more to say about compact operators presently. One can give direct proofs of these later results, but the details are tedious and a little messy, whereas when one uses the homomorphism property of the Fredholm index, which we are now going to introduce, they drop out very nicely.

The index and the essential spectrum, which we shall also introduce, are indispensible items in the operator theorist's tool-kit. Nevertheless, many of the proofs in Fredholm theory are elementary (although often neither trivial nor obvious).

Let X, Y be Banach spaces and $u \in B(X, Y)$. We say u is *Fredholm* if $\ker(u)$ is finite-dimensional and $u(X)$ is finite-codimensional in Y. We define the *nullity* of u to be $\dim(\ker(u))$ and denote it by $\mathrm{nul}(u)$. The *defect* of u is the codimension of $u(X)$ in Y, and is denoted by $\mathrm{def}(u)$. The *index* of u is defined to be

$$\mathrm{ind}(u) = \mathrm{nul}(u) - \mathrm{def}(u).$$

The index is a very simple prototype of the application of algebraic topological methods to this subject. The connecting homomorphism in the K-theory of Banach algebras (to be introduced in Chapter 7) can be thought of as a generalised Fredholm index.

Note that because there is a finite-dimensional (and therefore closed) vector subspace Z of Y, such that $u(X) \oplus Z = Y$, it is a consequence of the following theorem that $u(X)$ is closed in Y.

1.4.7. Theorem. *Let X, Y be Banach spaces and $u \in B(X, Y)$. Suppose that there is a closed vector subspace Z of Y such that $u(X) \oplus Z = Y$. Then $u(X)$ is closed in Y.*

Proof. The bounded linear map

$$X/\ker(u) \to Y, \ x + \ker(u) \mapsto u(x),$$

has the same range as u and is injective, so we may suppose without loss of generality that u is injective.

The map

$$v: X \oplus Z \to Y, \ (x, z) \mapsto u(x) + z,$$

is a continuous linear isomorphism between Banach spaces, so by the open mapping theorem, v^{-1} is also continuous. If $x \in X$, then $\|x\| = \|v^{-1}u(x)\| \leq \|v^{-1}\|\|u(x)\|$, so $\|u(x)\| \geq \|v^{-1}\|^{-1}\|x\|$. Thus, u is bounded below, and therefore $u(X)$ is closed in Y. $\qquad\square$

The following theorem is a fundamental result of Fredholm theory.

1.4.8. Theorem. *Let $X \xrightarrow{u} Y \xrightarrow{v} Z$ be Fredholm linear maps between Banach spaces X, Y, Z. Then vu is Fredholm and*

$$\mathrm{ind}(vu) = \mathrm{ind}(v) + \mathrm{ind}(u).$$

Proof. Set $Y_2 = \ker(v) \cap u(X)$ and choose suitable closed vector subspaces Y_1, Y_3, Y_4 of Y, such that $u(X) = Y_2 \oplus Y_3$, $\ker(v) = Y_1 \oplus Y_2$, and $Y = Y_1 \oplus u(X) \oplus Y_4$. Note that Y_1, Y_2, Y_4 are finite-dimensional.

The map

$$\ker(vu) \to Y_2, \ x \mapsto u(x),$$

is surjective and it has the same kernel as u, so the kernel of vu is finite-dimensional and $\mathrm{nul}(vu) = \mathrm{nul}(u) + \dim(Y_2)$.

Since $v(Y) = v(Y_3) \oplus v(Y_4)$ and $v(Y_3) = vu(X)$, therefore $v(Y) = vu(X) \oplus v(Y_4)$. Choose a finite-dimensional vector subspace Z' of Z such that $v(Y) \oplus Z' = Z$, so $Z = vu(X) \oplus v(Y_4) \oplus Z'$. Because $v(Y_4) \oplus Z'$ is finite-dimensional, $vu(X)$ is finite-codimensional in Z. Therefore, vu is a Fredholm operator.

The map

$$Y_4 \to v(Y_4), \ y \mapsto v(y),$$

is a linear isomorphism, so $\dim(Y_4) = \dim(v(Y_4))$. Hence, $\mathrm{def}(vu) = \dim(Y_4) + \dim(Z') = \dim(Y_4) + \mathrm{def}(v)$. Consequently, $\mathrm{nul}(vu) + \mathrm{def}(u) + \mathrm{def}(v) = \mathrm{nul}(u) + \dim(Y_4) + \mathrm{nul}(v) + \mathrm{def}(v) = \mathrm{nul}(u) + \mathrm{nul}(v) + \mathrm{def}(vu)$, and therefore, $\mathrm{ind}(vu) = \mathrm{nul}(vu) - \mathrm{def}(vu) = \mathrm{nul}(u) + \mathrm{nul}(v) - \mathrm{def}(u) - \mathrm{def}(v) = \mathrm{ind}(u) + \mathrm{ind}(v)$. $\qquad\square$

We give an immediate easy application of the index:

1.4.9. Theorem. *Let u be a compact operator on a Banach space X, and let $\lambda \in \mathbf{C} \setminus \{0\}$.*

(1) The operator $u - \lambda$ is Fredholm of index zero.
(2) If p denotes the (finite) ascent of $u - \lambda$, then

$$X = \ker(u - \lambda)^p \oplus (u - \lambda)^p(X).$$

Proof. That $u - \lambda$ is Fredholm follows from Theorem 1.4.5, and the ascent and descent of $u - \lambda$ are finite by Theorem 1.4.6. If we suppose that m, n are integers greater than $\max\{\operatorname{ascent}(u - \lambda), \operatorname{descent}(u - \lambda)\}$, then we have $\operatorname{nul}(u - \lambda)^m = \operatorname{nul}(u - \lambda)^n$ and $\operatorname{def}(u - \lambda)^m = \operatorname{def}(u - \lambda)^n$, so $\operatorname{ind}((u - \lambda)^m) = \operatorname{ind}((u - \lambda)^n)$, and therefore $m \operatorname{ind}(u - \lambda) = n \operatorname{ind}(u - \lambda)$ by Theorem 1.4.8. It follows that $\operatorname{ind}(u - \lambda) = 0$. Thus, Condition (1) is proved.

If $x \in \ker(u - \lambda)^p \cap (u - \lambda)^p(X)$, then there is an element $y \in X$ such that $x = (u - \lambda)^p(y)$ and $(u - \lambda)^{2p}(y) = 0$. Since $\ker(u - \lambda)^p = \ker(u - \lambda)^{2p}$, it follows that $(u - \lambda)^p(y) = 0$; that is, $x = 0$. Moreover, since $\operatorname{nul}(u - \lambda)^p = \operatorname{def}(u - \lambda)^p$, because $\operatorname{ind}(u - \lambda)^p = 0$, it follows that $X = \ker(u - \lambda)^p \oplus (u - \lambda)^p(X)$. $\qquad\square$

1.4.10. Corollary (Fredholm Alternative). *The operator $u - \lambda$ is injective if and only if it is surjective.*

Proof. Since the index of $u - \lambda$ is zero, the nullity is zero if and only if the defect is zero; that is, $u - \lambda$ is injective if and only if it is surjective. $\quad\square$

1.4.2. Remark. If Y, Z are complementary vector subspaces of a vector space X, and u, v are linear maps on Y, Z, respectively, we denote by $u \oplus v$ the linear map on X given by

$$(u \oplus v)(y + z) = u(y) + v(z) \quad (y \in Y, \ z \in Z).$$

Clearly, $u \oplus v$ is invertible if and only if u and v are invertible.

If X is a Banach space and $w \in B(X)$, we write $\sigma(w)$ for $\sigma_{B(X)}(w)$. If Y, Z are closed complementary vector subspaces of X, and if $u \in B(Y)$, $v \in B(Z)$, $w \in B(X)$, and $w = u \oplus v$, then $\sigma(w) = \sigma(u) \cup \sigma(v)$, by the preceding observation.

1.4.11. Theorem. *Let u be a compact operator on a Banach space X. Then $\sigma(u)$ is countable, and each non-zero point of $\sigma(u)$ is an eigenvalue of u and an isolated point of $\sigma(u)$.*

Proof. If λ is a non-zero point of $\sigma(u)$, then by the Fredholm alternative, Corollary 1.4.10, $u - \lambda$ is not injective, and therefore λ is an eigenvalue of u. The operator $u - \lambda$ has finite ascent, p say, and by Theorem 1.4.9

we can write $X = Y \oplus Z$, where $Y = \ker(u - \lambda)^p$ and $Z = (u - \lambda)^p(X)$. The spaces Y, Z are closed and invariant for u (that is, $u(Y) \subseteq Y$ and $u(Z) \subseteq Z$). Hence, $u - \lambda = (u_Y - \lambda \operatorname{id}_Y) \oplus (u_Z - \lambda \operatorname{id}_Z)$, where u_Y, u_Z are the restrictions of u to Y, Z, respectively. Since $(u_Y - \lambda \operatorname{id}_Y)^p = 0$, the spectrum $\sigma(u_Y)$ is the singleton set $\{\lambda\}$. Also, the operator u_Z is compact and $\ker(u_Z - \lambda \operatorname{id}_Z)^p = 0$, so $(u_Z - \lambda \operatorname{id}_Z)^p$ is invertible (as it is injective and Fredholm of index zero), and therefore $u_Z - \lambda \operatorname{id}_Z$ is invertible. Hence, $\lambda \notin \sigma(u_Z)$. This implies that $\sigma(u) \setminus \{\lambda\} = \sigma(u_Z)$, so λ is an isolated point of $\sigma(u)$ because $\sigma(u_Z)$ is closed in $\sigma(u)$.

Countability of $\sigma(u)$ follows by elementary topology. $\square$

1.4.2. Example. Let us interpret our resullts now in terms of integral equations. Let $I = [0, 1]$ and suppose $k \in C(I^2)$. Consider the integral equation

$$\int_0^1 k(s, t) f(t) \, dt - \lambda f(s) = g(s).$$

Here λ is a non-zero scalar, $g \in C(I)$ is a known function, and $f \in C(I)$ is the unknown. If u is the compact integral operator corresponding to the kernel k, as in Example 1.4.1, then we can rewrite our equation as

$$(u - \lambda)(f) = g.$$

The non-zero spectrum of u is of the form $\{\lambda_n \mid 1 \leq n \leq N\}$, where N is an integer or ∞. If $\lambda \neq \lambda_n$ for all n, then the integral equation has a unique solution: $f = (u - \lambda)^{-1}(g)$. If on the other hand $\lambda = \lambda_n$ say, then the homogeneous equation

$$\int_0^1 k(s, t) f(t) \, dt - \lambda f(s) = 0$$

has a non-zero solution by the Fredholm alternative (Corollary 1.4.10), and by Theorem 1.4.5 the solution set is finite-dimensional.

Observe that if $N = \infty$, then $\lim_{n \to \infty} \lambda_n = 0$ by Theorem 1.4.11.

1.4.3. Example. One should not be misled by Theorem 1.4.11—the spectral behaviour of compact operators is not typical of all operators. To illustrate this, let H be a separable Hilbert space with an orthonormal basis $(e_n)_{n=1}^\infty$. If (λ_n) is a bounded sequence of scalars, define $u \in B(H)$ by setting $u(x) = \sum_{n=1}^\infty \lambda_n \alpha_n e_n$ when $x = \sum_{n=1}^\infty \alpha_n e_n$. We call u the *diagonal* operator with *diagonal* (λ_n) with respect to the basis (e_n). It is readily verified that $\|u\| = \sup_n |\lambda_n|$, and that u is invertible if and only if $\inf_n |\lambda_n| > 0$, and in this case u^{-1} is the diagonal operator with respect to (e_n) with diagonal (λ_n^{-1}). These observations imply that $\sigma(u)$ is the closure of the set $\{\lambda_n \mid n = 1, 2, \ldots\}$.

Suppose that a non-empty compact set K in $\mathbf{C}$ is given and choose a dense sequence (λ_n) in K. If u is the corresponding diagonal operator, then $\sigma(u) = K$. Thus, the spectrum is an arbitrary non-empty compact set in general.

We need to consider now a few elementary results (some of which extend what we said in Remark 1.4.2). These will be used immediately for the proof of Theorem 1.4.15.

A linear map $p\colon X \to X$ on a vector space X is *idempotent* if $p^2 = p$. In this case $X = p(X) \oplus \ker(p)$, since $\ker(p) = (1-p)(X)$. In the reverse direction, if $X = Y \oplus Z$, where Y and Z are vector subspaces of X, then there is a unique idempotent p on X such that $p(X) = Y$ and $\ker(p) = Z$. We call p the *projection of X on Y along Z*.

1.4.12. Theorem. *Let Y, Z be closed complementary vector subspaces of a Banach space X. Then the projection p of X on Y along Z is bounded.*

Proof. Let (x_n) be a sequence in X converging to 0 and suppose that $(p(x_n))$ converges to a point y of X. By the closed graph theorem, p will have been shown to be bounded if we show that $y = 0$. Now $y \in Y$, since $p(x_n) \in Y$ and Y is closed in X, and $-y \in Z$, since $x_n - p(x_n) \in Z$ and $-y = \lim_{n \to \infty}(x_n - p(x_n))$. Hence, $y \in Y \cap Z$, and therefore $y = 0$. $\square$

1.4.13. Corollary. *Let $v \in B(Y)$ and $w \in B(Z)$, and suppose that $u = v \oplus w$. Then $u \in B(X)$.*

Proof. We have to show u is continuous. Let p be the projection of X onto Y along Z. Suppose that (x_n) is a sequence in X converging to a point x. Then $(u(x_n)) = (vp(x_n) + w(1-p)(x_n))$ converges to $u(x) = vp(x) + w(1-p)(x)$ by continuity of v, w, and p. $\square$

1.4.14. Theorem. *Let $u\colon X \to Y$ be a linear map between normed vector spaces X and Y and suppose that X is finite-dimensional. Then u is bounded.*

Proof. Define a new norm on X by setting

$$\|x\|' = \max(\|x\|, \|u(x)\|).$$

Then $\|.\|'$ is equivalent to the original norm on X (because all norms on finite-dimensional vector spaces are equivalent). This shows that u is bounded. $\square$

If u is a linear map between vector spaces, a *pseudo-inverse* of u is a linear map $v\colon Y \to X$ such that $uvu = u$. Observe that uv and vu are idempotents, that $\ker(vu) = \ker(u)$, and that $uv(Y) = u(X)$.

1.4.15. Theorem. *Let X, Y be Banach spaces and let $u \in B(X, Y)$ be a Fredholm operator. Then u admits a pseudo-inverse v that is Fredholm and is such that $1 - uv$ and $1 - vu$ are of finite rank. Moreover, if $\mathrm{ind}(u) = 0$, we may choose v to be invertible.*

Proof. Choose a closed vector subspace X_1 of X such that $\ker(u) \oplus X_1 = X$ and a finite-dimensional vector subspace Y_1 of Y such that $u(X) \oplus Y_1 = Y$. The restriction

$$u_1 \colon X_1 \to u(X), \quad x \mapsto u(x),$$

is a continuous linear isomorphism, so by the open mapping theorem its inverse $v_1 \colon u(X) \to X_1$ is also continuous. Let $v \colon Y \to X$ be the linear map defined as follows: On $u(X)$, $v = v_1$; and on Y_1

$$v = \begin{cases} 0, & \text{if } \mathrm{ind}(u) \neq 0 \\ w, & \text{if } \mathrm{ind}(u) = 0, \end{cases}$$

where w is a linear isomorphism of Y_1 onto $\ker(u)$ (such an isomorphism exists if $\mathrm{ind}(u) = 0$). It is easily checked that v is continuous and that $uvu = u$. Now $\ker(v) \subseteq Y_1$ and $X_1 \subseteq v(Y)$, so v has finite nullity and defect and is therefore a Fredholm operator.

Because $(1 - vu)(X) = \ker(vu) = \ker(u)$, the idempotent $1 - vu$ is of finite rank. Also, $uv(Y) = u(X)$, so $(1 - uv)(Y) \subseteq (1 - uv)(Y_1)$, and therefore $1 - uv$ is of finite rank.

If we now suppose that $\mathrm{ind}(u) = 0$, then $v(Y) = X$, and $\ker(v) \subseteq Y_1$, so $\ker(v) = 0$. Hence, v is invertible. $\qquad\qquad\qquad\qquad\qquad\qquad \square$

The following characterisation of Fredholm operators is extremely useful. Note incidentally that all operators on a finite-dimensional vector space are Fredholm, so that in this case Fredholm theory is degenerate. Thus, we shall be interested only in infinite-dimensional spaces for these operators.

1.4.16. Theorem (Atkinson). *Let X be an infinite-dimensional Banach space and let $u \in B(X)$. Then u is Fredholm if and only if $u + K(X)$ is invertible in the quotient algebra $B(X)/K(X)$.*

Proof. Let π be the quotient homomorphism from $B(X)$ to $B(X)/K(X)$. If u is Fredholm, then by Theorem 1.4.15 there is a Fredholm operator v in $B(X)$ such that $1 - vu$ and $1 - uv$ are of finite rank and therefore compact. Hence, $0 = \pi(1 - uv) = 1 - \pi(u)\pi(v)$ and $0 = \pi(1 - vu) = 1 - \pi(v)\pi(u)$, so $\pi(u)$ is invertible in $B(X)/K(X)$.

Conversely, suppose $\pi(u)$ is invertible, with inverse $\pi(v)$. Then $uv = 1 + w_1$ and $vu = 1 + w_2$, where $w_1, w_2 \in K(X)$. Clearly $\ker(u) \subseteq \ker(1 + w_2)$, and $\ker(1 + w_2)$ is finite-dimensional by Theorem 1.4.5, so $\mathrm{nul}(u) < +\infty$. Also, $(1 + w_1)(X) = uv(X) \subseteq u(X)$, and $(1 + w_1)(X)$ has finite codimension in X by Theorem 1.4.5. Consequently, $\mathrm{def}(u) < \infty$. Thus, u is Fredholm.$\square$

1.4.17. Theorem. *Let X be an infinite-dimensional Banach space and let Φ denote the set of Fredholm operators on X. Then Φ is open in $B(X)$ and the index function*

$$\text{ind}: \Phi \to \mathbf{Z}, \ u \mapsto \text{ind}(u),$$

is continuous.

Proof. If, as usual, π denotes the quotient homomorphism from $B(X)$ to $B(X)/K(X)$, then $\Phi = \pi^{-1}(\text{Inv}(B(X)/K(X)))$ by the Atkinson characterisation, Theorem 1.4.16. By Theorem 1.2.3, the set of invertible elements in $B(X)/K(X)$ is open, and therefore Φ is open in $B(X)$ by continuity of π.

Let $u \in \Phi$ and choose $v \in \Phi$, a pseudo-inverse of u, such that $1 - vu$ and $1 - uv \in K(X)$ (this is possible by Theorem 1.4.15). Suppose that $w \in \Phi$ and $\|u - w\| < \|v\|^{-1}$. Then $\|uv - wv\| < 1$, so $s = 1 + wv - uv$ is invertible in $B(X)$ by Theorem 1.2.2. Now $u + wvu = uvu + su$, so $wvu = su$ (as $u = uvu$) and therefore $\text{ind}(w) + \text{ind}(v) + \text{ind}(u) = \text{ind}(s) + \text{ind}(u)$. But $\text{ind}(s) = 0$, because s is invertible, so $\text{ind}(w) = -\text{ind}(v)$. Thus, the index map is locally constant and therefore continuous. $\square$

1.4.18. Theorem. *Let X be an infinite-dimensional Banach space, and suppose that $w \in K(X)$, that $u \in B(X)$, and that u is Fredholm. Then*

$$\text{ind}(u + w) = \text{ind}(u).$$

Proof. By Theorem 1.4.17, the function

$$\alpha: [0, 1] \to \mathbf{Z}, \ t \mapsto \text{ind}(u + tw),$$

is continuous, and therefore $\alpha[0, 1]$ is connected in the discrete space $\mathbf{Z}$. Hence, $\alpha[0, 1]$ is a singleton set, so $\text{ind}(u) = \alpha(0) = \alpha(1) = \text{ind}(u + w)$. $\square$

1.4.3. *Remark.* Let u be a Fredholm operator on an infinite-dimensional Banach space X. If u is the sum of an invertible operator and a compact operator, then by Theorem 1.4.18 $\text{ind}(u) = 0$, since invertible operators are of course of index zero. The converse is also true; that is, if $\text{ind}(u) = 0$, then u is the sum of an invertible operator and a compact operator. For by Theorem 1.4.15 there is an invertible pseudo-inverse v for u, and if we denote by π the quotient map from $B(X)$ to $B(X)/K(X)$, the equation $u = uvu$ implies that $\pi(u) = \pi(u)\pi(v)\pi(u)$, and since $\pi(u)$ is invertible by Theorem 1.4.16, it follows that $\pi(u) = \pi(v^{-1})$. Hence, $u - v^{-1}$ is compact, so u is the sum of an invertible and a compact operator. Incidentally, it is easy to give examples of operators that are of index zero and not invertible (for instance, if p is a finite-rank non-zero idempotent, then $1 - p$ is Fredholm of index zero, and non-invertible).

Again suppose X to be an infinite-dimensional Banach space and suppose that $u \in B(X)$. We define the *essential spectrum* of u to be

$$\sigma_e(u) = \{\lambda \in \mathbf{C} \mid u - \lambda \text{ is not Fredholm}\}.$$

Let C denote the quotient algebra $B(X)/K(X)$. This algebra is called the *Calkin* algebra on X. If π is the quotient map from $B(X)$ to C, it is clear from the Atkinson characterisation (Theorem 1.4.16) that $\sigma_e(u) = \sigma_C(\pi u)$. Thus, $\sigma_e(u)$ is a non-empty compact set. Obviously, $\sigma_e(u) \subseteq \sigma(u)$.

1.4.4. Example. Suppose that H is a Hilbert space with an orthonormal basis $(e_n)_{n=1}^\infty$. The *unilateral shift* on this basis is the operator u in $B(H)$ such that $u(e_n) = e_{n+1}$ for all n. Observe that $\text{nul}(u) = 0$ and $\text{def}(u) = 1$, so u is a Fredholm operator and $\text{ind}(u) = -1$.

If instead we suppose that $(f_n)_{n \in \mathbf{Z}}$ is an orthonormal basis for H, the *bilateral shift* on this basis is the operator v such that $v(f_n) = f_{n+1}$ for all $n \in \mathbf{Z}$. This operator is invertible, so $\text{ind}(v) = 0$. Hence, u and v are not similar (two elements a, b of a unital algebra are *similar* if there is an invertible element c such that $a = c^{-1}bc$).

It follows from Theorem 1.4.16 that if $\pi \colon B(H) \to B(H)/K(H)$ is the quotient homomorphism, then $\pi(u)$ is invertible. It is natural to ask if one can write $\pi(u) = \pi(w)$ for some invertible operator w in $B(H)$. If this were the case, then $\text{ind}(u) = \text{ind}(w)$, since $u - w \in K(H)$. This is, however, impossible, since $\text{ind}(u) = -1$, and $\text{ind}(w) = 0$. An interesting consequence is that $\pi(u)$ provides an example of an invertible element that cannot be written as an exponential, for if $\pi(u) = e^w$ for some w in the Calkin algebra, then $w = \pi(w')$ for some $w' \in B(H)$, and therefore $\pi(u) = e^{\pi(w')} = \pi(e^{w'})$. But $e^{w'}$ is invertible in $B(H)$, which contradicts what we have just shown. Thus, $\pi(u)$ has no logarithm in the Calkin algebra.

We shall have more to say about shifts in the next chapter.

We shall see further examples and applications concerning compact and Fredholm operators in later chapters. We turn in Chapter 2 to the case where the algebras have involutions and the operators have adjoints. This is the self-adjoint theory, and it is in this setting that some of the deepest results concerning algebras and operators have been proved.

1. Exercises

1. Let $(A_\lambda)_{\lambda \in \Lambda}$ denote a family of Banach algebras. The *direct sum* $A = \oplus_\lambda A_\lambda$ is the set of all $(a_\lambda) \in \prod_\lambda A_\lambda$ such that $\|(a_\lambda)\| = \sup_\lambda \|a_\lambda\|$ is finite. Show that this is a Banach algebra under the pointwise-defined operations

$$(a_\lambda) + (b_\lambda) = (a_\lambda + b_\lambda)$$
$$\mu(a_\lambda) = (\mu a_\lambda)$$
$$(a_\lambda)(b_\lambda) = (a_\lambda b_\lambda),$$

and norm given by $(a_\lambda) \mapsto \|(a_\lambda)\|$. Show that A is unital or abelian if this is the case for all of the algebras A_λ.

The *restricted sum* $B = \oplus_\lambda^{c_0} A_\lambda$ is the set of all elements $(a_\lambda) \in A$ such that for each $\varepsilon > 0$ there exists a finite subset F of Λ for which $\|a_\lambda\| < \varepsilon$ if $\lambda \in \Lambda \setminus F$. Show that B is a closed ideal in A.

2. Let A be a Banach algebra and Ω a non-empty set. Denote by $\ell^\infty(\Omega, A)$ the set of all bounded maps f from Ω to A. Show that $\ell^\infty(\Omega, A)$ is a Banach algebra with the pointwise-defined operations and the sup-norm $\|f\| = \sup\{\|f(\omega)\| \mid \omega \in \Omega\}$. If Ω is a compact Hausdorff space, show that the set $C(\Omega, A)$ of all continuous functions from Ω to A is a closed subalgebra of $\ell^\infty(\Omega, A)$.

3. Give an example of a unital non-abelian Banach algebra A in which 0 and A are the only closed ideals.

4. Give an example of a non-modular maximal ideal in an abelian Banach algebra. (If A is the disc algebra, let $A_0 = \{f \in A \mid f(0) = 0\}$. Then A_0 is a closed subalgebra of A and admits an ideal of the type required.)

5. Let A be a unital abelian Banach algebra.
(a) Show that $\sigma(a+b) \subseteq \sigma(a) + \sigma(b)$ and $\sigma(ab) \subseteq \sigma(a)\sigma(b)$ for all $a, b \in A$. Show that this is not true for *all* Banach algebras.
(b) Show that if A contains an *idempotent* e (that is, $e = e^2$) other than 0 and 1, then $\Omega(A)$ is disconnected.
(c) Let $a_1, \ldots, a_n$ generate A as a Banach algebra. Show that $\Omega(A)$ is homeomorphic to a compact subset of $\mathbf{C}^n$. More precisely, set $\sigma(a_1, \ldots, a_n) = \{(\tau(a_1), \ldots, \tau(a_n)) \mid \tau \in \Omega(A)\}$. Show that the canonical map from $\Omega(A)$ to $\sigma(a_1, \ldots, a_n)$ is a homeomorphism.

6. Let A be a unital Banach algebra.
(a) If a is invertible in A, show that $\sigma(a^{-1}) = \{\lambda^{-1} \mid \lambda \in \sigma(a)\}$.
(b) For any element $a \in A$, show that $r(a^n) = (r(a))^n$.
(c) If A is abelian, show that the Gelfand representation is isometric if and only if $\|a^2\| = \|a\|^2$ for all $a \in A$.

7. Let A be a Banach algebra. Show that the spectral radius function $r\colon A \to \mathbf{R}$ is upper semi-continuous. (One can show that r is not in general continuous [Hal, Problem 104].)

8. Show that if B is a maximal abelian subalgebra of a unital Banach algebra A, then B is closed and contains the unit. Show that $\sigma_A(b) = \sigma_B(b)$ for all $b \in B$.

9. Let (Ω, μ) be a measure space. Show that the linear span of the idempotents is dense in $L^\infty(\Omega, \mu)$. Show that the spectrum of the Banach algebra $L^\infty(\Omega, \mu)$ is totally disconnected, by showing that if A is an arbitrary abelian Banach algebra in which the idempotents have dense linear span, its spectrum $\Omega(A)$ is totally disconnected.

10. Let $A = C^1[0,1]$, as in Example 1.2.6. Let $x \colon [0,1] \to \mathbf{C}$ be the inclusion. Show that x generates A as a Banach algebra. If $t \in [0,1]$, show that τ_t belongs to $\Omega(A)$, where τ_t is defined by $\tau_t(f) = f(t)$, and show that the map $[0,1] \to \Omega(A)$, $t \mapsto \tau_t$, is a homeomorphism. Deduce that $r(f) = \|f\|_\infty$ $(f \in A)$. Show that the Gelfand representation is not surjective for this example.

11. Let A be a unital Banach algebra and set

$$\zeta(a) = \inf_{\|b\|=1} \|ab\| \qquad (a \in A).$$

We say that an element a of A is a *left topological zero divisor* if there is a sequence of unit vectors (a_n) of A such that $\lim_{n\to\infty} aa_n = 0$. Equivalently, $\zeta(a) = 0$.

(a) Show that left topological zero divisors are not invertible.

(b) Show that $|\zeta(a) - \zeta(b)| \leq \|a - b\|$ for all $a, b \in A$. Hence, ζ is a continuous function.

(c) If a is a boundary point of the set $\mathrm{Inv}(A)$ in A, show that there is a sequence of invertible elements (v_n) converging to a such that $\lim_{n\to\infty} \|v_n^{-1}\|^{-1} = 0$. Using the continuity of ζ, deduce that $\zeta(a) = 0$. Thus, boundary points of $\mathrm{Inv}(A)$ are left topological zero divisors. In particular, if λ is a boundary point of the spectrum of an element a of A, then $\lambda - a$ is a left topological zero divisor.

(d) Let Ω be a compact Hausdorff space and let $A = C(\Omega)$. Show that in this case the topological zero divisors are precisely the non-invertible elements (if f is non-invertible, then 0 is a boundary point of the spectrum of $\bar{f}f$).

(e) Give an example of a unital Banach algebra and a non-invertible element that is not a left topological zero divisor.

12. A derivation on an algebra A is a linear map $d \colon A \to A$ such that $d(ab) = adb + d(a)b$. Show that the Leibnitz formula,

$$d^n(ab) = \sum_{r=0}^{n} \binom{n}{r} d^r(a) d^{n-r}(b) \qquad (n = 1, 2, \ldots),$$

holds.

13. Suppose that d is a bounded derivation on a unital Banach algebra A and $\lambda \in \mathbf{C} \setminus \{0\}$ such that $da = \lambda a$. Show that a is nilpotent, that is, that $a^n = 0$ for some positive integer n (use the boundedness of $\sigma(d)$).

14. Suppose that d is a bounded derivation on a unital Banach algebra A, and that $a \in A$ and $d^2a = 0$. Show that da is quasinilpotent. (Hint: Show that $d^{n+1}(a^n) = 0$ and hence, $d^n(a^n) = n!(da)^n$.) For $a \in A$, the map $b \mapsto [a, b] = ab - ba$ is a bounded derivation on A. Therefore, the Kleinecke–Shirokov theorem holds: If $[a, [a, b]] = 0$, then $[a, b]$ is quasinilpotent.

15. Let H be a Hilbert space with an orthonormal basis $(e_n)_{n=1}^{\infty}$, and let u be an operator in $B(H)$ diagonal with respect to (e_n) with diagonal the sequence (λ_n). Show that u is compact if and only if $\lim_{n \to \infty} \lambda_n = 0$.

16. Let X be a Banach space. If $p \in B(X)$ is a compact idempotent, show that its rank is finite.

17. Let $u \colon X \to Y$ be a compact operator between Banach spaces. Show that if the range of u is closed, then it is finite-dimensional. (Hint: Show that the well-defined operator

$$X/\ker(u) \to u(X), \quad x + \ker(u) \mapsto u(x),$$

is an invertible compact operator.)

18. Let X, Y be Banach spaces and suppose that $u \in B(X, Y)$ has compact transpose u^*. Show that u is compact using the fact that u^{**} is compact.

19. Let $u \colon X \to Y$ and $u' \colon X' \to Y'$ be bounded operators between Banach spaces. Show that the linear map

$$u \oplus u' \colon X \oplus X' \to Y \oplus Y', \quad (x, x') \mapsto (u(x), u'(x')),$$

is bounded with norm $\max\{\|u\|, \|u'\|\}$. Show that if u and u' are Fredholm operators, so is $u \oplus u'$, and $\mathrm{ind}(u \oplus u') = \mathrm{ind}(u) + \mathrm{ind}(u')$.

20. If X is an infinite-dimensional Banach space and $u \in B(X)$, show that

$$\bigcap_{v \in K(X)} \sigma(u + v) = \sigma(u) \setminus \{\lambda \in \mathbf{C} \mid u - \lambda \text{ is Fredholm of index zero}\}.$$

1. Addenda

Let G be a locally compact abelian group. If μ is Haar measure on G, we write $L^1(G)$ for $L^1(G, \mu)$. If $f, g \in L^1(G)$, then there is an element $f * g \in L^1(G)$ such that

$$(f * g)(x) = \int f(x - y)g(y)\, d\mu(y)$$

for almost all x in G. The product $f * g$ is the *convolution* of f and g. Under the multiplication operation given by $(f, g) \mapsto f * g$, $L^1(G)$ is an abelian Banach algebra, called the *group algebra* of G. It has a unit if and only if G is discrete.

Let $\hat{G}$ be the dual group of G, that is, the set of continuous homomorphisms γ from G to $\mathbf{T}$. This is endowed with a suitable topology making it a locally compact group. For $f \in L^1(G)$ and $\gamma \in \hat{G}$, define

$$\hat{f}(\gamma) = \int f(x)\overline{\gamma(x)}\, d\mu(x).$$

Then the function

$$\tau_\gamma \colon L^1(G) \to \mathbf{C}, \ f \mapsto \hat{f}(\gamma),$$

is a character on $L^1(G)$, and all characters on $L^1(G)$ are of this form. The map

$$\hat{G} \to \Omega(L^1(G)), \ \gamma \mapsto \tau_\gamma,$$

is a homeomorphism.

Reference: [Cnw 2].

Suppose u is a non-zero compact operator on an infinite-dimensional Banach space X. Then there is a non-trivial closed vector subspace Y of X such that $v(Y) \subseteq Y$ for all operators $v \in B(X)$ commuting with u. This is a special case of a theorem of Lomonosov [TL, Theorem 7.15].

Let X be an infinite-dimensional Banach space. An operator $u \in B(X)$ is a *Riesz* operator if its essential spectrum is the zeroset, $\sigma_e(u) = \{0\}$. The spectral theory of these operators is similar to that of compact operators. Obviously, the sum of a quasinilpotent operator and a compact operator is a Riesz operator. The converse is true for Hilbert spaces and is also known for some other Banach spaces.

For certain Banach algebras and certain of their closed ideals, one can develop a Fredholm theory that is analogous to the classical Fredholm theory relative to $B(X)$ and $K(X)$ for Banach spaces X.

References: [BMSW], [Wes].

CHAPTER 2

C*-Algebras and
Hilbert Space Operators

In this chapter we commence our study of C*-algebras and of operators on Hilbert spaces. Hilbert spaces are very well-behaved compared with general Banach spaces, and the same is even more true of C*-algebras as compared with general Banach algebras. The main results of this chapter are a theorem of Gelfand, which asserts that (up to isomorphism) all abelian C*-algebras are of the form $C_0(\Omega)$, where Ω is a locally compact Hausdorff space, and the spectral theorem. This theorem enables us to "synthesize" a normal operator from linear combinations of projections where the coefficients lie in the spectrum. It is a very powerful result.

2.1. C*-Algebras

We begin by defining a number of concepts that make sense in any algebra with an involution.

An *involution* on an algebra A is a conjugate-linear map $a \mapsto a^*$ on A, such that $a^{**} = a$ and $(ab)^* = b^*a^*$ for all $a, b \in A$. The pair $(A, *)$ is called an *involutive* algebra, or a **-algebra*. If S is a subset of A, we set $S^* = \{a^* \mid a \in S\}$, and if $S^* = S$ we say S is *self-adjoint*. A self-adjoint subalgebra B of A is a **-subalgebra* of A and is a *-algebra when endowed with the involution got by restriction. Because the intersection of a family of *-subalgebras of A is itself one, there is for every subset S of A a smallest *-algebra B of A containing S, called the *-algebra *generated* by S.

If I is self-adjoint ideal of A, then the quotient algebra A/I is a *-algebra with the involution given by $(a + I)^* = a^* + I$ $(a \in A)$.

We define an involution on $\tilde{A}$ extending that of A by setting $(a, \lambda)^* = (a^*, \bar{\lambda})$. Thus, $\tilde{A}$ is a *-algebra, and A is a self-adjoint ideal in $\tilde{A}$.

An element a in A is *self-adjoint* or *hermitian* if $a = a^*$. For each $a \in A$ there exist unique hermitian elements $b, c \in A$ such that $a = b + ic$ ($b = \frac{1}{2}(a + a^*)$ and $c = \frac{1}{2i}(a - a^*)$). The elements a^*a and aa^* are hermitian. The set of hermitian elements of A is denoted by A_{sa}.

We say a is *normal* if $a^*a = aa^*$. In this case the $*$-algebra it generates is abelian and is in fact the linear span of all $a^m a^{*n}$, where $m, n \in \mathbf{N}$ and $n + m > 0$.

An element p is a *projection* if $p = p^* = p^2$.

If A is unital, then $1^* = 1$ ($1^* = (11^*)^* = 1$). If $a \in \text{Inv}(A)$, then $(a^*)^{-1} = (a^{-1})^*$. Hence, for any $a \in A$,

$$\sigma(a^*) = \sigma(a)^* = \{\bar{\lambda} \in \mathbf{C} \mid \lambda \in \sigma(a)\}.$$

An element u in A is a *unitary* if $u^*u = uu^* = 1$. If $u^*u = 1$, then u is an *isometry*, and if $uu^* = 1$, then u is a *co-isometry*.

If $\varphi \colon A \to B$ is a homomorphism of $*$-algebras A and B and φ preserves adjoints, that is, $\varphi(a^*) = (\varphi(a))^*$ ($a \in A$), then φ is a *$*$-homomorphism*. If in addition φ is a bijection, it is a *$*$-isomorphism*. If $\varphi \colon A \to B$ is a $*$-homomorphism, then $\ker(\varphi)$ is a self-adjoint ideal in A and $\varphi(A)$ is a $*$-subalgebra of B.

An *automorphism* of a $*$-algebra A is a $*$-isomorphism $\varphi \colon A \to A$. If A is unital and u is a unitary in A, then

$$\text{Ad}\, u \colon A \to A, \quad a \mapsto uau^*,$$

is an automorphism of A. Such automorphisms are called *inner*. We say elements a, b of A are *unitarily equivalent* if there exists a unitary u of A such that $b = uau^*$. Since the unitaries form a group, this is an equivalence relation on A. Note that $\sigma(a) = \sigma(b)$ if a and b are unitarily equivalent.

A *Banach $*$-algebra* is a $*$-algebra A together with a complete submultiplicative norm such that $\|a^*\| = \|a\|$ ($a \in A$). If, in addition, A has a unit such that $\|1\| = 1$, we call A a *unital Banach $*$-algebra*.

A *C*-algebra* is a Banach $*$-algebra such that

$$\|a^*a\| = \|a\|^2 \qquad (a \in A). \tag{1}$$

A closed $*$-subalgebra of a C*-algebra is obviously also a C*-algebra. We shall therefore call a closed $*$-subalgebra of a C*-algebra a *C*-subalgebra*.

If a C*-algebra has a unit 1, then automatically $\|1\| = 1$, because $\|1\| = \|1^*1\| = \|1\|^2$. Similarly, if p is a non-zero projection, then $\|p\| = 1$.

If u is a unitary of A, then $\|u\| = 1$, since $\|u\|^2 = \|u^*u\| = \|1\| = 1$. Hence, $\sigma(u) \subseteq \mathbf{T}$, for if $\lambda \in \sigma(u)$, then $\lambda^{-1} \in \sigma(u^{-1}) = \sigma(u^*)$, so $|\lambda|$ and $|\lambda^{-1}| \leq 1$; that is, $|\lambda| = 1$.

The seemingly mild requirement on a C*-algebra in Eq. (1) is in fact very strong—far more is known about the nature and structure of these algebras than perhaps of any other non-trivial class of algebras. Because of the existence of the involution, C*-algebra theory can be thought of as "infinite-dimensional real analysis." For instance, the study of linear functionals on C*-algebras (and of traces, *cf.* Section 6.2) is "non-commutative measure theory."

2.1.1. Example. The scalar field C is a unital C*-algebra with involution given by complex conjugation $\lambda \mapsto \bar{\lambda}$.

2.1.2. Example. If Ω is a locally compact Hausdorff space, then $C_0(\Omega)$ is a C*-algebra with involution $f \mapsto \bar{f}$.
   Similarly, all of the following algebras are C*-algebras with involution given by $f \mapsto \bar{f}$:
(a) $\ell^\infty(S)$ where S is a set;
(b) $L^\infty(\Omega, \mu)$ where (Ω, μ) is a measure space;
(c) $C_b(\Omega)$ where Ω is a topological space;
(d) $B_\infty(\Omega)$ where Ω is a measurable space.

2.1.3. Example. If H is a Hilbert space, then $B(H)$ is a C*-algebra. We shall see that every C*-algebra can be thought of as a C*-subalgebra of some $B(H)$ (Gelfand–Naimark theorem). We defer to Section 2.3 a fuller consideration of this example.

2.1.4. Example. If $(A_\lambda)_{\lambda \in \Lambda}$ is a family of C*-algebras, then the direct sum $\oplus_\lambda A_\lambda$ is a C*-algebra with the pointwise-defined involution, and the restricted sum $\oplus_\lambda^{co} A_\lambda$ is a closed self-adjoint ideal (*cf.* Exercise 1.1).

2.1.5. Example. If Ω is a non-empty set and A is a C*-algebra, then $\ell^\infty(\Omega, A)$ is a C*-algebra with the pointwise-defined involution. This of course generalises Example 2.1.2 (a). If Ω is a locally compact Hausdorff space, we say a continuous function $f \colon \Omega \to A$ *vanishes at infinity* if, for each $\varepsilon > 0$, the set $\{\omega \in \Omega \mid \|f(\omega)\| \geq \varepsilon\}$ is compact. Denote by $C_0(\Omega, A)$ the set of all such functions. This is a C*-subalgebra of $\ell^\infty(\Omega, A)$.

   The following easy result has a surprising and important corollary:

2.1.1. Theorem. *If a is a self-adjoint element of a C*-algebra A, then $r(a) = \|a\|$.*

Proof. Clearly, $\|a^2\| = \|a\|^2$, and therefore by induction $\|a^{2^n}\| = \|a\|^{2^n}$, so $r(a) = \lim_{n \to \infty} \|a^n\|^{1/n} = \lim_{n \to \infty} \|a^{2^n}\|^{1/2^n} = \|a\|$. $\square$

2.1.2. Corollary. *There is at most one norm on a $*$-algebra making it a C*-algebra.*

Proof. If $\|.\|_1$ and $\|.\|_2$ are norms on a *-algebra A making it a C*-algebra, then

$$\|a\|_j^2 = \|a^*a\|_j = r(a^*a) = \sup_{\lambda \in \sigma(a^*a)} |\lambda| \qquad (j=1,2),$$

so $\|a\|_1 = \|a\|_2$. □

2.1.3. Lemma. *Let A be a Banach algebra endowed with an involution such that $\|a\|^2 \leq \|a^*a\|$ ($a \in A$). Then A is a C*-algebra.*

Proof. The inequalities $\|a\|^2 \leq \|a^*a\| \leq \|a^*\|\|a\|$ imply that $\|a\| \leq \|a^*\|$ for all a. Hence, $\|a\| = \|a^*\|$, and therefore $\|a\|^2 = \|a^*a\|$. □

We associate to each C*-algebra A a certain unital C*-algebra $M(A)$ which contains A as an ideal. This algebra is of great importance in more advanced aspects of the theory, especially in certain approaches to K-theory.

A *double centraliser* for a C*-algebra A is a pair (L, R) of bounded linear maps on A, such that for all $a, b \in A$

$$L(ab) = L(a)b, \quad R(ab) = aR(b) \quad \text{and} \quad R(a)b = aL(b).$$

For example, if $c \in A$ and L_c, R_c are the linear maps on A defined by $L_c(a) = ca$ and $R_c(a) = ac$, then (L_c, R_c) is a double centraliser on A. It is easily checked that for all $c \in A$

$$\|c\| = \sup_{\|b\| \leq 1} \|cb\| = \sup_{\|b\| \leq 1} \|bc\|,$$

and therefore $\|L_c\| = \|R_c\| = \|c\|$.

2.1.4. Lemma. *If (L, R) is a double centraliser on a C*-algebra A, then $\|L\| = \|R\|$.*

Proof. Since $\|aL(b)\| = \|R(a)b\| \leq \|R\|\|a\|\|b\|$, we have

$$\|L(b)\| = \sup_{\|a\| \leq 1} \|aL(b)\| \leq \|R\|\|b\|,$$

and therefore $\|L\| \leq \|R\|$. Also, $\|R(a)b\| = \|aL(b)\| \leq \|L\|\|a\|\|b\|$ implies

$$\|R(a)\| = \sup_{\|b\| \leq 1} \|R(a)b\| \leq \|L\|\|a\|,$$

and therefore $\|R\| \leq \|L\|$. Thus, $\|L\| = \|R\|$. □

If A is a C*-algebra, we denote the set of its double centralisers by $M(A)$. We define the norm of the double centraliser (L, R) to be $\|L\| = \|R\|$. It is easy to check $M(A)$ is a closed vector subspace of $B(A) \oplus B(A)$.

If (L_1, R_1) and $(L_2, R_2) \in M(A)$, we define their product to be

$$(L_1, R_1)(L_2, R_2) = (L_1 L_2, R_2 R_1).$$

Straightforward computations show that this product is again a double centraliser of A and that $M(A)$ is an algebra under this multiplication.

If $L \colon A \to A$, define $L^* \colon A \to A$ by setting $L^*(a) = (L(a^*))^*$. Then L^* is linear and the map $L \mapsto L^*$ is an isometric conjugate-linear map from $B(A)$ to itself such that $L^{**} = L$ and $(L_1 L_2)^* = L_1^* L_2^*$. If (L, R) is a double centraliser on A, so is $(L, R)^* = (R^*, L^*)$. It is easily verified that the map $(L, R) \mapsto (L, R)^*$ is an involution on $M(A)$.

2.1.5. Theorem. *If A is a C*-algebra, then $M(A)$ is a C*-algebra under the multiplication, involution, and norm defined above.*

Proof. The only thing that is not completely straightforward that has to be checked is that if $T = (L, R)$ is a double centraliser, then $\|T^*T\| = \|T\|^2$. If $\|a\| \leq 1$, then $\|L(a)\|^2 = \|(L(a))^* L(a)\| = \|L^*(a^*)L(a)\| = \|a^* R^* L(a)\| \leq \|R^* L\| = \|T^*T\|$, so

$$\|T\|^2 = \sup_{\|a\| \leq 1} \|L(a)\|^2 \leq \|T^*T\| \leq \|T\|^2,$$

and therefore $\|T^*T\| = \|T\|^2$. $\square$

The algebra $M(A)$ is called *multiplier algebra* of A.

The map
$$A \to M(A), \quad a \mapsto (L_a, R_a),$$

is an isometric *-homomorphism, and therefore we can, and do, identify A as a C*-subalgebra of $M(A)$. In fact A is an ideal of $M(A)$. Note that $M(A)$ is unital (the double centraliser $(\mathrm{id}_A, \mathrm{id}_A)$ is the unit), so $A = M(A)$ if and only if A is unital.

We have already seen in Chapter 1 that for every Banach algebra A, its unitisation $\tilde{A}$ is a Banach algebra with the norm $\|(a, \lambda)\| = \|a\| + |\lambda|$. If A is a Banach *-algebra, then so is $\tilde{A}$ with this norm. However, if A is a C*-algebra, there is a problem here, since this norm does not make $\tilde{A}$ a C*-algebra in general. For instance, if $A = \mathbf{C}$ and $(a, \lambda) = (-2, 1)$, we have $\|(a, \lambda)\|^2 = 9$, but $\|(a, \lambda)^*(a, \lambda)\| = 1$.

We can, however, endow $\tilde{A}$ with a norm making it a C*-algebra:

2.1.6. Theorem. *If A is a C*-algebra, then there is a (necessarily unique) norm on its unitisation $\tilde{A}$ making it into a C*-algebra, and extending the norm of A.*

Proof. Uniqueness of the norm is given by Corollary 2.1.2. The proof of existence falls into two cases, depending on whether A is unital or non-unital.

Suppose first that A has a unit e. Then the map φ from $\tilde{A}$ to the direct sum of the C*-algebras A and $\mathbf{C}$ defined by $\varphi(a, \lambda) = (a + \lambda e, \lambda)$ is a *-isomorphism. Hence, one gets a norm on $\tilde{A}$ making it a C*-algebra by setting $\|(a, \lambda)\| = \|\varphi(a, \lambda)\|$.

Now suppose A has no unit. If 1 is the unit of $M(A)$, then $A \cap \mathbf{C}1 = 0$. The map φ from $\tilde{A}$ onto the C*-subalgebra $A \oplus \mathbf{C}1$ of $M(A)$ defined by setting $\varphi(a, \lambda) = a + \lambda 1$ is a *-isomorphism, so we get a norm on $\tilde{A}$ making it a C*-algebra by setting $\|(a, \lambda)\| = \|\varphi(a, \lambda)\|$.   □

If A is a C*-algebra, we shall always understand the norm of $\tilde{A}$ to be the one making it a C*-algebra.

Note that when A is non-unital, $M(A)$ is in general very much bigger than $\tilde{A}$. For instance, it is shown in Section 3.1 that if $A = C_0(\Omega)$, where Ω is a locally compact Hausdorff space, then $M(A) = C_b(\Omega)$.

If $\varphi: A \to B$ is a *-homomorphism between *-algebras A and B, then it extends uniquely to a unital *-homomorphism $\tilde{\varphi}: \tilde{A} \to \tilde{B}$.

2.1.7. Theorem. *A *-homomorphism $\varphi: A \to B$ from a Banach *-algebra A to a C*-algebra B is necessarily norm-decreasing.*

Proof. We may suppose that A, B, and φ are unital (by going to $\tilde{A}$, $\tilde{B}$, and $\tilde{\varphi}$ if necessary). If $a \in A$, then $\sigma(\varphi a) \subseteq \sigma(a)$, so $\|\varphi a\|^2 = \|\varphi(a)^* \varphi(a)\| = \|\varphi(a^* a)\| = r(\varphi(a^* a)) \leq r(a^* a) \leq \|a^* a\| \leq \|a\|^2$. Hence, $\|\varphi(a)\| \leq \|a\|$.   □

2.1.8. Theorem. *If a is a hermitian element of a C*-algebra A, then $\sigma(a) \subseteq \mathbf{R}$.*

Proof. We may suppose that A is unital. Since e^{ia} is unitary, $\sigma(e^{ia}) \subseteq \mathbf{T}$. If $\lambda \in \sigma(a)$ and $b = \sum_{n=1}^{\infty} i^n (a - \lambda)^{n-1} / n!$ then $e^{ia} - e^{i\lambda} = (e^{i(a-\lambda)} - 1)e^{i\lambda} = (a - \lambda)be^{i\lambda}$. Since b commutes with a, and since $a - \lambda$ is non-invertible, $e^{ia} - e^{i\lambda}$ is non-invertible. Hence, $e^{i\lambda} \in \mathbf{T}$, and therefore $\lambda \in \mathbf{R}$. Thus, $\sigma(a) \subseteq \mathbf{R}$.   □

2.1.9. Theorem. *If τ is a character on a C*-algebra A, then it preserves adjoints.*

Proof. If $a \in A$, then $a = b + ic$ where b, c are hermitian elements of A. The numbers $\tau(b)$ and $\tau(c)$ are real because they are in $\sigma(b)$ and $\sigma(c)$ respectively, so $\tau(a^*) = \tau(b - ic) = \tau(b) - i\tau(c) = (\tau(b) + i\tau(c))^- = \tau(a)^-$. □

The character space of a unital abelian Banach algebra is non-empty, so this is true in particular for unital abelian C*-algebras. However, there are non-unital, non-zero, abelian Banach algebras for which the character space is empty. Fortunately, this cannot happen in the case of C*-algebras. Let A be a non-unital, non-zero, abelian C*-algebra. Then A contains a non-zero hermitian element, a say. Since $r(a) = \|a\|$ by Theorem 2.1.1, it follows that there is a character τ on $\tilde{A}$ such that $|\tau(a)| = \|a\| \neq 0$. Hence, the restriction of τ to A is a non-zero homomorphism from A to $\mathbf{C}$, that is, a character on A.

We shall now completely determine the abelian C*-algebras. This result can be thought of as a preliminary form of the spectral theorem. It allows us to construct the functional calculus, a very useful tool in the analysis of non-abelian C*-algebras.

2.1.10. Theorem (Gelfand). *If A is a non-zero abelian C*-algebra, then the Gelfand representation*

$$\varphi \colon A \to C_0(\Omega(A)), \quad a \mapsto \hat{a},$$

is an isometric $*$-isomorphism.

Proof. That φ is a norm-decreasing homomorphism, such that $\|\varphi(a)\| = r(a)$, is given by Theorem 1.3.6. If $\tau \in \Omega(A)$, then $\varphi(a^*)(\tau) = \tau(a^*) = \tau(a)^- = \varphi(a)^*(\tau)$, so φ is a $*$-homomorphism. Moreover, φ is isometric, since $\|\varphi(a)\|^2 = \|\varphi(a)^*\varphi(a)\| = \|\varphi(a^*a)\| = r(a^*a) = \|a^*a\| = \|a\|^2$. Clearly, then, $\varphi(A)$ is a closed $*$-subalgebra of $C_0(\Omega)$ separating the points of $\Omega(A)$, and having the property that for any $\tau \in \Omega(A)$ there is an element $a \in A$ such that $\varphi(a)(\tau) \neq 0$. The Stone–Weierstrass theorem implies, therefore, that $\varphi(A) = C_0(\Omega(A))$. $\square$

Let S be a subset of a C*-algebra A. The C*-algebra *generated* by S is the smallest C*-subalgebra of A containing S. If $S = \{a\}$, we denote by $C^*(a)$ the C*-subalgebra generated by S. If a is a normal, then $C^*(a)$ is abelian. Similarly, if A is unital and a normal, then the C*-subalgebra generated by 1 and a is abelian.

Observe that $r(a) = \|a\|$ if a is a normal element of a C*-algebra (apply Theorem 2.1.10 to $C^*(a)$).

The following result is important.

2.1.11. Theorem. *Let B be a C*-subalgebra of a unital C*-algebra A containing the unit of A. Then*

$$\sigma_B(b) = \sigma_A(b) \qquad (b \in B).$$

Proof. First suppose that b is a hermitian element of B. Since in this case $\sigma_A(b)$ is contained in $\mathbf{R}$, it has no holes, and therefore by Theorem 1.2.8, $\sigma_A(b) = \sigma_B(b)$. Therefore, b is invertible in B if and only if it is invertible in A.

Now suppose that b is an arbitrary element of B, that is invertible in A, so there is an element $a \in A$ such that $ba = ab = 1$. Then $a^*b^* = b^*a^* = 1$, so $bb^*a^*a = 1 \Rightarrow bb^*$ is invertible in A and therefore in B. Hence, there is an element $c \in B$ such that $bb^*c = 1$. Consequently, $b^*c = a$, so $a \in B$, which implies that b is invertible in B. Thus, for any element of B, its invertibility in A is equivalent to its invertibility in B. The theorem follows. □

If A is a unital C*-algebra and $a \in A_{sa}$, then e^{ia} is a unitary, but not all unitaries are of this form—it will be seen later that the Calkin algebra on a Hilbert space is a C*-algebra and the image of the unilateral shift in this algebra provides an example of a unitary that has no logarithm (cf. Example 1.4.4). Using Theorem 2.1.10, we can give some useful conditions that ensure a unitary *does* have a logarithm.

2.1.12. Theorem. Let u be a unitary in a unital C*-algebra A. If $\sigma(u) \neq \mathbf{T}$, then there exists $a \in A_{sa}$ such that $u = e^{ia}$.

(If $\|1 - u\| < 2$, then $\sigma(u) \neq \mathbf{T}$.)

Proof. By replacing u by λu for some $\lambda \in \mathbf{T}$ if necessary, we may suppose that $-1 \notin \sigma(u)$. Since u is normal, we may also suppose that A is abelian (replacing A by the C*-subalgebra generated by 1 and u if need be). Let $\varphi \colon A \to C(\Omega)$ be the Gelfand representation, let $f = \varphi(u)$, and as usual denote by $ln \colon \mathbf{C} \setminus (-\infty, 0] \to \mathbf{C}$ the principal branch of the logarithm function. Then $g = ln \circ f$ is a well-defined element of $C_0(\Omega)$, and $e^g = f$. Since $|f(\omega)| = 1$ for all $\omega \in \Omega$, the real part of g vanishes, so $g = ih$ where $h = \bar{h} \in C_0(\Omega)$. Let $a = \varphi^{-1}(h)$. Then $a \in A_{sa}$ and $u = e^{ia}$ because $\varphi(u) = e^{ih} = e^{\varphi(ia)} = \varphi(e^{ia})$.

The parenthetical observation in the statement of the theorem follows from the equations

$$\|1 - u\| = r(1 - u) = \sup\{|1 - \lambda| \mid \lambda \in \sigma(u)\},$$

which imply that $-1 \notin \sigma(u)$ when $\|1 - u\| < 2$. □

We are now going to set up the functional calculus, for which we need to make two easy observations:

If $\theta \colon \Omega \to \Omega'$ is a continuous map between compact Hausdorff spaces Ω and Ω', then the *transpose* map

$$\theta^t \colon C(\Omega') \to C(\Omega), \quad f \mapsto f\theta,$$

is a unital *-homomorphism. Moreover, if θ is a homeomorphism, then θ^t is a *-isomorphism.

Our second observation is that a *-isomorphism of C*-algebras is necessarily isometric. This is an immediate consequence of Theorem 2.1.7.

2.1.13. Theorem. *Let a be a normal element of a unital C*-algebra A, and suppose that z is the inclusion map of $\sigma(a)$ in $\mathbf{C}$. Then there is a unique unital $*$-homomorphism $\varphi\colon C(\sigma(a)) \to A$ such that $\varphi(z) = a$. Moreover, φ is isometric and $\mathrm{im}(\varphi)$ is the C*-subalgebra of A generated by 1 and a.*

Proof. Denote by B the (abelian) C*-algebra generated by 1 and a, and let $\psi\colon B \to C(\Omega(B))$ be the Gelfand representation. Then ψ is a $*$-isomorphism by Theorem 2.1.10, and so is $\hat{a}^t\colon C(\sigma(a)) \to C(\Omega(B))$, since $\hat{a}\colon \Omega(B) \to \sigma(a)$ is a homeomorphism. Let $\varphi\colon C(\sigma(a)) \to A$ be the composition $\psi^{-1}\hat{a}^t$, so φ is a $*$-homomorphism. Then $\varphi(z) = a$, since $\varphi(z) = \psi^{-1}(\hat{a}^t(z)) = \psi^{-1}(\hat{a}) = a$, and obviously φ is unital. From the Stone–Weierstrass theorem, we know that $C(\sigma(a))$ is generated by 1 and z; φ is therefore the unique unital $*$-homomorphism from $C(\sigma(a))$ to A such that $\varphi(z) = 1$.

It is clear that φ is isometric and $\mathrm{im}(\varphi) = B$. □

As in Theorem 2.1.13, let a be a normal element of a unital C*-algebra A, and let z be the inclusion map of $C(\sigma(a))$ in $\mathbf{C}$. We call the unique unital $*$-homomorphism $\varphi\colon C(\sigma(a)) \to A$ such that $\varphi(z) = a$ the *functional calculus* at a. If p is a polynomial, then $\varphi(p) = p(a)$, so for $f \in C(\sigma(a))$ we may write $f(a)$ for $\varphi(a)$. Note that $f(a)$ is normal.

Let B be the image of φ, so B is the C*-algebra generated by 1 and a. If $\tau \in \Omega(B)$, then $f(\tau(a)) = \tau(f(a))$, since the maps $f \mapsto f(\tau(a))$ and $f \mapsto \tau(f(a))$ from $C(\sigma(a))$ to $\mathbf{C}$ are $*$-homomorphisms agreeing on the generators 1 and z and hence are equal.

2.1.14. Theorem (Spectral Mapping). *Let a be a normal element of a unital C*-algebra A, and let $f \in C(\sigma(a))$. Then*

$$\sigma(f(a)) = f(\sigma(a)).$$

Moreover, if $g \in C(\sigma(f(a)))$, then

$$(g \circ f)(a) = g(f(a)).$$

Proof. Let B be the C*-subalgebra generated by 1 and a. Then $\sigma(f(a)) = \{\tau(f(a)) \mid \tau \in \Omega(B)\} = \{f(\tau(a)) \mid \tau \in \Omega(B)\} = f(\sigma(a))$.

If C denotes the C*-subalgebra generated by 1 and $f(a)$, then $C \subseteq B$ and for any $\tau \in \Omega(B)$ its restriction τ_C is a character on C. We therefore have $\tau((g \circ f)(a)) = g(f(\tau(a))) = g(\tau_C(f(a))) = \tau_C(g(f(a))) = \tau(g(f(a)))$. Hence, $(g \circ f)(a) = g(f(a))$. □

We close this section by showing that if Ω is a compact Hausdorff space, then the character space of $C(\Omega)$ is Ω.

2.1.15. Theorem. Let Ω be a compact Hausdorff space, and for each $\omega \in \Omega$ let δ_ω be the character on $C(\Omega)$ given by evaluation at ω; that is, $\delta_\omega(f) = f(\omega)$. Then the map

$$\Omega \to \Omega(C(\Omega)), \quad \omega \mapsto \delta_\omega,$$

is a homeomorphism.

Proof. This map is continuous because if $(\omega_\lambda)_{\lambda \in \Lambda}$ is a net in Ω converging to a point ω, then $\lim_{\lambda \in \Lambda} f(\omega_\lambda) = f(\omega)$ for all $f \in C(\Omega)$, so the net $(\delta_{\omega_\lambda})$ is weak* convergent to δ_ω. The map is also injective, because if ω, ω' are distinct points of Ω, then by Urysohn's lemma there is a function $f \in C(\Omega)$ such that $f(\omega) = 0$ and $f(\omega') = 1$, and therefore $\delta_\omega \neq \delta_{\omega'}$.

Now we show surjectivity of the map. Let $\tau \in \Omega(C(\Omega))$. Then $M = \ker(\tau)$ is a proper C*-algebra of $C(\Omega)$. Also, M separates the points of Ω, for if ω, ω' are distinct points of Ω, then as we have just seen there is a function $f \in C(\Omega)$ such that $f(\omega) \neq f(\omega')$, so $g = f - \tau(f)$ is a function in M such that $g(\omega) \neq g(\omega')$. It follows from the Stone–Weierstrass theorem that there is a point $\omega \in \Omega$ such that $f(\omega) = 0$ for all $f \in M$. Hence, $(f - \tau(f))(\omega) = 0$, so $f(\omega) = \tau(f)$, for all $f \in C(\Omega)$. Therefore, $\tau = \delta_\omega$. Thus, the map is a continuous bijection between compact Hausdorff spaces and therefore is a homeomorphism. $\square$

2.2. Positive Elements of C*-Algebras

In this section we introduce a partial order relation on the hermitian elements of a C*-algebra. The principal results are the existence of a unique positive square root for each positive element and Theorem 2.2.4, which asserts that elements of the form a^*a are positive.

2.2.1. Remark. Let $A = C_0(\Omega)$, where Ω is a locally compact Hausdorff space. Then A_{sa} is the set of real-valued functions in A and there is a natural partial order on A_{sa} given by $f \leq g$ if and only if $f(\omega) \leq g(\omega)$ for all $\omega \in \Omega$. An element $f \in A$ is positive, that is, $f \geq 0$, if and only if f is of the form $f = \bar{g}g$ for some $g \in A$, and in this case f has a unique positive square root in A, namely the function $\omega \mapsto \sqrt{f(\omega)}$. Note that if $f = \bar{f}$ we can also express the positivity condition in terms of the norm: If $t \in \mathbf{R}$, then f is positive if $\|f - t\| \leq t$, and in the reverse direction if $\|f\| \leq t$ and $f \geq 0$, then $\|f - t\| \leq t$. We shall presently define a partial order on an arbitrary C*-algebra that generalises that of $C_0(\Omega)$, and we shall obtain similar, and many other, results.

Let A be a unital algebra and B a subalgebra such that $B + \mathbf{C}1 = A$. Then $\sigma_B(b) \cup \{0\} = \sigma_A(b) \cup \{0\}$ for all $b \in B$. If B is non-unital, this is seen by observing that the map $\tilde{B} \to A$, $(b, \lambda) \mapsto b + \lambda 1$, is an isomorphism.

If B has a unit e not equal to the unit 1 of A, then for any $b \in B$ and $\lambda \in \mathbf{C} \setminus \{0\}$ invertibility of $b + \lambda$ in A is equivalent to invertibility of $b + \lambda e$ in B, so $\sigma_A(b) = \sigma_B(b) \cup \{0\}$.

From these observations and Theorem 2.1.11, it is clear that for any C*-subalgebra B of a C*-algebra A we have $\sigma_B(b) \cup \{0\} = \sigma_A(b) \cup \{0\}$ for all $b \in B$.

An element a of a C*-algebra A is *positive* if a is hermitian and $\sigma(a) \subseteq \mathbf{R}^+$. We write $a \geq 0$ to mean that a is positive, and denote by A^+ the set of positive elements of A. By the preceding observation $B^+ = B \cap A^+$ for any C*-subalgebra B of A.

If S is a non-empty set, then an element $f \in \ell^\infty(S)$ is positive in the C*-algebra sense if and only if $f(x) \geq 0$ for all $x \in S$, because $\sigma(f)$ is the closure of the range of f. Hence, if Ω is any locally compact Hausdorff space, then $f \in C_0(\Omega)$ is positive if and only if $f(\omega) \geq 0$ for all $\omega \in \Omega$.

If a is a hermitian element of a C*-algebra A observe that $C^*(a)$ is the closure of the set of polynomials in a with zero constant term.

2.2.1. Theorem. Let A be a C*-algebra and $a \in A^+$. Then there exists a unique element $b \in A^+$ such that $b^2 = a$.

Proof. That there exists $b \in C^*(a)$ such that $b \geq 0$ and $b^2 = a$ follows from the Gelfand representation, since we may use it to identify $C^*(a)$ with $C_0(\Omega)$, where Ω is the character space of $C^*(a)$, and then apply Remark 2.2.1.

Suppose that c is another element of A^+ such that $c^2 = a$. As c commutes with a it must commute with b, since b is the limit of a sequence of polynomials in a. Let B be the (necessarily abelian) C*-subalgebra of A generated by b and c, and let $\varphi \colon B \to C_0(\Omega)$ be the Gelfand representation of B. Then $\varphi(b)$ and $\varphi(c)$ are positive square roots of $\varphi(a)$ in $C_0(\Omega)$, so by another application of Remark 2.2.1, $\varphi(b) = \varphi(c)$, and therefore $b = c$. $\square$

If A is a C*-algebra and a is a positive element, we denote by $a^{1/2}$ the unique positive element b such that $b^2 = a$.

If c is a hermitian element, then c^2 is positive, and we set $|c| = (c^2)^{1/2}$, $c^+ = \frac{1}{2}(|c| + c)$, and $c^- = \frac{1}{2}(|c| - c)$. Using the Gelfand representation of $C^*(c)$, it is easy to check that $|c|, c^+$ and c^- are positive elements of A such that $c = c^+ - c^-$ and $c^+ c^- = 0$.

2.2.2. Remark. If a is a hermitian element of the closed unit ball of a unital C*-algebra A, then $1 - a^2 \in A^+$ and the elements

$$u = a + i\sqrt{1 - a^2} \qquad \text{and} \qquad v = a - i\sqrt{1 - a^2}$$

are unitaries such that $a = \frac{1}{2}(u + v)$. Therefore, the unitaries linearly span A, a result that is frequently useful.

2.2.2. Lemma. *Suppose that A is a unital C*-algebra, a is a hermitian element of A and $t \in \mathbf{R}$. Then, $a \geq 0$ if $\|a-t\| \leq t$. In the reverse direction, if $\|a\| \leq t$ and $a \geq 0$, then $\|a - t\| \leq t$.*

Proof. We may suppose that A is the (abelian) C*-subalgebra generated by 1 and a, so by the Gelfand representation $A = C(\sigma(a))$. The result now follows from Remark 2.1.1.   □

It is immediate from Lemma 2.2.2 that A^+ is closed in A.

2.2.3. Lemma. *The sum of two positive elements in a C*-algebra is a positive element.*

Proof. Let A be a C*-algebra and a, b positive elements. To show that $a+b \geq 0$ we may suppose that A is unital. By Lemma 2.2.2, $\|a-\|a\|\| \leq \|a\|$ and $\|b-\|b\|\| \leq \|b\|$, so $\|a+b-\|a\|-\|b\|\| \leq \|a-\|a\|\|+\|b-\|b\|\| \leq \|a\|+\|b\|$. By Lemma 2.2.2 again, $a + b \geq 0$.   □

2.2.4. Theorem. *If a is an arbitrary element of a C*-algebra A, then a^*a is positive.*

Proof. First we show that $a = 0$ if $-a^*a \in A^+$. Since $\sigma(-aa^*) \setminus \{0\} = \sigma(-a^*a) \setminus \{0\}$ by Remark 1.2.1, $-aa^* \in A^+$ because $-a^*a \in A^+$. Write $a = b + ic$, where $b, c \in A_{sa}$. Then $a^*a + aa^* = 2b^2 + 2c^2$, so $a^*a = 2b^2 + 2c^2 - aa^* \in A^+$. Hence, $\sigma(a^*a) = \mathbf{R}^+ \cap (-\mathbf{R}^+) = \{0\}$, and therefore $\|a\|^2 = \|a^*a\| = r(a^*a) = 0$.

Now suppose a is an arbitrary element of A, and we shall show that a^*a is positive. If $b = a^*a$, then b is hermitian, and therefore we can write $b = b^+ - b^-$. If $c = ab^-$, then $-c^*c = -b^-a^*ab^- = -b^-(b^+ - b^-)b^- = (b^-)^3 \in A^+$, so $c = 0$ by the first part of this proof. Hence, $b^- = 0$, so $a^*a = b^+ \in A^+$.   □

If A is a C*-algebra, we make A_{sa} a poset by defining $a \leq b$ to mean $b - a \in A^+$. The relation $\leq$ is translation-invariant; that is, $a \leq b \Rightarrow a + c \leq b + c$ for all $a, b, c \in A_{sa}$. Also, $a \leq b \Rightarrow ta \leq tb$ for all $t \in \mathbf{R}^+$, and $a \leq b \Leftrightarrow -a \geq -b$.

Using Theorem 2.2.4 we can extend our definition of $|a|$: for arbitrary a set $|a| = (a^*a)^{1/2}$.

We summarise some elementary facts about A^+ in the following result.

2.2.5. Theorem. *Let A be a C*-algebra.*

(1) *The set A^+ is equal to $\{a^*a \mid a \in A\}$.*
(2) *If $a, b \in A_{sa}$ and $c \in A$, then $a \leq b \Rightarrow c^*ac \leq c^*bc$.*
(3) *If $0 \leq a \leq b$, then $\|a\| \leq \|b\|$.*
(4) *If A is unital and a, b are positive invertible elements, then $a \leq b \Rightarrow 0 \leq b^{-1} \leq a^{-1}$.*

Proof. Conditions (1) and (2) are implied by Theorem 2.2.4 and the existence of positive square roots for positive elements. To prove Condition (3) we may suppose that A is unital. The inequality $b \leq \|b\|$ is given by the Gelfand representation applied to the C*-algebra generated by 1 and b. Hence, $a \leq \|b\|$. Applying the Gelfand representation again, this time to the C*-algebra generated by 1 and a, we obtain the inequality $\|a\| \leq \|b\|$.

To prove Condition (4) we first observe that if $c \geq 1$, then c is invertible and $c^{-1} \leq 1$. This is given by the Gelfand representation applied to the C*-subalgebra generated by 1 and c. Now $a \leq b \Rightarrow 1 = a^{-1/2}aa^{-1/2} \leq a^{-1/2}ba^{-1/2} \Rightarrow (a^{-1/2}ba^{-1/2})^{-1} \leq 1$, that is, $a^{1/2}b^{-1}a^{1/2} \leq 1$. Hence, $b^{-1} \leq (a^{1/2})^{-1}(a^{1/2})^{-1} = a^{-1}$. $\qquad \square$

2.2.6. Theorem. *If a, b are positive elements of a C*-algebra A, then the inequality $a \leq b$ implies the inequality $a^{1/2} \leq b^{1/2}$.*

Proof. We show $a^2 \leq b^2 \Rightarrow a \leq b$ and this will prove the theorem. We may suppose that A is unital. Let $t > 0$ and let c, d be the real and imaginary hermitian parts of the element $(t + b + a)(t + b - a)$. Then

$$c = \tfrac{1}{2}((t + b + a)(t + b - a)) + (t + b - a)(t + b + a))$$
$$= t^2 + 2tb + b^2 - a^2$$
$$\geq t^2.$$

Consequently, c is both invertible and positive. Since $1 + ic^{-1/2}dc^{-1/2} = c^{-1/2}(c + id)c^{-1/2}$ is invertible, therefore $c + id$ is invertible. It follows that $t + b - a$ is left invertible, and therefore invertible, because it is hermitian. Consequently, $-t \notin \sigma(b - a)$. Hence, $\sigma(b - a) \subseteq \mathbf{R}^+$, so $b - a$ is positive, that is, $a \leq b$. $\qquad \square$

It is not true that $0 \leq a \leq b \Rightarrow a^2 \leq b^2$ in arbitrary C*-algebras. For example, take $A = M_2(\mathbf{C})$. This is a C*-algebra where the involution is given by

$$\begin{pmatrix} \alpha & \beta \\ \gamma & \delta \end{pmatrix}^* = \begin{pmatrix} \bar{\alpha} & \bar{\gamma} \\ \bar{\beta} & \bar{\delta} \end{pmatrix}.$$

Let p and q be the projections

$$p = \begin{pmatrix} 1 & 0 \\ 0 & 0 \end{pmatrix} \qquad \text{and} \qquad q = \tfrac{1}{2}\begin{pmatrix} 1 & 1 \\ 1 & 1 \end{pmatrix}.$$

Then $p \leq p + q$, but $p^2 = p \not\leq (p + q)^2 = p + q + pq + qp$, since the matrix

$$q + pq + qp = \tfrac{1}{2}\begin{pmatrix} 3 & 2 \\ 2 & 1 \end{pmatrix}$$

has a negative eigenvalue.

It can be shown that the implication $0 \leq a \leq b \Rightarrow a^2 \leq b^2$ holds only in abelian C*-algebras [Ped, Proposition 1.3.9].

2.3. Operators and Sesquilinear Forms

In this section (and the next) we shall interpret and apply many of the ideas of Chapter 1 and the first two sections of this chapter in the context of operators on Hilbert spaces. We shall also prove the invaluable polar decomposition theorem. An important concern in the present section is the correspondence of operators and sesquilinear forms. This is interesting in its own right, but it also has wide applicability—for example, we shall use it in the proof of the spectral theorem.

We begin by showing that operators on Hilbert spaces have adjoints.

2.3.1. Theorem. *Let H_1 and H_2 be Hilbert spaces.*

(1) *If $u \in B(H_1, H_2)$, then there is a unique element $u^* \in B(H_2, H_1)$ such that*

$$\langle u(x_1), x_2 \rangle = \langle x_1, u^*(x_2) \rangle \qquad (x_1 \in H_1, \ x_2 \in H_2).$$

(2) *The map $u \mapsto u^*$ is conjugate-linear and $u^{**} = u$. Also*

$$\|u\| = \|u^*\| = \|u^*u\|^{1/2}.$$

Proof. If $u \in B(H_1, H_2)$ and $x_2 \in H_2$, then the function

$$H_1 \to \mathbf{C}, \ x_1 \mapsto \langle u(x_1), x_2 \rangle,$$

is continuous and linear, so by the Riesz representation theorem for linear functionals on Hilbert spaces there is a unique element $u^*(x_2) \in H_1$ such that $\langle u(x_1), x_2 \rangle = \langle x_1, u^*(x_2) \rangle$ $(x_1 \in H_1)$. Moreover,

$$\|u^*(x_2)\| = \sup_{\|x_1\| \le 1} |\langle u(x_1), x_2 \rangle| \le \|u\| \|x_2\|.$$

The map $u^* \colon H_2 \to H_1$, $x_2 \mapsto u^*(x_2)$, is linear and $\|u^*\| \le \|u\|$. Thus, u^* satisfies Condition (1) (uniqueness of u is obvious).

If $x_1 \in H_1$ and $\|x_1\| \le 1$, then $\langle u(x_1), u(x_1) \rangle = \langle x_1, u^*u(x_1) \rangle \le \|u^*u\|$, so

$$\|u\|^2 = \sup_{\|x_1\| \le 1} \|u(x_1)\|^2 \le \|u^*u\| \le \|u\|^2.$$

Hence, $\|u\| = \|u^*u\|^{1/2}$. The other assertions in Condition (2) of the theorem have routine verifications. $\qquad \square$

If $u \colon H_1 \to H_2$ is a continuous linear map between Hilbert spaces, we call the map $u^* \colon H_2 \to H_1$ the *adjoint* of u. Note that $\ker(u^*) = (\mathrm{im}(u))^\perp$, where $\mathrm{im}(u)$ is the range of u, and hence, $(\mathrm{im}(u^*))^- = \ker(u)^\perp$.

If $H_1 \overset{u}{\to} H_2 \overset{v}{\to} H_3$ are continuous linear maps between Hilbert spaces, then $(vu)^* = u^*v^*$.

If H is a Hilbert space, then $B(H)$ is a C*-algebra under the involution $u \mapsto u^*$, where u^* is the adjoint of u.

It follows in particular that $M_n(\mathbf{C}) = B(\mathbf{C}^n)$ is a C*-algebra. Observe that the involution on $M_n(\mathbf{C})$ is given by $(\lambda_{ij})_{ij}^* = (\bar{\lambda}_{ji})_{ij}$.

If H is a vector space, a map $\sigma\colon H^2 \to \mathbf{C}$ is a *sesquilinear form* if it is linear in the first variable and conjugate-linear in the second. For such a form the *polarisation identity*

$$\sigma(x,y) = \tfrac{1}{4} \sum_{k=0}^{3} i^k \sigma(x + i^k y, x + i^k y)$$

holds. Thus, sesquilinear forms σ and σ' on H are equal if and only if $\sigma(x,x) = \sigma'(x,x)$ for all $x \in H$. Sesquilinear forms are taken up in more detail later in this section.

If H is a Hilbert space and $u \in B(H)$, then $(x,y) \mapsto \langle u(x), y \rangle$ is a sesquilinear form on H. Hence, if $u, v \in B(H)$, then $u = v$ if and only if $\langle u(x), x \rangle = \langle v(x), x \rangle$ for all $x \in H$.

If $u^*u = \mathrm{id}$ and $uu^* = \mathrm{id}$, we say u is a *unitary operator*. This is equivalent to u being isometric and surjective. Observe that u is isometric $\Leftrightarrow u^*u = \mathrm{id}$.

2.3.1. Example. Let $(e_n)_{n=1}^\infty$ be an orthonormal basis for a Hilbert space H, and suppose that u is an operator diagonal with respect to (e_n), with diagonal sequence (λ_n). Then u^* is also diagonal with respect to (e_n) and its diagonal sequence is $(\bar{\lambda}_n)$. This follows from the observation that $\langle u^*(e_n), e_m \rangle = \langle e_n, u(e_m) \rangle = \langle e_n, \lambda_m e_m \rangle = \bar{\lambda}_m \delta_{nm}$, where δ_{nm} is the Kronecker delta symbol, which implies that $u^*(e_n) = \bar{\lambda}_n e_n$. Since all operators diagonal with respect to the same basis commute, $uu^* = u^*u$; that is, u is normal.

2.3.2. Example. Let (e_n) and H be as in the preceding example, but this time let u denote the unilateral shift on this basis, so $u(e_n) = e_{n+1}$ for all $n \geq 1$. The adjoint u^* is the backward shift: $u^*(e_n) = e_{n-1}$ if $n > 1$ and $u^*(e_1) = 0$. It follows that $u^*u = 1$. It is easily seen that u has no eigenvalues. In contrast, u^* has very many, for if $|\lambda| < 1$, then λ is an eigenvalue: Set $x = \sum_{n=1}^\infty \lambda^n e_n$ and observe that $x \in H$ because $\sum_{n=1}^\infty |\lambda|^{2n} < \infty$, and that $x \neq 0$ and $u^*(x) = \lambda x$. It follows from this, and the fact that $\|u^*\| = \|u\| = 1$, that $\sigma(u) = \sigma(u^*) = \mathbf{D}$.

Incidentally, if $(f_n)_{n=1}^\infty$ is an orthonormal basis for another Hilbert space K and v is the unilateral shift on (f_n), so $v(f_n) = f_{n+1}$, then $v = wuw^*$, where $w\colon H \to K$ is the unitary operator such that $w(e_n) = f_n$ for all $n \geq 1$. From the abstract point of view, the operators u and v are therefore the same, so one can speak of "the" unilateral shift.

If K is a closed vector subspace of a Hilbert space H, we call the projection p of H on K along $K^\perp$ the *(orthogonal) projection* on K. This is self-adjoint. If $u \in B(H)$, then K is invariant for u (that is, $u(K) \subseteq K$) if and only if $pup = up$. We say that K is *reducing* for u if both K and $K^\perp$ are invariant for u. This is equivalent to p commuting with u, because $K^\perp$ is invariant for u if and only if K is invariant for u^*.

The following result on projections will be used frequently and tacitly.

2.3.2. Theorem. *Let p, q be projections on a Hilbert space H. Then the following conditions are equivalent:*

(1) $p \leq q$.
(2) $pq = p$.
(3) $qp = p$.
(4) $p(H) \subseteq q(H)$.
(5) $\|p(x)\| \leq \|q(x)\|$ $(x \in H)$.
(6) $q - p$ is a projection.

Proof. Equivalence of Conditions (2),(3), and (4) is clear, as are the implications (2) $\Rightarrow$ (6) $\Rightarrow$ (1). We show (1) $\Rightarrow$ (5) $\Rightarrow$ (2), and this will prove the theorem.

If we assume Condition (1) holds, $\|q(x)\|^2 - \|p(x)\|^2 = \langle (q - p)(x), x \rangle = \|(q - p)^{1/2}(x)\|^2 \geq 0$, so Condition (5) holds.

If now we assume Condition (5) holds, $\|p(1 - q)(x)\| \leq \|(q - q^2)(x)\| = 0$, and therefore $p = pq$; that is, Condition (2) holds. $\square$

Let $u: H_1 \to H_2$ be a continuous linear map between Hilbert spaces. Since $(u(H_1))^\perp = \ker(u^*)$, the operator u is Fredholm if and only if $u(H_1)$ is closed in H_2 and the spaces $\ker(u)$ and $\ker(u^*)$ are finite-dimensional. In this case $\mathrm{ind}(u) = \dim(\ker(u)) - \dim(\ker(u^*))$, and the adjoint of u is also Fredholm and such that $\mathrm{ind}(u^*) = -\mathrm{ind}(u)$. (To see that u^* has closed range, recall from Theorem 1.4.15 that there is a continuous linear map $v: H_2 \to H_1$ such that $u = uvu$. Hence, $u^* = u^*v^*u^*$, so u^*v^* is an idempotent and $u^*(H_2) = u^*v^*(H_1)$. Thus, $u^*(H_2)$ is closed in H_1.)

An operator u on a Hilbert space H is normal if and only if $\|u(x)\| = \|u^*(x)\|$ $(x \in H)$, since $\langle (uu^* - u^*u)(x), x \rangle = \|u^*(x)\|^2 - \|u(x)\|^2$. Thus, $\ker(u) = \ker(u^*)$ if u is normal, and therefore a normal Fredholm operator has index zero.

A continuous linear map $u: H_1 \to H_2$ between Hilbert spaces H_1, H_2 is a *partial isometry* if u is isometric on $\ker(u)^\perp$, that is, $\|u(x)\| = \|x\|$ for all $x \in \ker(u)^\perp$.

2.3.3. Theorem. *Let H_1, H_2 be Hilbert spaces and $u \in B(H_1, H_2)$. Then the following conditions are equivalent:*

(1) $u = uu^*u$.

(2) u^*u is a projection.
(3) uu^* is a projection.
(4) u is a partial isometry.

Proof. The implication $(1) \Rightarrow (2)$ is obvious. To show the converse suppose that u^*u is a projection. Then $\|u(x)\|^2 = \langle u(x), u(x) \rangle = \langle u^*u(x), x \rangle = \|u^*u(x)\|^2$ for all $x \in H_1$, so $u(1 - u^*u) = 0$, and therefore $u = uu^*u$.

To show that $(2) \Rightarrow (3)$, suppose again that u^*u is a projection. Then $(uu^*)^3 = (uu^*)^2$, so $\sigma(uu^*) \subseteq \{0,1\}$. Hence, uu^* is a projection by the functional calculus. Thus, $(2) \Rightarrow (3)$, and clearly, then, $(3) \Rightarrow (2)$ by symmetry.

To show that $(1) \Rightarrow (4)$, suppose that $u = uu^*u$. Then u^*u is the projection onto $\ker(u)^\perp$, since $u^* = u^*uu^*$, and $\ker(u)^\perp = (u^*(H_2))^- = u^*u(H_1)$. Hence, if $x \in \ker(u)^\perp$, then $\|u(x)\|^2 = \langle u^*u(x), x \rangle = \langle x, x \rangle = \|x\|^2$. Thus, u is a partial isometry, so $(1) \Rightarrow (4)$.

Finally, we show $(4) \Rightarrow (2)$ (and this will prove the theorem). Suppose that u is a partial isometry. If p is the projection of H_1 on $\ker(u)^\perp$ and $x \in \ker(u)^\perp$, then $\langle u^*u(x), x \rangle = \|u(x)\|^2 = \langle x, x \rangle = \langle p(x), x \rangle$. If $x \in \ker(u)$, then $\langle u^*u(x), x \rangle = 0 = \langle p(x), x \rangle$. Thus, $\langle u^*u(x), x \rangle = \langle p(x), x \rangle$ for all $x \in H_1$. Hence, $u^*u = p$, so $(4) \Rightarrow (2)$. $\square$

Just as we can write a complex number as the product of a unitary (= number of modulus one) times a non-negative number, the following result asserts that we can write an operator as the product of a partial isometry times a positive operator.

2.3.4. Theorem (Polar Decomposition). *Let v be a continuous linear operator on a Hilbert space H. Then there is a unique partial isometry $u \in B(H)$ such that*

$$v = u|v| \quad \text{and} \quad \ker(u) = \ker(v).$$

Moreover, $u^*v = |v|$.

Proof. If $x \in H$, $\||v|(x)\|^2 = \langle |v|(x), |v|(x) \rangle = \langle |v|^2(x), x \rangle = \langle v^*v(x), x \rangle = \langle v(x), v(x) \rangle = \|v(x)\|^2$. Hence, the map

$$u_0 \colon |v|(H) \to H, \quad |v|(x) \mapsto v(x),$$

is well-defined and isometric. It is also linear. Therefore, it has a unique linear isometric extension (also denoted u_0) to $(|v|(H))^-$. Define u in $B(H)$ by setting

$$u = \begin{cases} u_0, & \text{on } \overline{|v|(H)} \\ 0, & \text{on } \overline{|v|(H)}^\perp. \end{cases}$$

Then $u|v| = v$, and u is isometric on $\ker(u)^\perp$, because $\ker(u) = \overline{|v|(H)}^\perp$. Thus, u is a partial isometry and $\ker(u) = \ker(|v|)$. Now $\langle u^*v(x), |v|(y) \rangle =$

$\langle v(x), v(y) \rangle = \langle v^*v(x), y \rangle = \langle |v|(x), |v|(y) \rangle \Rightarrow \langle u^*v(x), z \rangle = \langle |v|(x), z \rangle$ for all $z \in \overline{|v|(H)}$, and therefore for all $z \in H$. Thus, $u^*v = |v|$. It follows that $\ker(|v|) = \ker(v)$, so $\ker(u) = \ker(v)$.

Now suppose that $w \in B(H)$ is another partial isometry such that $v = w|v|$ and $\ker(w) = \ker(v)$. Then w is equal to u on $\overline{|v|(H)}$ and on $\overline{|v|(H)}^\perp = \ker(v) = \ker(w) = \ker(u)$. Thus, $w = u$. $\square$

Before we turn to the correspondence between sesquilinear forms and operators, we present a very brief survey of the basic definitions and facts pertaining to sesquilinear forms, since these are not always covered in books on general functional analysis.

The sesquilinear form σ on a vector space H is said to be *hermitian* if $\sigma(y, x) = \sigma(x, y)^-$ for all $x, y \in H$. It follows from the polarisation identity that a sesquilinear form σ is hermitian if and only if $\sigma(x, x) \in \mathbf{R}$ $(x \in H)$. A sesquilinear form σ is *positive* if $\sigma(x, x) \geq 0$ for all $x \in H$. Thus, positive sesquilinear forms are hermitian.

The inequality

$$|\sigma(x, y)| \leq \sqrt{\sigma(x, x)}\sqrt{\sigma(y, y)} \qquad (x, y \in H),$$

which holds for any positive sesquilinear form σ, is called the *Cauchy–Schwarz inequality*. It implies that the function $p\colon x \mapsto \sqrt{\sigma(x, x)}$ is a semi-norm on H; that is, p satisfies the axioms of a norm except that the implication $p(x) = 0 \Rightarrow x = 0$ may not hold.

A sesquilinear form σ on a normed vector space H is *bounded* if there is a positive number M such that

$$|\sigma(x, y)| \leq M\|x\|\|y\| \qquad (x, y \in H).$$

The *norm* $\|\sigma\|$ of σ is the infimum of all such numbers M. Obviously, $|\sigma(x, y)| \leq \|\sigma\|\|x\|\|y\|$. A sesquilinear form is continuous if and only if it is bounded.

The proofs of these facts are elementary and are the same as for the corresponding results on inner products.

2.3.5. Theorem. *If u is an operator on a Hilbert space H, then the sesquilinear form*

$$\sigma_u\colon H^2 \to \mathbf{C}, \ (x, y) \mapsto \langle u(x), y \rangle,$$

is hermitian if and only if u is hermitian, and positive if and only if u is positive.

Proof. We show only the implication, σ_u is positive $\Rightarrow u$ is positive, since the other assertions are easy exercises (if u is positive, use the existence of a positive square root for u to show the converse of the result we are now going to prove).

Suppose that σ_u is positive. Then it is hermitian and therefore u is hermitian. To see that $\sigma(u) \subseteq \mathbf{R}^+$, we show that $u - \lambda$ is invertible if $\lambda < 0$. In this case if $x \in H$, then

$$
\begin{aligned}
\|(u - \lambda)(x)\|^2 &= \langle (u - \lambda)(x), (u - \lambda)(x) \rangle \\
&= \|u(x)\|^2 + |\lambda|^2 \|x\|^2 - 2\lambda \langle u(x), x \rangle \\
&\geq |\lambda|^2 \|x\|^2.
\end{aligned}
$$

Thus, $\|(u - \lambda)(x)\| \geq |\lambda| \|x\|$, so $u - \lambda$ is bounded below. Hence, $(u - \lambda)(H)$ is closed in H and $\ker(u - \lambda) = 0$. Therefore, $(u - \lambda)(H) = \ker(u^* - \bar{\lambda})^\perp = \ker(u - \lambda)^\perp = 0^\perp = H$. Hence, $u - \lambda$ is invertible. $\square$

By the preceding theorem, if u is a operator on a Hilbert space H, then u is hermitian if and only if $\langle u(x), x \rangle \in \mathbf{R}$ $(x \in H)$, and u is positive if and only if $\langle u(x), x \rangle \geq 0$ $(x \in H)$.

2.3.6. Theorem. *Let σ be a bounded sesquilinear form on a Hilbert space H. Then there is a unique operator u on H such that*

$$
\sigma(x, y) = \langle u(x), y \rangle \qquad (x, y \in H).
$$

Moreover, $\|u\| = \|\sigma\|$.

Proof. Uniqueness of u is obvious.

For each $y \in H$, the function $H \to \mathbf{C}$, $x \mapsto \sigma(x, y)$, is continuous and linear, so by the Riesz representation theorem there is a unique element $v(y) \in H$ such that $\sigma(x, y) = \langle x, v(y) \rangle$ $(x \in H)$. Also,

$$
\|v(y)\| = \sup_{\|x\| \leq 1} |\sigma(x, y)| \leq \|\sigma\| \|y\|.
$$

The map $v \colon H \to H$, $y \mapsto v(y)$, is linear and $\|v\| \leq \|\sigma\|$. If $u = v^*$, then $\sigma(x, y) = \langle u(x), y \rangle$ $(x, y \in H)$, and also the inequality $|\sigma(x, y)| \leq \|u\| \|x\| \|y\|$ which holds for all x, y, implies that $\|\sigma\| \leq \|u\|$. Hence, $\|\sigma\| = \|u\|$. $\square$

2.4. Compact Hilbert Space Operators

In this section we analyse two closely related classes of compact operators, the Hilbert–Schmidt and the trace-class operators. Some of the details are a little technical, but the results are useful to us for the analysis of von Neumann algebras, as well as being important in applications and having intrinsic interest. We begin by looking at general compact operators on a Hilbert space and we strengthen some of the results of Section 1.4 in this case.

We shall need to view Hilbert spaces as dual spaces. Let H be a Hilbert space and $H_* = H$ as an additive group, but define a new scalar multiplication on H_* by setting $\lambda.x = \bar{\lambda}x$, and a new inner product by setting $\langle x, y \rangle_* = \langle y, x \rangle$. Then H_* is a Hilbert space, and obviously the norm induced by the new inner product is the same as that induced by the old one. If $x \in H$, define $v(x) \in (H_*)^*$ by setting $v(x)(y) = \langle y, x \rangle_* = \langle x, y \rangle$. It is a direct consequence of the Riesz representation theorem that the map

$$v: H \to (H_*)^*, \quad x \mapsto v(x),$$

is an isometric linear isomorphism, which we use to identify these Banach spaces. The weak* topology on H is called the *weak* topology. A net $(x_\lambda)_{\lambda \in \Lambda}$ converges to a point x in H in the weak topology if and only if $\langle x, y \rangle = \lim_\lambda \langle x_\lambda, y \rangle$ $(y \in H)$. Consequently, the weak topology is weaker than the norm topology, and a bounded linear map between Hilbert spaces is necessarily weakly continuous. The importance to us of the weak topology is the fact that the closed unit ball of H is weakly compact (Banach–Alaoglu theorem).

2.4.1. Theorem. *Let* $u: H_1 \to H_2$ *be a compact linear map between Hilbert spaces* H_1 *and* H_2. *Then the image of the closed unit ball of* H_1 *under* u *is compact.*

Proof. Let S be the closed unit ball of H_1. It is weakly compact, and u is weakly continuous, so $u(S)$ is weakly compact and therefore weakly closed. Hence, $u(S)$ is norm-closed, since the weak topology is weaker than the norm topology. Since u is a compact operator, this implies that $u(S)$ is norm-compact. $\square$

2.4.2. Theorem. *Let* u *be a compact operator on a Hilbert space* H. *Then both* $|u|$ *and* u^* *are compact.*

Proof. Suppose that u has polar decomposition $u = w|u|$ say. Then $|u| = w^*u$, so $|u|$ is compact, and $u^* = |u|w^*$, so u^* is compact. $\square$

2.4.3. Corollary. *If* H *is any Hilbert space, then* $K(H)$ *is self-adjoint.*

Thus, $K(H)$ is a C*-algebra, since (as we saw in Chapter 1) $K(H)$ is a closed ideal in $B(H)$.

An operator u on a Hilbert space H is *diagonalisable* if H admits an orthonormal basis consisting of eigenvectors of u. Diagonalisable operators are necessarily normal, but not all normal operators are diagonalisable. For instance, the bilateral shift is normal (it is a unitary), but it has no eigenvalues.

2.4.4. Theorem. *If u is a compact normal operator on a Hilbert space H, then it is diagonalisable.*

Proof. By Zorn's lemma there is a maximal orthonormal set E of eigenvectors of u. If K is the closed linear span of E, then $H = K \oplus K^{\perp}$, and K reduces u. The restriction $u_{K^{\perp}} \colon K^{\perp} \to K^{\perp}$ is compact and normal. An eigenvector of $u_{K^{\perp}}$ is one for u also, so by maximality of E, the operator $u_{K^{\perp}}$ has no eigenvectors, and therefore $\sigma(u_{K^{\perp}}) = \{0\}$ by Theorem 1.4.11. Hence, $\|u_{K^{\perp}}\| = r(u_{K^{\perp}})$ (by normality) $= 0$, so $K^{\perp} = 0$. Thus, $K = H$ and E is an orthonormal basis of eigenvectors of u, so u is diagonalisable.□

If H is a Hilbert space, we denote by $F(H)$ the set of finite-rank operators on H. It is easy to check that $F(H)$ is a self-adjoint ideal of $B(H)$.

2.4.5. Theorem. *If H is a Hilbert space, then $F(H)$ is dense in $K(H)$.*

Proof. Since $F(H)^{-}$ and $K(H)$ are both self-adjoint, it suffices to show that if u is a hermitian element of $K(H)$, then $u \in F(H)^{-}$. Let E be an orthonormal basis of H consisting of eigenvectors of u, and let $\varepsilon > 0$. By Theorem 1.4.11 the set S of eigenvalues λ of u such that $|\lambda| \geq \varepsilon$ is finite. From Theorem 1.4.5 it is therefore clear that the set S' of elements of E corresponding to elements of S is finite. Now define a finite-rank diagonal operator v on H by setting $v(x) = \lambda x$ if $x \in S'$ and λ is the eigenvalue corresponding to x, and setting $v(x) = 0$ if $x \in E \setminus S'$. It is easily checked that $\|v - u\| \leq \sup_{\lambda \in \sigma(u) \setminus S} |\lambda| \leq \varepsilon$. This shows that $u \in F(H)^{-}$.   □

If x, y are elements of a Hilbert space H we define the operator $x \otimes y$ on H by

$$(x \otimes y)(z) = \langle z, y \rangle x.$$

Clearly, $\|x \otimes y\| = \|x\| \|y\|$. The rank of $x \otimes y$ is one if x and y are non-zero. If $x, x', y, y' \in H$ and $u \in B(H)$, then the following equalities are readily verified:

$$(x \otimes x')(y \otimes y') = \langle y, x' \rangle (x \otimes y')$$
$$(x \otimes y)^{*} = y \otimes x$$
$$u(x \otimes y) = u(x) \otimes y$$
$$(x \otimes y)u = x \otimes u^{*}(y).$$

The operator $x \otimes x$ is a rank-one projection if and only if $\langle x, x \rangle = 1$, that is, x is a unit vector. Conversely, every rank-one projection is of the form $x \otimes x$ for some unit vector x. Indeed, if $e_1, \ldots, e_n$ is an orthonormal set in H, then the operator $\sum_{j=1}^{n} e_j \otimes e_j$ is the orthogonal projection of H onto the vector subspace $\mathbf{C}e_1 + \cdots + \mathbf{C}e_n$.

If $u \in B(H)$ is a rank-one operator and x a non-zero element of its range, then $u = x \otimes y$ for some $y \in H$. For if $z \in H$, then $u(z) = \tau(z)x$ for some scalar $\tau(z) \in \mathbf{C}$. It is readily verified that the map $z \mapsto \tau(z)$ is a bounded linear functional on H, and therefore, by the Riesz representation theorem, there exists $y \in H$ such that $\tau(z) = \langle z, y \rangle$ for all $z \in H$. Therefore, $u = x \otimes y$.

2.4.6. Theorem. *If H is a Hilbert space, then $F(H)$ is linearly spanned by the rank-one projections.*

Proof. Let $u \in F(H)$ and we shall show it is a linear combination of rank-one projections. The real and imaginary parts of u are in $F(H)$, since $F(H)$ is self-adjoint, so we may suppose that u is hermitian. Now $u = u^+ - u^-$, and by the polar decomposition $|u| \in F(H)$, so u^+ and u^- belong to $F(H)$. Hence, we may assume that $u \geq 0$. The range $u(H)$ is finite-dimensional, and therefore it is a Hilbert space with an orthonormal basis, $e_1, \ldots, e_n$ say. Let $p = \sum_{j=1}^{n} e_j \otimes e_j$, so p is the projection of H onto $u(H)$. Then $u = pu = u^{1/2}pu^{1/2} \Rightarrow u = \sum_{j=1}^{n} x_j \otimes x_j$, where $x_j = u^{1/2}(e_j)$. Now $x_j = \lambda_j f_j$ for some unit vector f_j and scalar λ_j, so $u = \sum_{j=1}^{n} |\lambda_j|^2 f_j \otimes f_j$, and since the operators $f_j \otimes f_j$ are rank-one projections we are done.   □

2.4.7. Theorem. *If H is a Hilbert space and I a non-zero ideal in $B(H)$, then I contains $F(H)$.*

Proof. Let u be a non-zero operator in I. Then for some $x \in H$ we have $u(x) \neq 0$. If p is a rank-one projection, then $p = y \otimes y$ for some unit vector $y \in H$, and clearly there exists $v \in B(H)$ such that $vu(x) = y$ (take $v = (y \otimes u(x))/\|u(x)\|^2$, for instance). Hence, $p = vu(x \otimes x)u^*v^*$, so $p \in I$ as $u \in I$. Thus, I contains all the rank-one projections and therefore by Theorem 2.4.6 it contains $F(H)$.   □

If $u: H \to H'$ is a unitary between Hilbert spaces H and H', then the map

$$\mathrm{Ad}\, u: K(H) \to K(H'), \quad v \mapsto uvu^*,$$

is a *-isomorphism. In fact, all *-isomorphisms between $K(H)$ and $K(H')$ are obtained in this way:

2.4.8. Theorem. *Let H and H' be Hilbert spaces and suppose that the map $\varphi: K(H) \to K(H')$ is a *-isomorphism. Then there exists a unitary $u: H \to H'$ such that $\varphi = \mathrm{Ad}\, u$.*

Proof. Let E be an orthonormal basis for H, and for $e \in E$ let $p_e = e \otimes e$. Then p_e is a rank-one projection and $p_e K(H)p_e = \mathbf{C}p_e$. Hence, $q_e = \varphi(p_e)$ is a projection on H' such that $q_e K(H')q_e = \varphi(p_e K(H)p_e) = \varphi(\mathbf{C}p_e) = \mathbf{C}q_e$. It is easily inferred from this that q_e is also of rank one. Thus, we may write $q_e = \tilde{e} \otimes \tilde{e}$ for a unit vector $\tilde{e}$ in H'. If e, f are distinct elements of E, then $\langle \tilde{f}, \tilde{e} \rangle \tilde{e} \otimes \tilde{f} = q_e q_f = \varphi(p_e p_f) = \langle f, e \rangle \varphi(e \otimes f) = 0$, and therefore $\tilde{e}$ and $\tilde{f}$ are orthogonal. Thus, $\tilde{E} = \{\tilde{e} \mid e \in E\}$ is an orthonormal set in H'. We claim it is maximal; that is, it is an orthonormal basis for H'. For if we suppose the contrary, then there is a unit vector x of H' orthogonal to $\tilde{E}$. Reasoning as above, but this time using φ^{-1} instead of φ, there is an element y of H such that $\varphi^{-1}(x \otimes x) = y \otimes y$, and the set $E \cup \{y\}$ is orthonormal in H. This contradicts the fact that E is an orthonormal basis. This argument therefore shows that $\tilde{E}$ is an orthonormal basis for H' as claimed.

For $e, f \in E$ let $q_{ef} = \varphi(e \otimes f)$. Then $q_{ee} = q_e$, and

$$q_{ef} q_{gh} = \langle g, f \rangle q_{eh} \qquad (e, f, g, h \in E),$$

since

$$(e \otimes f)(g \otimes h) = \langle g, f \rangle e \otimes h.$$

Because $q_{ef} = q_e q_{ef}$, the range of q_{ef} is $\mathbf{C}\tilde{e}$. Hence, q_{ef} can be written in the form $\tilde{e} \otimes y$ for some unit vector $y \in H'$. Since $q_{fe} = q_{ef}^* = y \otimes \tilde{e}$, and q_{fe} has range $\mathbf{C}\tilde{f}$, we have $y = \bar{\lambda}_{ef}\tilde{f}$ for some scalar λ_{ef} of modulus one. Thus, $q_{ef} = \lambda_{ef}\tilde{e} \otimes \tilde{f}$. Since $q_{eg} = q_{ef} q_{fg}$,

$$\lambda_{eg}\tilde{e} \otimes \tilde{g} = \lambda_{ef}(\tilde{e} \otimes \tilde{f})\lambda_{fg}(\tilde{f} \otimes \tilde{g})$$
$$= \lambda_{ef}\lambda_{fg}\tilde{e} \otimes \tilde{g}.$$

Therefore, $\lambda_{eg} = \lambda_{ef}\lambda_{fg}$. Observe also that $\bar{\lambda}_{eg} = \lambda_{ge}$, since $q_{eg}^* = q_{ge}$. Thus, if we fix an element, f say, in E and set $\mu_e = \lambda_{ef}$ for all $e \in E$, we get $\lambda_{eg} = \mu_e \bar{\mu}_g$.

Let $u: H \to H'$ be the unitary such that $u(e) = \mu_e \tilde{e}$ for all $e \in E$. Then $\operatorname{Ad} u(e \otimes g) = u(e) \otimes u(g) = \mu_e \tilde{e} \otimes \mu_g \tilde{g} = \lambda_{eg}\tilde{e} \otimes \tilde{g} = \varphi(e \otimes g)$. From this it follows that $\operatorname{Ad} u$ and φ are equal at $x \otimes y$ for x, y in the linear span of E, and hence for all x, y in H, since E has closed linear span H. Thus, $\operatorname{Ad} u$ and φ are equal on all the rank-one operators on H, and since these have closed linear span $K(H)$, we have $\operatorname{Ad} u = \varphi$. $\qquad\qquad\square$

We make a few observations now which we shall need in the proof of the next theorem, and which are also of independent interest.

Let Ω be a locally compact Hausdorff space. For $\omega \in \Omega$, denote by τ_ω the character on $C_0(\Omega)$ given by evaluation at ω: $\tau_\omega(f) = f(\omega)$. If $\omega_1, \ldots, \omega_n$ are distinct points of Ω, then $\tau_{\omega_1}, \ldots, \tau_{\omega_n}$ are linearly independent. For if $\lambda_1 \tau_{\omega_1} + \cdots + \lambda_n \tau_{\omega_n} = 0$ and we fix i, then by Urysohn's lemma we may choose $f \in C_0(\Omega)$ such that $f(\omega_i) = 1$ and $f(\omega_j) = 0$ for $j \neq i$. Hence, $0 = \sum_{j=1}^{n} \lambda_j f(\omega_j) = \lambda_i$.

It follows that if $C_0(\Omega)$ is finite-dimensional, then Ω is finite.

From this observation we show that the projections linearly span an abelian finite-dimensional C*-algebra. We may suppose the algebra is of the form $C_0(\Omega)$ by the Gelfand representation. Then Ω is finite and therefore discrete, so the characteristic functions of the singleton sets span $C_0(\Omega)$.

Suppose now that A is an arbitrary finite-dimensional C*-algebra. It is linearly spanned by its self-adjoint elements, and they in turn are linear combinations of projections by what we have just shown, so it follows that A is the linear span of its projections.

If p is a finite-rank projection on a Hilbert space H, then the C*-algebra $A = pB(H)p$ is finite-dimensional. To see this, write $p = \sum_{j=1}^{n} e_j \otimes e_j$, where $e_1, \ldots, e_n \in H$. If $u \in B(H)$, then

$$pup = \sum_{j,k=1}^{n} (e_j \otimes e_j)u(e_k \otimes e_k) = \sum_{j,k=1}^{n} \langle u(e_k), e_j \rangle e_j \otimes e_k.$$

Hence, A is in the linear span of the operators $e_j \otimes e_k$ ($j, k = 1, \ldots, n$), and therefore $\dim(A) < \infty$.

A closed vector subspace K of H is *invariant* for a subset $A \subseteq B(H)$ if it is invariant for every operator in A. If A is a C*-subalgebra of $B(H)$, it is said to be *irreducible*, or to act *irreducibly* on H, if the only closed vector subspaces of H that are invariant for A are 0 and H. The concept of irreducibility is of great importance in the representation theory of C*-algebras which we shall be taking up in Chapter 5. The following theorem gives a nice connection between irreducibility and the ideal of compact operators, and will be needed in succeeding chapters.

2.4.9. Theorem. *Let A be a C*-algebra acting irreducibly on a Hilbert space H and having non-zero intersection with $K(H)$. Then $K(H) \subseteq A$.*

Proof. The intersection $A \cap K(H)$ is a non-zero self-adjoint set, so it contains a non-zero self-adjoint element, u say. Now $r(u) = \|u\| > 0$, so $\sigma(u)$ contains non-zero elements. Hence, by Theorem 1.4.11 u admits a non-zero eigenvalue, λ say. By the same theorem, the non-zero points of $\sigma(u)$ are isolated, so if f is the characteristic function of $\{\lambda\}$ on $\sigma(u)$, then f is continuous, and $p = f(u)$ is a projection in A. Moreover, p is non-zero because f is non-zero. If z is the inclusion function of $\sigma(u)$ in $\mathbf{C}$, then $(z - \lambda)f = 0$, so $(u - \lambda)p = 0$, and therefore $p(H) \subseteq \ker(u - \lambda)$. By Theorem 1.4.5 the space $\ker(u - \lambda)$ is finite-dimensional, so p is therefore of finite rank.

Let q be a non-zero projection in A of minimal finite rank. Then the C*-algebra qAq is finite-dimensional; therefore, it is the linear span of its projections, by the remarks preceding this theorem. However, the minimal rank assumption on q implies that the only projections in qAq can be 0 and q, so $qAq = \mathbf{C}q$. Now let y be a non-zero element of $q(H)$. If K is the closure of the set of vectors $u(y)$ ($u \in A$), then K is a vector subspace of H invariant for A, and is non-zero since it contains $y = q(y)$. It follows from the irreducibility of A that $K = H$. Hence, if x is an arbitrary element of $q(H)$, then $x = \lim_{n\to\infty} u_n(y)$ for some sequence (u_n) in A. Therefore, $x = \lim_{n\to\infty} qu_nq(y)$, because $x = q(x)$ and $y = q(y)$. But $qu_nq = \lambda_n q$ for some $\lambda_n \in \mathbf{C}$, because $qAq = \mathbf{C}q$, so $x \in \mathbf{C}y$. This shows that $q(H) = \mathbf{C}y$, and therefore $q = y \otimes y$.

Now suppose that x is an arbitrary unit vector of H. As before, there are operators $u_n \in A$ such that $x = \lim_{n\to\infty} u_n(y)$, so

$$x \otimes x = \lim_{n\to\infty} u_n(y) \otimes u_n(y) = \lim_{n\to\infty} u_n(y \otimes y)u_n^* = \lim_{n\to\infty} u_n q u_n^*.$$

Hence, $x \otimes x \in A$. Therefore, all rank-one projections are in A, so $F(H) \subseteq A$ by Theorem 2.4.6, and therefore $K(H) \subseteq A$, by Theorem 2.4.5.   □

Before we introduce the Hilbert–Schmidt operators, it is convenient to make a few observations about summable families. Let $(x_\lambda)_{\lambda\in\Lambda}$ be a family of elements of a Banach space X. Let Λ' denote the set of all non-empty finite subsets of Λ, and for each $F \in \Lambda'$, set $x_F = \sum_{\lambda\in F} x_\lambda$. Then $(x_F)_{F\in\Lambda'}$ is a net where $F \leq G$ in Λ' if $F \subseteq G$. We say $(x_\lambda)_{\lambda\in\Lambda}$ is *summable* to an element $x \in X$ if the net $(x_F)_{F\in\Lambda'}$ converges to x, and in this case we write $x = \sum_{\lambda\in\Lambda} x_\lambda$.

If all x_λ are in $\mathbf{R}^+$, then the family $(x_\lambda)_{\lambda\in\Lambda}$ is summable if and only if $\sup_F \sum_{\lambda\in F} x_\lambda < +\infty$, and in this case

$$\sum_{\lambda\in\Lambda} x_\lambda = \sup_{F\in\Lambda'} \sum_{\lambda\in F} x_\lambda.$$

We thus can use the right-hand side of this expression to define $\sum_{\lambda\in\Lambda} x_\lambda$ whether $(x_\lambda)_{\lambda\in\Lambda}$ is summable or not, provided all x_λ are in $\mathbf{R}^+$.

Let u be an operator on a Hilbert space H, and suppose that E is an orthonormal basis for H. We define the *Hilbert–Schmidt norm* of u to be

$$\|u\|_2 = \left(\sum_{x\in E} \|u(x)\|^2 \right)^{1/2}.$$

This definition is independent of the choice of basis. To see this let E' be another orthonormal basis for H. Then for each finite non-empty set F of E,

$$\sum_{x\in F} \|u(x)\|^2 = \sum_{x\in F} \sum_{y\in E'} |\langle u(x), y\rangle|^2$$

$$= \sum_{y\in E'} \sum_{x\in F} |\langle u(x), y\rangle|^2$$

$$\leq \sum_{y\in E'} \|u^*(y)\|^2,$$

so

$$\sum_{x \in E} \|u(x)\|^2 \le \sum_{y \in E'} \|u^*(y)\|^2.$$

By symmetry, therefore,

$$\sum_{x \in E} \|u(x)\|^2 = \sum_{x \in E} \|u^*(x)\|^2 = \sum_{y \in E'} \|u(y)\|^2.$$

This shows not only that the expression for $\|u\|_2$ is independent of the choice of basis, but also that $\|u^*\|_2 = \|u\|_2$.

An operator u is a *Hilbert–Schmidt operator* if $\|u\|_2 < +\infty$. We denote the class of all Hilbert–Schmidt operators on H by $L^2(H)$.

2.4.1. Example. Let $(e_n)_{n=1}^\infty$ be an orthonormal basis for a Hilbert space H and let u be an operator on H diagonal with respect to (e_n), with diagonal sequence (λ_n). Then u is a Hilbert–Schmidt operator if and only if $\sum_{n=1}^\infty |\lambda_n|^2 < \infty$, since $\|u\|_2 = \sqrt{\sum_{n=1}^\infty |\lambda_n|^2}$.

More generally, if u is an arbitrary operator in $B(H)$ and $(\alpha_{n,m})$ is its matrix with respect to the basis (e_n), so that $\alpha_{n,m} = \langle u(e_m), e_n \rangle$, then from the definition

$$\|u\|_2 = \sqrt{\sum_{m=1}^\infty \sum_{n=1}^\infty |\alpha_{n,m}|^2},$$

and, therefore, u is Hilbert–Schmidt if and only if $\sum_m \sum_n |\alpha_{n,m}|^2 < \infty$.

2.4.2. Example. Let $L^2(\mathbf{T})$ and $L^2(\mathbf{T}^2)$ denote the Lebesgue L^2-spaces of $\mathbf{T}$ and $\mathbf{T}^2$ with the usual measures, normalised arc length m (that is, m is the Haar measure of $\mathbf{T}$), and the corresponding product measure $m \times m$. By elementary measure theory, $C(\mathbf{T})$ and $C(\mathbf{T}^2)$ are L^2-dense in $L^2(\mathbf{T})$ and $L^2(\mathbf{T}^2)$, respectively. Define $e_n \in C(\mathbf{T})$ by $e_n(\lambda) = \lambda^n$, and $e_{nm} \in C(\mathbf{T}^2)$ by $e_{nm}(\lambda, \mu) = \lambda^n \mu^m$. These sequences are orthonormal in the corresponding L^2-spaces. By the Stone–Weierstrass theorem, the sup-norm closed linear span of (e_n) in $C(\mathbf{T})$ is $C(\mathbf{T})$ itself, since this closed span is a C*-subalgebra separating the points of $\mathbf{T}$ and containing the constants. By similar reasoning the sup-norm closed linear span of (e_{nm}) is $C(\mathbf{T}^2)$. Thus, (e_n) and (e_{nm}) have L^2-norm dense linear span in, and are therefore orthonormal bases of, $L^2(\mathbf{T})$ and $L^2(\mathbf{T}^2)$, respectively.

Let k be an element of $L^2(\mathbf{T}^2)$. Then for almost all $\lambda \in \mathbf{T}$,

$$\int |k(\lambda, \mu) f(\mu)| \, dm\mu < \infty,$$

since

$$\iint |k(\lambda,\mu)f(\mu)|\, d(m\times m)(\lambda,\mu)$$
$$\leq (\iint |k(\lambda,\mu)|^2\, d(m\times m)(\lambda,\mu))^{1/2}(\iint |f(\mu)|^2\, d(m\times m)(\lambda,\mu))^{1/2}$$
$$= \|k\|_2\|f\|_2.$$

Define the *integral* operator $u = u_k$ on $L^2(\mathbf{T})$ by

$$(uf)(\lambda) = \int k(\lambda,\mu)f(\mu)\, dm\mu$$

for almost all λ. That $u(f) \in L^2(\mathbf{T})$ follows by another application of the Cauchy–Schwarz inequality,

$$\int |(uf)(\lambda)|^2\, dm\lambda = \int |\int k(\lambda,\mu)f(\mu)\, dm\mu|^2\, dm\lambda$$
$$\leq \int (\int |k(\lambda,\mu)|^2\, dm\mu)(\int |f(\mu)|^2\, dm\mu)\, dm\lambda$$
$$= \|k\|_2^2\|f\|_2^2.$$

Hence, u is bounded with norm $\|u\| \leq \|k\|_2$.

Now we compute $\|u\|_2$. From the definition,

$$\|u\|_2^2 = \sum_{n\in\mathbf{Z}} \|u(e_n)\|^2$$
$$= \sum_{n,m\in\mathbf{Z}} |\langle u(e_n), e_m\rangle|^2$$
$$= \sum_{n,m\in\mathbf{Z}} |\int (u(e_n))(\lambda)\overline{e_m(\lambda)}\, dm\lambda|^2$$
$$= \sum_{n,m\in\mathbf{Z}} |\iint k(\lambda,\mu)e_n(\mu)\overline{e_m(\lambda)}\, dm\mu\, dm\lambda|^2$$
$$= \sum_{n,m\in\mathbf{Z}} |\langle k, e_{m,-n}\rangle|^2.$$

Thus, $\|u\|_2 = \|k\|_2$, and therefore u is a Hilbert–Schmidt operator.

2.4.10. Theorem. *Let u,v be operators on a Hilbert space H, and $\lambda \in \mathbf{C}$. Then*

(1) $\|u + v\|_2 \leq \|u\|_2 + \|v\|_2$ and $\|\lambda u\|_2 = |\lambda|\|u\|_2$;
(2) $\|u\| \leq \|u\|_2$;
(3) $\|uv\|_2 \leq \|u\|\|v\|_2$ and $\|uv\|_2 \leq \|u\|_2\|v\|$.

Proof. If F is is any finite set of orthonormal vectors of H, then

$$\sqrt{\sum_{x \in F} \|u(x) + v(x)\|^2} \leq \sqrt{\sum_{x \in F} (\|u(x)\| + \|v(x)\|)^2}$$

$$\leq \sqrt{\sum_{x \in F} \|u(x)\|^2} + \sqrt{\sum_{x \in F} \|v(x)\|^2}.$$

It follows that $\|u + v\|_2 \leq \|u\|_2 + \|v\|_2$. The equality $\|\lambda u\|_2 = |\lambda| \|u\|_2$ is trivial.

If x is a unit vector of H, there is an orthonormal basis E containing x. Hence, $\|u(x)\|^2 \leq \sum_{y \in E} \|u(y)\|^2 = \|u\|_2^2$, so $\|u\| \leq \|u\|_2$.

If E is an arbitrary orthonormal basis of H, then

$$\|uv\|_2^2 = \sum_{x \in E} \|uv(x)\|^2 \leq \|u\|^2 \sum_{x \in E} \|v(x)\|^2 = \|u\|^2 \|v\|_2^2.$$

Hence, $\|uv\|_2 \leq \|u\| \|v\|_2$. Therefore, $\|uv\|_2 = \|v^*u^*\|_2 \leq \|v^*\| \|u^*\|_2 = \|u\|_2 \|v\|$. $\qquad\square$

2.4.11. Corollary. *The set $L^2(H)$ is a self-adjoint ideal of $B(H)$, and a normed $*$-algebra (that is, a normed algebra with an isometric involution), where the norm is given by $u \mapsto \|u\|_2$.*

Note that if $x, y \in H$, then $\|x \otimes y\|_2 = \|x\| \|y\|$, so $x \otimes y \in L^2(H)$. Hence, $F(H) \subseteq L^2(H)$.

2.4.12. Lemma. *Let u_1, u_2 be Hilbert–Schmidt operators on a Hilbert space H. If E is an orthonormal basis of H and $v = u_1^* u_2$, then the family $(\langle v(x), x \rangle)_{x \in E}$ is absolutely summable, that is, $\sum_{x \in E} |\langle v(x), x \rangle| < +\infty$, and*

$$\sum_{x \in E} \langle v(x), x \rangle = \tfrac{1}{4} \sum_{k=0}^{3} i^k \|u_2 + i^k u_1\|_2^2.$$

Proof. If F is a finite non-empty subset of E, then

$$\sum_{x \in F} |\langle v(x), x \rangle| = \sum_{x \in F} |\langle u_2(x), u_1(x) \rangle|$$

$$\leq \sum_{x \in F} \|u_2(x)\| \|u_1(x)\|$$

$$\leq \sqrt{\sum_{x \in F} \|u_2(x)\|^2} \sqrt{\sum_{x \in F} \|u_1(x)\|^2}.$$

Hence, $(\langle v(x), x \rangle)_{x \in E}$ is absolutely summable. Also,

$$\langle v(x), x \rangle = \langle u_2(x), u_1(x) \rangle = \tfrac{1}{4} \sum_{k=0}^{3} i^k \|u_2(x) + i^k u_1(x)\|^2$$

by the polarisation identity, so

$$\sum_{x \in E} \langle v(x), x \rangle = \tfrac{1}{4} \sum_{k=0}^{3} i^k \sum_{x \in E} \|(u_2 + i^k u_1)(x)\|^2 = \tfrac{1}{4} \sum_{k=0}^{3} i^k \|u_2 + i^k u_1\|_2^2,$$

which is the required result. $\square$

If u is an operator on a Hilbert space H, we define its *trace-class norm* to be $\|u\|_1 = \||u|^{1/2}\|_2^2$. If E is an orthonormal basis of H, then

$$\|u\|_1 = \sum_{x \in E} \langle |u|(x), x \rangle.$$

If $\|u\|_1 < +\infty$, we call u a *trace-class operator*. The connection between trace-class operators and Hilbert–Schmidt operators is given in the following result.

2.4.13. Theorem.   *Let v be an operator on a Hilbert space H. The following conditions are equivalent:*

(1) v *is trace-class.*
(2) $|v|$ *is trace-class.*
(3) $|v|^{1/2}$ *is a Hilbert–Schmidt operator.*
(4) *There exist Hilbert–Schmidt operators u_1, u_2 on H such that $v = u_1 u_2$.*

Proof.   The implications $(1) \Rightarrow (2) \Rightarrow (3) \Rightarrow (4)$ are easy (for $(3) \Rightarrow (4)$ use the polar decomposition of v), so we prove $(4) \Rightarrow (1)$ only.

Assume that $v = u_1 u_2$, where $u_1, u_2 \in L^2(H)$. If $v = w|v|$ is the polar decomposition of v, then $|v| = w^* v = (w^* u_1) u_2$. If E is any orthonormal basis of H, then by the "polarisation identity" of the preceding lemma, $\sum_{x \in E} \langle |v|(x), x \rangle < +\infty$, so $\|v\|_1 < +\infty$. $\square$

It is clear from Theorem 2.4.13 that if v is a trace-class operator and u is an arbitrary operator on H, then uv and vu are also trace-class operators.

We define the *trace* of a trace-class operator v to be

$$\mathrm{tr}(v) = \sum_{x \in E} \langle v(x), x \rangle,$$

where E is any orthonormal basis of H. By Lemma 2.4.12 the definition of tr is independent of the choice of orthonormal basis.

2.4.14. Theorem. *Let u and v be operators on a Hilbert space H. Then*

$$\mathrm{tr}(uv) = \mathrm{tr}(vu)$$

if either
(1) *u and v are both Hilbert–Schmidt operators,*
or
(2) *v is trace-class.*

Proof. In Case (1),

$$\mathrm{tr}(uv) = \tfrac{1}{4} \sum_{k=0}^{3} i^k \|v + i^k u^*\|_2^2$$

$$= \tfrac{1}{4} \sum_{k=0}^{3} i^k \|(v + i^k u^*)^*\|_2^2$$

$$= \tfrac{1}{4} \sum_{k=0}^{3} i^k \|u + i^k v^*\|_2^2$$

$$= \mathrm{tr}(vu).$$

In Case (2) $v = u_1 u_2$ for some $u_1, u_2 \in L^2(H)$, so $\mathrm{tr}(uv) = \mathrm{tr}((uu_1)u_2)$ $= \mathrm{tr}(u_2(uu_1))$ (by Case (1)) $= \mathrm{tr}(u_1(u_2 u))$ (same reason) $= \mathrm{tr}(vu)$. □

There are similar results for the trace-class norm as for the Hilbert–Schmidt norm, but the proofs require more work:

2.4.15. Theorem. *Let u, v be operators on a Hilbert space H and $\lambda \in \mathbf{C}$.*
(1) $\|u + v\|_1 \leq \|u\|_1 + \|v\|_1$ *and* $\|\lambda u\|_1 = |\lambda| \|u\|_1$.
(2) $\|u\| \leq \|u\|_1 = \|u^*\|_1$.
(3) $\|uv\|_1 \leq \|u\| \|v\|_1$ *and* $\|uv\|_1 \leq \|u\|_1 \|v\|$.

Proof. Beginning with Condition (2) we have $\|u\|_1 = \||u|^{1/2}\|_2^2 \geq \||u|^{1/2}\|^2$ $= \||u|\| = \|u\|$. If $u = w|u|$ is the polar decomposition of u, then $uu^* = w|u|^2 w^*$, so $|u^*|^2 = (w|u|w^*)^2$, and therefore $|u^*| = w|u|w^*$. Hence, $\|u^*\|_1 = \mathrm{tr}(|u^*|) = \mathrm{tr}(w|u|w^*) = \mathrm{tr}(w^*u) = \mathrm{tr}(|u|) = \|u\|_1$. This proves Condition (2).

Next, we show that Condition (3) holds. Let $vu = w'|vu|$ be the polar decomposition of vu and $w'' = w'^* vw$. Then $|vu| = w'^* vu = w''|u|$. Hence, $|vu|^2 = |u|w''^* w''|u| \leq |u|^2 \|w''\|^2 \leq |u|^2 \|v\|^2$, so $|vu| \leq |u| \|v\|$ by Theorem 2.2.6. Consequently, if E is an orthonormal basis for H,

$$\|vu\|_1 = \sum_{x \in E} \langle |vu|(x), x \rangle$$

$$\leq \sum_{x \in E} \langle |u|(x), x \rangle \|v\|$$

$$= \|u\|_1 \|v\|.$$

Also, $\|uv\|_1 = \|v^*u^*\|_1 \le \|v\|\|u\|_1$. This proves Condition (3).

Finally, we show Condition (1). The equality $\|\lambda u\|_1 = |\lambda|\|u\|_1$ follows from the corresponding statement for the norm $\|.\|_2$. Suppose that u and v are trace-class operators, and let $u = w|u|$, $v = w'|v|$, and $u + v = w''|u+v|$ be the respective polar decompositions. Then

$$|u+v| = w''^*(u+v) = w''^*w|u| + w''^*w'|v|.$$

If E is an orthonormal basis of H,

$$
\begin{aligned}
\|u+v\|_1 &= \sum_{x \in E} \langle |u+v|(x), x \rangle \\
&= |\sum_{x \in E} \langle w''^*w|u|(x), x \rangle + \sum_{x \in E} \langle w''^*w'|v|(x), x \rangle| \\
&\le \sum_{x \in E} |\langle |u|^{1/2}(x), |u|^{1/2}w^*w''(x) \rangle| + \sum_{x \in E} |\langle |v|^{1/2}(x), |v|^{1/2}w'^*w''(x) \rangle| \\
&\le (\sum_{x \in E} \||u|^{1/2}(x)\|^2)^{1/2} (\sum_{x \in E} \||u|^{1/2}w^*w''(x)\|^2)^{1/2} \\
&\quad + (\sum_{x \in E} \||v|^{1/2}(x)\|^2)^{1/2} (\sum_{x \in E} \||v|^{1/2}w'^*w''(x)\|^2)^{1/2} \\
&= \|u\|_1^{1/2} \||u|^{1/2}w^*w''\|_2 + \|v\|_1^{1/2} \||v|^{1/2}w'^*w''\|_2 \\
&\le \|u\|_1^{1/2}\|u\|_1^{1/2} + \|v\|_1^{1/2}\|v\|_1^{1/2} \\
&= \|u\|_1 + \|v\|_1,
\end{aligned}
$$

so $\|u+v\|_1 \le \|u\|_1 + \|v\|_1$. $\qquad\qquad\qquad\qquad\qquad\qquad\qquad\qquad\qquad$ □

If H is a Hilbert space, we denote the set of trace-class operators on H by $L^1(H)$. From the preceding theorem it is clear that $L^1(H)$ is a self-adjoint ideal of $B(H)$, and the function $u \mapsto \|u\|_1$ is a norm on $L^1(H)$ making it a normed $*$-algebra.

2.4.16. Theorem. Let H be a Hilbert space. The function

$$\mathrm{tr}: L^1(H) \to \mathbf{C}, \quad u \mapsto \mathrm{tr}(u),$$

is linear, and

$$|\mathrm{tr}(vu)| \le \|v\|\|u\|_1 \qquad (v \in B(H), u \in L^1(H)).$$

Proof. Linearity of the trace is clear. To show the inequality let $u = w|u|$ be the polar decomposition of u and let E be an orthonormal basis of H.

Then

$$|\operatorname{tr}(vu)| = |\sum_{x \in E} \langle vu(x), x \rangle|$$

$$= |\sum_{x \in E} \langle |u|^{1/2}(x), |u|^{1/2} w^* v^*(x) \rangle|$$

$$\leq \sum_{x \in E} \||u|^{1/2}(x)\| \||u|^{1/2} w^* v^*(x)\|$$

$$\leq (\sum_{x \in E} \||u|^{1/2}(x)\|^2)^{1/2} (\sum_{x \in E} \||u|^{1/2} w^* v^*(x)\|^2)^{1/2}$$

$$= \|u\|_1^{1/2} \||u|^{1/2} w^* v^*\|_2$$

$$\leq \|u\|_1^{1/2} \||u|^{1/2}\|_2 \|v\|$$

$$= \|u\|_1 \|v\|,$$

so $|\operatorname{tr}(vu)| \leq \|u\|_1 \|v\|$.   □

If $x, y \in H$, then $\|x \otimes y\|_1 = \|x\| \|y\|$ and $\operatorname{tr}(x \otimes y) = \langle x, y \rangle$. The inclusions $F(H) \subseteq L^1(H) \subseteq L^2(H)$ hold.

2.4.17. Theorem. *Let H be a Hilbert space. Then for $i = 1, 2$, the ideal $L^i(H)$ is contained in $K(H)$, and $F(H)$ is dense in $L^i(H)$ in the norm $\|.\|_i$.*

Proof. An easy exercise.   □

2.5. The Spectral Theorem

The normal operators form one of the best understood and most tractable of classes of operators. The principal reason for this is the spectral theorem, a powerful structure theorem that answers many (not all) questions about these operators. In this section we actually prove a more general result than the spectral theorem for normal operators (Theorem 2.5.6), and we get this extra useful generality without any increase in difficulty of the proofs. Indeed, the more general situation illustrates nicely the connection between spectral measures and representations of abelian C*-algebras.

Let Ω be a compact Hausdorff space and H a Hilbert space. A *spectral measure* E relative to (Ω, H) is a map from the σ-algebra of all Borel sets of Ω to the set of projections in $B(H)$ such that
  (1) $E(\emptyset) = 0$, $E(\Omega) = 1$;
  (2) $E(S_1 \cap S_2) = E(S_1)E(S_2)$ for all Borel sets S_1, S_2 of Ω;
  (3) for all $x, y \in H$, the function $E_{x,y}: S \mapsto \langle E(S)x, y \rangle$, is a regular Borel complex measure on Ω.

Denote by $M(\Omega)$ the Banach space of all regular Borel complex measures on Ω, and by $B_\infty(\Omega)$ the C*-algebra of all bounded Borel-measurable complex-valued functions on Ω.

2.5.1. *Example*. Let Ω be a compact Hausdorff space and let μ be a positive regular Borel measure on Ω. Define $M_\varphi \in B(L^2(\Omega, \mu))$ by

$$M_\varphi(f) = \varphi f \quad (f \in L^2(\Omega, \mu)).$$

That M_φ is bounded is given by

$$\|M_\varphi(f)\|_2^2 = \int |\varphi(\omega)f(\omega)|^2 \, d\mu\omega \leq \|\varphi\|_\infty^2 \int |f(\omega)|^2 \, d\mu\omega,$$

which implies that $\|M_\varphi\| \leq \|\varphi\|_\infty$. The operator M_φ is called a *multiplication* operator. The map

$$L^\infty(\Omega, \mu) \to B(L^2(\Omega, \mu)), \quad \varphi \mapsto M_\varphi,$$

is a *-homomorphism of C*-algebras. In particular, the adjoint of M_φ is $M_{\bar{\varphi}}$, and M_φ is normal. In fact, these operators are typical of all normal operators (see Section 4.4).

If S is a Borel set of Ω, then χ_S (the characteristic function of S) is a projection in $L^\infty(\Omega, \mu)$, so $E(S) = M_{\chi_S}$ is a projection in $B(L^2(\Omega, \mu))$. The map $E\colon S \mapsto E(S)$ is a spectral measure relative to the pair $(\Omega, L^2(\Omega, \mu))$.

Since the multiplication operators are a very important class we linger with this example a little longer to show that if $\varphi \in L^\infty(\Omega, \mu)$, then $\|M_\varphi\| = \|\varphi\|_\infty$. For if this is false, then there exists a positive number ε such that $\|\varphi\|_\infty - \varepsilon > \|M_\varphi\|$ and, therefore, there is a Borel set S of Ω such that $\mu(S) > 0$ and $|\varphi(\omega)| > \|M_\varphi\| + \varepsilon$ for all $\omega \in S$. Since μ is regular,

$$\mu(S) = \sup\{\mu(K) \mid K \text{ is compact and } K \subseteq S\},$$

so we may suppose that S is compact. Then $\mu(S) < \infty$, again by regularity of μ. However,

$$\begin{aligned}
\|M_\varphi\|^2 \mu(S) &\geq \|M_\varphi(\chi_S)\|_2^2 \\
&= \int |\varphi(\omega)\chi_S(\omega)|^2 \, d\mu\omega \\
&\geq \int (\|M_\varphi\| + \varepsilon)^2 \chi_S(\omega) \, d\mu\omega \\
&= (\|M_\varphi\| + \varepsilon)^2 \mu(S),
\end{aligned}$$

and therefore after dividing by $\mu(S)$, we get $\|M_\varphi\| \geq \|M_\varphi\| + \varepsilon$, a contradiction. This shows that $\|M_\varphi\| = \|\varphi\|_\infty$ as claimed.

This result means that the map $\varphi \mapsto M_\varphi$, is in fact an isometric *-isomorphism of $L^\infty(\Omega, \mu)$ onto a C*-subalgebra of $B(L^2(\Omega, \mu))$. We therefore have $\sigma(M_\varphi) = \sigma(\varphi)$ (the spectrum of φ in $L^\infty(\Omega, \mu)$).

2.5.1. Lemma. *Let Ω be a compact Hausdorff space, let H be a Hilbert space, and suppose that $\mu_{x,y} \in M(\Omega)$ for all $x, y \in H$. Suppose also that for each Borel set S of Ω the function*

$$\sigma_S: H^2 \to \mathbf{C}, \ (x, y) \mapsto \mu_{x,y}(S),$$

is a sesquilinear form. Then for each $f \in B_\infty(\Omega)$ the function

$$\sigma_f: H^2 \to \mathbf{C}, \ (x, y) \mapsto \int f \, d\mu_{x,y},$$

is a sesquilinear form.

Proof. Suppose first that f is simple, so we can write $f = \sum_{j=1}^n \lambda_j \chi_{S_j}$, where $S_1, \ldots, S_n$ are pairwise disjoint Borel sets of Ω, and $\lambda_1, \ldots, \lambda_n$ are complex numbers. Then

$$\int f \, d\mu_{x,y} = \sum_{j=1}^n \lambda_j \int \chi_{S_j} \, d\mu_{x,y} = \sum_{j=1}^n \lambda_j \mu_{x,y}(S_j).$$

The set of sesquilinear forms on H is a vector space with the pointwise-defined operations, and we have just shown that σ_f is a linear combination of the σ_{S_j}, so σ_f is a sesquilinear form.

Now suppose that f is an arbitrary element of $B_\infty(\Omega)$. Then f is the uniform limit of a sequence (f_n), where each f_n is a simple function in $B_\infty(\Omega)$. Hence, $\int |f_n - f| \, d|\mu_{x,y}| \leq \|f_n - f\|_\infty |\mu_{x,y}|(\Omega)$, so $\int f \, d\mu_{x,y} = \lim_{n \to \infty} \int f_n \, d\mu_{x,y}$ for each $x, y \in H$. It follows immediately that σ_f is a sesquilinear form on H. $\qquad\square$

2.5.2. Theorem. *Let Ω be a compact Hausdorff space, H a Hilbert space, and E a spectral measure relative to (Ω, H). Then for each $f \in B_\infty(\Omega)$ the function*

$$\sigma_f: H^2 \to \mathbf{C}, \ (x, y) \mapsto \int f \, dE_{x,y},$$

is a bounded sesquilinear form on H, and $\|\sigma_f\| \leq \|f\|_\infty$.

Proof. That σ_f is a sesquilinear form follows from the preceding lemma, so we need only show $\|\sigma_f\| \leq \|f\|_\infty$. Suppose that $\Omega = S_1 \cup \ldots \cup S_n$, where $S_1, \ldots, S_n$ are pairwise disjoint Borel sets of Ω. Then

$$\sum_{j=1}^n |\langle E(S_j)(x), y \rangle| = \sum_{j=1}^n |\langle E(S_j)(x), E(S_j)(y) \rangle|$$

$$\leq \left(\sum_{j=1}^n \|E(S_j)(x)\|^2 \right)^{1/2} \left(\sum_{j=1}^n \|E(S_j)(y)\|^2 \right)^{1/2}$$

$$= \|E(\Omega)(x)\| \|E(\Omega)(y)\|$$

$$= \|x\| \|y\|.$$

Hence, $\|E_{x,y}\| \leq \|x\|\|y\|$. Therefore,

$$|\int f \, dE_{x,y}| \leq \|f\|_\infty \|E_{x,y}\| \leq \|f\|_\infty \|x\|\|y\|,$$

so $\|\sigma_f\| \leq \|f\|_\infty$. $\qquad\qquad\qquad\qquad\qquad\qquad\qquad\qquad\qquad\qquad\square$

2.5.3. Theorem. *Let Ω be a compact Hausdorff space, H a Hilbert space, and E a spectral measure relative to (Ω, H). Then for each $f \in B_\infty(\Omega)$ there is a unique bounded operator u on H such that*

$$\langle u(x), y \rangle = \int f \, dE_{x,y} \qquad (x, y \in H).$$

Proof. Immediate from the preceding theorem and Theorem 2.3.6. $\qquad\square$

We write $\int f \, dE$ for u and call it the *integral* of f with respect to E. Note that $\int \chi_S \, dE = E(S)$ for each Borel set S.

2.5.4. Theorem. *With the same assumptions on Ω, H, and E as in the preceding theorem, the map*

$$\varphi: B_\infty(\Omega) \to B(H), \ f \mapsto \int f \, dE,$$

is a unital $*$-homomorphism.

Proof. Linearity is routine and boundedness follows from Theorems 2.5.2 and 2.3.6. To show that $\varphi(fg) = \varphi(f)\varphi(g)$ and $\varphi(\bar{f}) = (\varphi(f))^-$, we need only show these results when f, g are simple, because the simple elements of $B_\infty(\Omega)$ are dense. Hence, we may reduce further and suppose that $f = \chi_S$ and $g = \chi_{S'}$ by linearity of φ and the fact that all simple elements of $B_\infty(\Omega)$ are linear combinations of such characteristic functions. Then $\varphi(fg) = \int \chi_S \chi_{S'} \, dE = E(S \cap S') = E(S)E(S') = \int \chi_S \, dE \int \chi_{S'} \, dE = \varphi(f)\varphi(g)$. Also, $\varphi(\bar{f}) = \varphi(f) = E(S) = (\varphi(f))^-$. $\qquad\square$

2.5.5. Theorem. *Let Ω be a compact Hausdorff space and H a Hilbert space, and suppose that $\varphi: C(\Omega) \to B(H)$ is a unital $*$-homomorphism. Then there is a unique spectral measure E relative to (Ω, H) such that*

$$\varphi(f) = \int f \, dE \qquad (f \in C(\Omega)).$$

Moreover, if $u \in B(H)$, then u commutes with $\varphi(f)$ for all $f \in C(\Omega)$ if and only if u commutes with $E(S)$ for all Borel sets S of Ω.

Proof. If $x, y \in H$, then the function

$$\tau_{x,y}: C(\Omega) \to \mathbf{C}, \quad f \mapsto \langle \varphi(f)(x), y \rangle,$$

is linear and $\|\tau_{x,y}\| \leq \|x\| \|y\|$. By the Riesz–Kakutani theorem, there is a unique measure $\mu_{x,y}$ in $M(\Omega)$ such that $\tau_{x,y}(f) = \int f \, d\mu_{x,y}$ for all $f \in C(\Omega)$. Also, $\|\mu_{x,y}\| = \|\tau_{x,y}\|$. Since the function

$$H^2 \to \mathbf{C}, \quad (x, y) \mapsto \langle \varphi(f)(x), y \rangle,$$

is sesquilinear, the maps from H to $M(\Omega)$ given by

$$x \mapsto \mu_{x,y} \quad \text{and} \quad y \mapsto \mu_{x,y}$$

are, respectively, linear and conjugate-linear. Hence, for each $f \in B_\infty(\Omega)$ the function

$$H^2 \to \mathbf{C}, \quad (x, y) \mapsto \int f \, d\mu_{x,y},$$

is a sesquilinear form, by Lemma 2.5.1. Also,

$$\left| \int f \, d\mu_{x,y} \right| \leq \|f\|_\infty \|\mu_{x,y}\| \leq \|f\|_\infty \|x\| \|y\|,$$

so this sesquilinear form is bounded and its norm is not greater than $\|f\|_\infty$. By Theorem 2.3.6, there is a unique operator, $\psi(f)$ say, in $B(H)$ such that

$$\langle \psi(f)(x), y \rangle = \int f \, d\mu_{x,y} \qquad (x, y \in H).$$

Moreover, $\|\psi(f)\| \leq \|f\|_\infty$.

Now suppose that $f \in C(\Omega)$. Then

$$\langle \psi(f)(x), y \rangle = \int f \, d\mu_{x,y} = \tau_{x,y}(f) = \langle \varphi(f)(x), y \rangle \qquad (x, y \in H),$$

so $\psi(f) = \varphi(f)$.

It is straightforward to check that the map

$$\psi: B_\infty(\Omega) \to B(H), \quad f \mapsto \psi(f),$$

is linear and we already know it is norm-decreasing. We show now that ψ is a *-homomorphism.

If $f \in C(\Omega)$ and $\bar{f} = f$, then $\varphi(f)$ is hermitian, so $\int f \, d\mu_{x,x} = \langle \varphi(f)(x), x \rangle$ is a real number. Thus, the measure $\mu_{x,x}$ is real, that is, $\bar{\mu}_{x,x} = \mu_{x,x}$, and therefore if f is an arbitrary function in $B_\infty(\Omega)$ such that

$\bar{f} = f$, then $\langle \psi(f)(x), x \rangle = \int f \, d\mu_{x,x}$ is real. Since x is arbitrary, this shows that $\psi(f)$ is hermitian. Therefore, ψ preserves the involutions.

Let $f \in B_\infty(\Omega)$ and $x \in H$.

Assertion: If the equation

$$\langle \psi(fg)(x), x \rangle = \langle \psi(f)\psi(g)(x), x \rangle \tag{1}$$

holds for all $g \in C(\Omega)$, then it also holds for all $g \in B_\infty(\Omega)$.

Observe that Eq. (1) is equivalent to

$$\int gf \, d\mu_{x,x} = \int g \, d\mu_{x,\psi(\bar{f})(x)}. \tag{2}$$

To prove the assertion, note that if Eq. (1) holds for all $g \in C(\Omega)$, then the regular measures $f d\mu_{x,x}$ and $\mu_{x,\psi(\bar{f})(x)}$ are equal because Eq. (2) holds for all $g \in C(\Omega)$. Hence, Eq. (2) holds for all $g \in B_\infty(\Omega)$; that is, Eq. (1) holds for all such g, as claimed.

Since φ is a $*$-homomorphism, Eq. (1) holds for all $f, g \in C(\Omega)$. Hence, by the assertion, Eq. (1) holds if $f \in C(\Omega)$ and $g \in B_\infty(\Omega)$. Replacing f, g with their conjugates, we get $\langle \psi(\bar{f}\bar{g})(x), x \rangle = \langle \psi(\bar{f})\psi(\bar{g})(x), x \rangle$. Taking conjugates of both sides of this equation and using the fact that ψ preserves the involutions, we get

$$\langle \psi(gf)(x), x \rangle = \langle \psi(g)\psi(f)(x), x \rangle, \tag{3}$$

for all $g \in B_\infty(\Omega)$ and all $f \in C(\Omega)$. Using the assertion again (with the roles of f and g interchanged), we get Eq. (3) holds for all $f, g \in B_\infty(\Omega)$. Since x was an arbitrary element of H, this implies that $\psi(gf) = \psi(g)\psi(f)$, so ψ is a homomorphism.

If S is a Borel set of Ω, we put $E(S) = \psi(\chi_S)$. Obviously, $E(S)$ is a projection on H, and it is easily verified that the map $E: S \mapsto E(S)$ from the σ-algebra of Borel sets of Ω to $B(H)$ is a spectral measure relative to (Ω, H)—we have $E_{x,y} = \mu_{x,y} \in M(\Omega)$, since $E_{x,y}(S) = \langle E(S)(x), y \rangle = \langle \psi(\chi_S)(x), y \rangle = \int \chi_S \, d\mu_{x,y}$.

If $f \in B_\infty(\Omega)$, then

$$\langle \left(\int f \, dE \right)(x), y \rangle = \int f \, dE_{x,y} = \int f \, d\mu_{x,y} = \langle \psi(f)(x), y \rangle,$$

so $\psi(f) = \int f \, dE$. In particular, $\varphi(f) = \int f \, dE$ for all $f \in C(\Omega)$.

To see uniqueness of E, suppose that E' is another spectral measure relative to (Ω, H) such that $\varphi(f) = \int f \, dE'$ for all $f \in C(\Omega)$. Then $\int f \, dE'_{x,y} = \langle \varphi(f)(x), y \rangle = \int f \, dE_{x,y}$. Hence, $E'_{x,y} = E_{x,y}$, and therefore $\langle E'(S)(x), y \rangle = \langle E(S)(x), y \rangle$, so $E = E'$.

Now suppose u is an operator on H commuting with all of the elements of the range of φ. Then if $f \in C(\Omega)$, $\int f \, d\mu_{u(x),y} = \langle \psi(f)u(x), y \rangle =$

$\langle u\psi(f)(x), y \rangle = \langle \psi(f)(x), u^*(y) \rangle = \int f \, d\mu_{x,u^*(y)}$. Hence $E_{u(x),y} = E_{x,u^*(y)}$, so $E(S)u = uE(S)$ for all Borel sets S.

Conversely, suppose now that u commutes with all the projections $E(S)$. Then

$$\langle E(S)u(x), y \rangle = \langle uE(S)(x), y \rangle = \langle E(S)(x), u^*(y) \rangle,$$

so $E_{u(x),y} = E_{x,u^*(y)}$. Hence, for every $f \in C(\Omega)$,

$$\int f \, dE_{u(x),y} = \int f \, dE_{x,u^*(y)};$$

that is, $\langle \varphi(f)u(x), y \rangle = \langle \varphi(f)(x), u^*(y) \rangle$, so $\varphi(f)u = u\varphi(f)$.   □

The next result (which is a special case of Theorem 2.5.5) is one of the most important in single operator theory, and is called the *spectral theorem*.

2.5.6. Theorem. *Let u be a normal operator on a Hilbert space H. Then there is a unique spectral measure E relative to $(\sigma(u), H)$ such that $u = \int z \, dE$, where z is the inclusion map of $\sigma(u)$ in $\mathbf{C}$.*

Proof. Let $\varphi: C(\sigma(u)) \to B(H)$ be the functional calculus at u. By the preceding theorem, there exists a unique spectral measure E relative to $(\sigma(u), H)$ such that $\varphi(f) = \int f \, dE$ for all $f \in C(\sigma(u))$. In particular, $u = \varphi(z) = \int z \, dE$. If E' is another spectral measure such that $u = \int z \, dE'$, then $\int f \, dE' = \int f \, dE = \varphi(f)$ for all $f \in C(\sigma(u))$, since 1 and z generate $C(\sigma(u))$. Therefore, $E = E'$.   □

The spectral measure E in Theorem 2.5.6 is called the *resolution of the identity* for u. Since $f(u) = \int f \, dE$ for all $f \in C(\sigma(u))$, we can unambiguously define $f(u) = \int f \, dE$ for all $f \in B_\infty(\sigma(u))$. The unital *-homomorphism

$$B_\infty(\sigma(u)) \to B(H), \quad f \mapsto f(u),$$

is called the *Borel functional calculus* at u.

If $v \in B(H)$ commutes with both u and u^*, then v commutes with $f(u)$ for all $f \in B_\infty(\sigma(u))$. For in this case v commutes with all polynomials in u and u^*, and since 1 and z generate $C(\sigma(u))$ by the Stone–Weierstrass theorem, v commutes with $f(u)$ for all $f \in C(\sigma(u))$. By Theorem 2.5.5, therefore, v commutes with $E(S)$ for all Borel sets S of $\sigma(u)$. It follows that $E_{x,v^*(y)} = E_{v(x),y}$ for all $x, y \in H$. Hence, if $f \in B_\infty(\sigma(u))$,

$$\langle (vf(u))(x), y \rangle = \int f \, dE_{x,v^*(y)}$$

$$= \int f \, dE_{v(x),y}$$

$$= \langle f(u)v(x), y \rangle.$$

Therefore, $vf(u) = f(u)v$.

Incidentally, if S is a Borel set of $\sigma(u)$, then $\chi_S(u) = E(S)$.

2.5.7. Theorem. *Let u be a normal operator on a Hilbert space H, and suppose that $g: \mathbf{C} \to \mathbf{C}$ is a continuous function. Then $(g \circ f)(u) = g(f(u))$ for all $f \in B_\infty(\sigma(u))$.*

Proof. The result is easily seen by first showing it for g a polynomial in z and $\bar{z}$, and then observing that an arbitrary continuous function $g: \mathbf{C} \to \mathbf{C}$ is a uniform limit of such polynomials on the compact disc $\Delta = \{\lambda \in \mathbf{C} \mid |\lambda| \leq \|f\|_\infty\}$, using the Stone–Weierstrass theorem applied to $C(\Delta)$. □

We give an application of this to writing a unitary as an exponential.

2.5.8. Theorem. *Let u be a unitary operator in $B(H)$, where H is a Hilbert space. Then there exists a hermitian operator v in $B(H)$ such that $u = e^{iv}$ and $\|v\| \leq 2\pi$.*

Proof. The function

$$f: [0, 2\pi) \to \mathbf{T}, \ t \mapsto e^{it},$$

is a continuous bijection with Borel measurable inverse g. Since $\sigma(u) \subseteq \mathbf{T}$, we can set $v = g(u)$. The operator v is self-adjoint because g is real-valued. Moreover, $\|v\| \leq \|g\|_\infty \leq 2\pi$. By Theorem 2.5.7, $(f \circ g)(u) = f(g(u)) = f(v) = e^{iv}$. But $(f \circ g)(\lambda) = \lambda$ for all $\lambda \in \mathbf{T}$, so $(f \circ g)(u) = u$. Therefore, $u = e^{iv}$. □

2. Exercises

1. Let A be a Banach algebra such that for all $a \in A$ the implication

$$Aa = 0 \text{ or } aA = 0 \Rightarrow a = 0$$

holds. Let L, R be linear mappings from A to itself such that for all $a, b \in A$,

$$L(ab) = L(a)b, \quad R(ab) = aR(b), \quad \text{and} \quad R(a)b = aL(b).$$

Show that L and R are necessarily continuous.

2. Let A be a unital C*-algebra.
(a) If a, b are positive elements of A, show that $\sigma(ab) \subseteq \mathbf{R}^+$.
(b) If a is an invertible element of A, show that $a = u|a|$ for a unique unitary u of A. Give an example of an element of $B(H)$ for some Hilbert space H that cannot be written as a product of a unitary times a positive operator.
(c) Show that if $a \in \mathrm{Inv}(A)$, then $\|a\| = \|a^{-1}\| = 1$ if and only if a is a unitary.

3. Let Ω be a locally compact Hausdorff space, and suppose that the C*-algebra $C_0(\Omega)$ is generated by a sequence of projections $(p_n)_{n=1}^\infty$. Show that the hermitian element $h = \sum_{n=1}^\infty p_n/3^n$ generates $C_0(\Omega)$.

4. We shall see in the next chapter that all closed ideals in C*-algebras are necessarily self-adjoint. Give an example of an ideal in the C*-algebra $C(\mathbf{D})$ that is not self-adjoint.

5. Let $\varphi: A \to B$ be an isometric linear map between unital C*-algebras A and B such that $\varphi(a^*) = \varphi(a)^*$ $(a \in A)$ and $\varphi(1) = 1$. Show that $\varphi(A^+) \subseteq B^+$.

6. Let A be a unital C*-algebra.
(a) If $r(a) < 1$ and $b = (\sum_{n=0}^\infty a^{*n}a^n)^{1/2}$, show that $b \geq 1$ and $\|bab^{-1}\| < 1$.
(b) For all $a \in A$, show that

$$r(a) = \inf_{b \in Inv(A)} \|bab^{-1}\| = \inf_{b \in A_{sa}} \|e^b a e^{-b}\|.$$

7. Let A be a unital C*-algebra.
(a) If $a, b \in A$, show that the map

$$f: \mathbf{C} \to A, \quad \lambda \mapsto e^{i\lambda b} a e^{-i\lambda b},$$

is differentible and that $f'(0) = i(ba - ab)$.
(b) Let X be a closed vector subspace of A which is unitarily invariant in the sense that $uXu^* \subseteq X$ for all unitaries u of A. Show that $ba - ab \in X$ if $a \in X$ and $b \in A$.
(c) Deduce that the closed linear span X of the projections in A has the property that $a \in X$ and $b \in A$ implies that $ba - ab \in X$.

8. Let a be a normal element of a C*-algebra A, and b an element commuting with a. Show that b^* also commutes with a (Fuglede's theorem). (Hint: Define $f(\lambda) = e^{i\lambda a^*}be^{-i\lambda a^*}$ in $\tilde{A}$ and deduce from Exercise 2.7 that this map is differentiable and $f'(0) = i(a^*b - ba^*)$. Since $e^{i\lambda a}$ and b commute, $f(\lambda) = e^{2ic(\lambda)}be^{-2ic(\lambda)}$, where $c(\lambda) = Re(\lambda a^*)$. Hence, $\|f(\lambda)\| = \|b\|$, so by Liouville's theorem, $f(\lambda)$ is constant.)

In the following exercises H is a Hilbert space:

9. If I is an ideal of $B(H)$, show that it is self-adjoint.

10. Let $u \in B(H)$.

(a) Show that u is a left topological zero divisor in $B(H)$ if and only if it is not bounded below (*cf.* Exercise 1.11).

(b) Define
$$\sigma_{ap}(u) = \{\lambda \in \mathbf{C} \mid u - \lambda \text{ is not bounded below}\}.$$

This set is called the *approximate point spectrum* of u because $\lambda \in \sigma_{ap}(u)$ if and only if there is a sequence (x_n) of unit vectors of H such that $\lim_{n \to \infty} \|(u - \lambda)(x_n)\| = 0$. Show that $\sigma_{ap}(u)$ is a closed subset of $\sigma(u)$ containing $\partial\sigma(u)$.

(c) Show that u is bounded below if and only if it is left-invertible in $B(H)$.

(d) Show that $\sigma(u) = \sigma_{ap}(u)$ if u is normal.

11. Let $u \in B(H)$ be a normal operator with spectral resolution of the identity E.

(a) Show that u admits an invariant closed vector subspace other than 0 and H if $\dim(H) > 1$.

(b) If λ is an isolated point of $\sigma(u)$, show that $E(\lambda) = \ker(u - \lambda)$ and that λ is an eigenvalue of u.

12. An operator u on H is *subnormal* if there is a Hilbert space K containing H as a closed vector subspace and there exists a normal operator v on K such that H is invariant for v, and u is the restriction of v. We call v a *normal extension* of u.

(a) Show that the unilateral shift is a non-normal subnormal operator.

(b) Show that if u is subnormal, then $u^*u \geq uu^*$.

(c) A normal extension $v \in B(K)$ of a subnormal operator $u \in B(H)$ is a *minimal* normal extension if the only closed vector subspace of K reducing v and containing H is K itself. Show that u admits a minimal normal extension. In the case that v is a minimal normal extension, show that K is the closed linear span of all $v^{*n}(x)$ ($n \in \mathbf{N}$, $x \in H$).

(d) Show that if $v \in B(K)$ and $v' \in B(K')$ are minimal normal extensions of u, then there exists a unitary operator $w \colon K \to K'$ such that $v' = wvw^*$ (so there is only one minimal normal extension).

2. Addenda

In the following, H is an infinite-dimensional separable Hilbert space.

If u is a self-adjoint operator on H, then there exists a self-adjoint diagonalisable operator v and a self-adjoint compact operator w on H, such that $u = v + w$ (Weyl–von Neumann). Similarly, if u is a normal operator on H, there exists a diagonalisable operator v and a compact operator w on H, such that $u = v + w$ (I. D. Berg).

An operator $u \in B(H)$ is *essentially normal* if $u^*u - uu^*$ is a compact operator. If u is the sum of a normal and a compact operator, then obviously it is essentially normal. The unilateral shift is essentially normal, but it is not the sum of a normal and a compact operator, since it has non-zero Fredholm index. It turns out that the index is the only obstruction to an essentially normal operator being the sum of a normal and a compact operator. More precisely, for u an essentially normal operator on H, the following conditions are equivalent:

(a) u is the sum of a normal operator and a compact operator.
(b) u is the sum of a diagonalisable operator and a compact operator.
(c) For all $\lambda \in \mathbf{C} \setminus \sigma_e(u)$, the operator $u - \lambda$ has zero Fredholm index.

An operator u is *essentially unitary* if $u^*u - 1$ and $uu^* - 1$ are compact operators.

If v is the unilateral shift and u is an essentially unitary operator on H of Fredholm index n, then there exists a compact operator w such that
(a) $u - w$ is unitary if $n = 0$;
(b) $u - w = v^{-n}$ if n is negative;
(c) $u - w = v^{*n}$ if n is positive.

Let u, v be essentially normal operators on H. The following conditions are equivalent:
(a) There exists a compact operator w on H such that $v - w$ is unitarily equivalent to u.
(b) The esssential spectra of u and v are the same set, K say, and for each $\lambda \in \mathbf{C} \setminus K$ the operators $u - \lambda$ and $v - \lambda$ have the same Fredholm index.

These surprising and elegant results on essentially normal operators are due to L. Brown, R. G. Douglas, and P. Fillmore:

We shall see in the next chapter that $B(H)/K(H)$ is a C*-algebra. If $\pi \colon B(H) \to B(H)/K(H)$ is the quotient homomorphism, then for $u \in B(H)$ the image $\pi(u)$ is normal if and only if u is essentially normal, and $\pi(u)$ is unitary if and only if u is essentially unitary.

Although the BDF results are expressed purely in terms of single operator theory, the proofs involve C*-algebras and homological algebraic techniques. The introduction of the latter into the subject of operator algebras has given a revolutionary impetus to its development.

Reference: [BDF].

Let u be a subnormal operator on H. If v is the minimal normal extension of u, then $\sigma(v) \subseteq \sigma(u)$ (P. Halmos). Hence, $r(u) = \|u\|$. A much deeper result is that u necessarily has an invariant closed vector subspace other than 0 and H (Scott Brown).

For the theory of subnormal operators see [Cnw 1].

CHAPTER 3

Ideals and Positive Functionals

In this chapter we show that every C*-algebra can be realised as a C*-subalgebra of $B(H)$ for some Hilbert space H. This is the Gelfand–Naimark theorem, and it is one of the fundamental results of the theory of C*-algebras. A key step in its proof is the GNS construction which sets up a correspondence between the positive linear functionals and some of the representations of the algebra. This correspondence will be exploited in many situations in the sequel. There are also deep connections between the positive linear functionals and the closed ideals and closed left ideals of the algebra.

We also look at hereditary C*-subalgebras. These are a sort of generalisation of ideals and are of great importance in the theory.

In the final section of this chapter we apply some of the results we have developed so far to an interesting and highly non-trivial class of operators, the Toeplitz operators.

3.1. Ideals in C*-Algebras

In this section we prove basic results on ideals and homomorphisms. First, we show the existence of approximate units in C*-algebras. Of course, if a C*-algebra is non-unital, one can simply adjoin a unit, as we have frequently done. This is not always appropriate, however—consider the problem of showing that closed ideals are self-adjoint (this is shown by using approximate units).

An *approximate unit* for a C*-algebra A is an increasing net $(u_\lambda)_{\lambda \in \Lambda}$ of positive elements in the closed unit ball of A such that $a = \lim_\lambda au_\lambda$ for all $a \in A$. Equivalently, $a = \lim_\lambda u_\lambda a$ for all $a \in A$.

3.1.1. Example. Let H be a Hilbert space with an orthonormal basis $(e_n)_{n=1}^{\infty}$. The C*-algebra $K(H)$ is of course non-unital, since $\dim(H) = \infty$. If p_n is the projection onto $\mathbf{C}e_1 + \cdots + \mathbf{C}e_n$, then the increasing sequence (p_n) is an approximate unit for $K(H)$. To see this we need only show that $u = \lim_{n \to \infty} p_n u$ if $u \in F(H)$, since $F(H)$ is dense in $K(H)$. Now if $u \in F(H)$, there exist $x_1, \ldots, x_m, y_1, \ldots, y_m$ in H such that $u = \sum_{k=1}^{m} x_k \otimes y_k$. Hence, $p_n u = \sum_{k=1}^{m} p_n(x_k) \otimes y_k$. Since $\lim_{n \to \infty} p_n(x) = x$ for all $x \in H$, therefore for each k,

$$\lim_{n \to \infty} \|p_n(x_k) \otimes y_k - x_k \otimes y_k\| = \lim_{n \to \infty} \|p_n(x_k) - x_k\| \|y_k\| = 0.$$

Hence, $\lim_{n \to \infty} p_n u = u$.

Let A be an arbitrary C*-algebra and denote by Λ the set of all positive elements a in A such that $\|a\| < 1$. This set is a poset under the partial order of A_{sa}. In fact, Λ is also upwards-directed; that is, if $a, b \in \Lambda$, then there exists $c \in \Lambda$ such that $a, b \leq c$. We show this: If $a \in A^+$, then $1 + a$ is of course invertible in $\tilde{A}$, and $a(1 + a)^{-1} = 1 - (1 + a)^{-1}$. We claim

$$a, b \in A^+ \text{ and } a \leq b \Rightarrow a(1 + a)^{-1} \leq b(1 + b)^{-1}. \tag{1}$$

Indeed, if $0 \leq a \leq b$, then $1 + a \leq 1 + b$ implies $(1 + a)^{-1} \geq (1 + b)^{-1}$, by Theorem 2.2.5, and therefore $1 - (1 + a)^{-1} \leq 1 - (1 + b)^{-1}$; that is, $a(1 + a)^{-1} \leq b(1 + b)^{-1}$, proving the claim. Observe that if $a \in A^+$, then $a(1 + a)^{-1}$ belongs to Λ (use the Gelfand representation applied to the C*-subalgebra generated by 1 and a). Suppose then that a, b are an arbitrary pair of elements of Λ. Put $a' = a(1 - a)^{-1}$, $b' = b(1 - b)^{-1}$ and $c = (a' + b')(1 + a' + b')^{-1}$. Then $c \in \Lambda$, and since $a' \leq a' + b'$, we have $a = a'(1 + a')^{-1} \leq c$, by (1). Similarly, $b \leq c$, and therefore Λ is upwards-directed, as asserted.

3.1.1. Theorem. *Every C*-algebra A admits an approximate unit. Indeed, if Λ is the upwards-directed set of all $a \in A^+$ such that $\|a\| < 1$ and $u_\lambda = \lambda$ for all $\lambda \in \Lambda$, then $(u_\lambda)_{\lambda \in \Lambda}$ is an approximate unit for A (called the canonical approximate unit).*

Proof. From the remarks preceding this theorem, $(u_\lambda)_{\lambda \in \Lambda}$ is an increasing net of positive elements in the closed unit ball of A. Therefore, we need only show that $a = \lim_\lambda u_\lambda a$ for each $a \in A$. Since Λ linearly spans A, we can reduce to the case where $a \in \Lambda$.

Suppose then that $a \in \Lambda$ and that $\varepsilon > 0$. Let $\varphi \colon C^*(a) \to C_0(\Omega)$ be the Gelfand representation. If $f = \varphi(a)$, then $K = \{\omega \in \Omega \mid |f(\omega)| \geq \varepsilon\}$ is compact, and therefore by Urysohn's lemma there is a continuous function $g \colon \Omega \to [0, 1]$ of compact support such that $g(\omega) = 1$ for all $\omega \in K$. Choose $\delta > 0$ such that $\delta < 1$ and $1 - \delta < \varepsilon$. Then $\|f - \delta gf\| \leq \varepsilon$. If $\lambda_0 =$

$\varphi^{-1}(\delta g)$, then $\lambda_0 \in \Lambda$ and $\|a - u_{\lambda_0} a\| \leq \varepsilon$. Now suppose that $\lambda \in \Lambda$ and $\lambda \geq \lambda_0$. Then $1 - u_\lambda \leq 1 - u_{\lambda_0}$, so $a(1 - u_\lambda)a \leq a(1 - u_{\lambda_0})a$. Hence, $\|a - u_\lambda a\|^2 = \|(1 - u_\lambda)^{1/2}(1 - u_\lambda)^{1/2} a\|^2 \leq \|(1 - u_\lambda)^{1/2} a\|^2 = \|a(1 - u_\lambda)a\| \leq \|a(1 - u_{\lambda_0})a\| \leq \|(1 - u_{\lambda_0})a\| \leq \varepsilon$. This shows that $a = \lim_\lambda u_\lambda a$. $\square$

3.1.1. Remark. If a C*-algebra A is separable, then it admits an approximate unit which is a sequence. For in this case there exist finite sets $F_1 \subseteq F_2 \subseteq \ldots \subseteq F_n \subseteq \ldots$ such that $F = \cup_{n=1}^\infty F_n$ is dense in A. Let $(u_\lambda)_{\lambda \in \Lambda}$ be any approximate unit for A. If $\varepsilon > 0$, and $F_n = \{a_1, \ldots, a_m\}$ say, then there exist $\lambda_1, \ldots, \lambda_m \in \Lambda$ such that $\|a_j - a_j u_\lambda\| < \varepsilon$ if $\lambda \geq \lambda_j$. Choose $\lambda_\varepsilon \in \Lambda$ such that $\lambda_\varepsilon \geq \lambda_1, \ldots, \lambda_m$. Then $\|a - a u_\lambda\| < \varepsilon$ for all $a \in F_n$ and all $\lambda \geq \lambda_\varepsilon$. Hence, if n is a positive integer and $\varepsilon = 1/n$, then there exists $\lambda_n = \lambda_\varepsilon \in \Lambda$ such that $\|a - a \lambda_n\| < 1/n$ for all $a \in F_n$. Also, we may obviously choose the λ_n such that $\lambda_n \leq \lambda_{n+1}$ for all n. Consequently, $\lim_{n \to \infty} \|a - a u_{\lambda_n}\| = 0$, for all $a \in F$, and since F is dense in A, this also holds for all $a \in A$. Therefore, $(u_{\lambda_n})_{n=1}^\infty$ is an approximate unit for A.

3.1.2. Theorem. *If L is a closed left ideal in a C*-algebra A, then there is an increasing net $(u_\lambda)_{\lambda \in \Lambda}$ of positive elements in the closed unit ball of L such that $a = \lim_\lambda a u_\lambda$ for all $a \in L$.*

Proof. Set $B = L \cap L^*$. Since B is a C*-algebra, it admits an approximate unit, $(u_\lambda)_{\lambda \in \Lambda}$ say, by Theorem 3.1.1. If $a \in L$, then $a^* a \in B$, so $0 = \lim_\lambda a^* a(1 - u_\lambda)$. Hence, $\lim_\lambda \|a - a u_\lambda\|^2 = \lim_\lambda \|(1 - u_\lambda)a^* a(1 - u_\lambda)\| \leq \lim_\lambda \|a^* a(1 - u_\lambda)\| = 0$, and therefore $\lim_\lambda \|a - a u_\lambda\| = 0$. $\square$

In the preceding proof we worked in the unitisation $\tilde{A}$ of A. We shall frequently do this tacitly.

3.1.3. Theorem. *If I is a closed ideal in a C*-algebra A, then I is self-adjoint and therefore a C*-subalgebra of A. If $(u_\lambda)_{\lambda \in \Lambda}$ is an approximate unit for I, then for each $a \in A$*

$$\|a + I\| = \lim_\lambda \|a - u_\lambda a\| = \lim_\lambda \|a - a u_\lambda\|.$$

Proof. By Theorem 3.1.2 there is an increasing net $(u_\lambda)_{\lambda \in \Lambda}$ of positive elements in the closed unit ball of I such that $a = \lim_\lambda a u_\lambda$ for all $a \in I$. Hence, $a^* = \lim_\lambda u_\lambda a^*$, so $a^* \in I$, because all of the elements u_λ belong to I. Therefore, I is self-adjoint.

Suppose that $(u_\lambda)_{\lambda \in \Lambda}$ is an arbitrary approximate unit of I, that $a \in A$, and that $\varepsilon > 0$. There is an element b of I such that $\|a + b\| < \|a + I\| + \varepsilon/2$. Since $b = \lim_\lambda u_\lambda b$, there exists $\lambda_0 \in \Lambda$ such that $\|b - u_\lambda b\| < \varepsilon/2$ for all $\lambda \geq \lambda_0$, and therefore

$$\begin{aligned}
\|a - u_\lambda a\| &\leq \|(1 - u_\lambda)(a + b)\| + \|b - u_\lambda b\| \\
&\leq \|a + b\| + \|b - u_\lambda b\| \\
&< \|a + I\| + \varepsilon/2 + \varepsilon/2.
\end{aligned}$$

It follows that $\|a + I\| = \lim_\lambda \|a - u_\lambda a\|$, and therefore also $\|a + I\| = \|a^* + I\| = \lim_\lambda \|a^* - u_\lambda a^*\| = \lim_\lambda \|a - au_\lambda\|$. $\square$

3.1.2. Remark. Let I be a closed ideal in a C*-algebra A, and J a closed ideal in I. Then J is also an ideal in A. To show this we need only show that ab and ba are in J if $a \in A$ and b is a positive element of J (since J is a C*-algebra, J^+ linearly spans J). If $(u_\lambda)_{\lambda \in \Lambda}$ is an approximate unit for I, then $b^{1/2} = \lim_\lambda u_\lambda b^{1/2}$ because $b^{1/2} \in I$. Hence, $ab = \lim_\lambda au_\lambda b^{1/2} b^{1/2}$, so $ab \in J$ because $b^{1/2} \in J$, $au_\lambda b^{1/2} \in I$, and J is an ideal in I. Therefore, $a^* b \in J$ also, so $ba \in J$, since J is self-adjoint.

3.1.4. Theorem. *If I is a closed ideal of a C*-algebra A, then the quotient A/I is a C*-algebra under its usual operations and the quotient norm.*

Proof. Let $(u_\lambda)_{\lambda \in \Lambda}$ be a approximate unit for I. If $a \in A$ and $b \in I$, then

$$
\begin{aligned}
\|a + I\|^2 &= \lim_\lambda \|a - au_\lambda\|^2 \text{ (by Theorem 3.1.3)} \\
&= \lim_\lambda \|(1 - u_\lambda)a^* a(1 - u_\lambda)\| \\
&\leq \sup_\lambda \|(1 - u_\lambda)(a^* a + b)(1 - u_\lambda)\| + \lim_\lambda \|(1 - u_\lambda)b(1 - u_\lambda)\| \\
&\leq \|a^* a + b\| + \lim_\lambda \|b - u_\lambda b\| \\
&= \|a^* a + b\|.
\end{aligned}
$$

Therefore, $\|a + I\|^2 \leq \|a^* a + I\|$. By Lemma 2.1.3, A/I is a C*-algebra. $\square$

3.1.5. Theorem. *If $\varphi: A \to B$ is an injective $*$-homomorphism between C*-algebras A and B, then φ is necessarily isometric.*

Proof. It suffices to show that $\|\varphi(a)\|^2 = \|a\|^2$, that is, $\|\varphi(a^* a)\| = \|a^* a\|$. Thus, we may suppose that A is abelian (restrict to $C(a^* a)$ if necessary), and that B is abelian (replace B by $\varphi(A)^-$ if required). Moreover, by extending $\varphi: A \to B$ to $\tilde{\varphi}: \tilde{A} \to \tilde{B}$ if necessary, we may further assume that A, B, and φ are unital.

If τ is a character on B, then $\tau \circ \varphi$ is one on A. Clearly the map

$$\varphi': \Omega(B) \to \Omega(A), \quad \tau \mapsto \tau \circ \varphi,$$

is continuous. Hence, $\varphi'(\Omega(B))$ is compact, because $\Omega(A)$ is compact, and therefore $\varphi'(\Omega(B))$ is closed in $\Omega(A)$. If $\varphi'(\Omega(B)) \neq \Omega(A)$, then by Urysohn's lemma there is a non-zero continuous function $f: \Omega(A) \to \mathbf{C}$ such that f vanishes on $\varphi'(\Omega(B))$. By the Gelfand representation, $f = \hat{a}$ for some element $a \in A$. Hence, for each $\tau \in \Omega(B)$, $\tau(\varphi(a)) = \hat{a}(\tau \circ \varphi) = 0$.

Therefore, $\varphi(a) = 0$, so $a = 0$. But this implies that f is zero, a contradiction. The only way to avoid this is to have $\varphi'(\Omega(B)) = \Omega(A)$. Hence, for each $a \in A$,

$$\|a\| = \|\hat{a}\|_\infty = \sup_{\tau \in \Omega(A)} |\tau(a)| = \sup_{\tau \in \Omega(B)} |\tau(\varphi(a))| = \|\varphi(a)\|.$$

Thus, φ is isometric. $\square$

3.1.6. Theorem. *If $\varphi: A \to B$ is a $*$-homomorphism between C*-algebras, then $\varphi(A)$ is a C*-subalgebra of B.*

Proof.   The map

$$A/\ker(\varphi) \to B, \quad a + \ker(\varphi) \mapsto \varphi(a),$$

is an injective $*$-homomorphism between C*-algebras and is therefore isometric. Its image is $\varphi(A)$, so this space is necessarily complete and therefore closed in B. $\square$

3.1.7. Theorem.   *Let B and I be respectively a C*-subalgebra and a closed ideal in a C*-algebra A. Then $B + I$ is a C*-subalgebra of A.*

Proof.   We show only that $B + I$ is complete, because the rest is trivial. Since I is complete we need only prove that the quotient $(B + I)/I$ is complete. The intersection $B \cap I$ is a closed ideal in B and the map φ from $B/(B \cap I)$ to A/I defined by setting $\varphi(b + B \cap I) = b + I$ ($b \in B$) is a $*$-homomorphism with range $(B + I)/I$. By Theorem 3.1.6, $(B + I)/I$ is complete, because it is a C*-algebra. $\square$

3.1.3. Remark.   The map

$$\varphi: B/(B \cap I) \to (B + I)/I, \quad b + B \cap I \mapsto b + I,$$

in the preceding proof is in fact clearly a $*$-isomorphism.

We return to the topic of multiplier algebras, because we can now say a little more about them using the results of this section.

Suppose that I is a closed ideal in a C*-algebra A. If $a \in A$, define L_a and R_a in $B(I)$ by setting $L_a(b) = ab$ and $R_a(b) = ba$. It is a straightforward exercise to verify that (L_a, R_a) is a double centraliser on I and that the map

$$\varphi: A \to M(I), \quad a \mapsto (L_a, R_a),$$

is a $*$-homomorphism. Recall that we identified I as a closed ideal in $M(I)$ by identifying a with (L_a, R_a) if $a \in I$. Hence, φ is an extension of the inclusion map $I \to M(I)$.

If $I_1, I_2, \ldots, I_n$ are sets in A, we define $I_1 I_2 \ldots I_n$ to be the closed linear span of all products $a_1 a_2 \ldots a_n$, where $a_j \in I_j$. If I, J are closed ideals in A, then $I \cap J = IJ$. The inclusion $IJ \subseteq I \cap J$ is obvious. To show the reverse inclusion we need only show that if a is a positive element of $I \cap J$, then $a \in IJ$. Suppose then that $a \in (I \cap J)^+$. Hence, $a^{1/2} \in I \cap J$. If $(u_\lambda)_{\lambda \in \Lambda}$ is an approximate unit for I, then $a = \lim_\lambda (u_\lambda a^{1/2}) a^{1/2}$, and since $u_\lambda a^{1/2} \in I$ for all $\lambda \in \Lambda$, we get $a \in IJ$, as required.

Let I be a closed ideal I in A. We say I is *essential* in A if $aI = 0 \Rightarrow a = 0$ (equivalently, $Ia = 0 \Rightarrow a = 0$). From the preceding observations it is easy to check that I is essential in A if and only if $I \cap J \neq 0$ for all non-zero closed ideals J in A.

Every C*-algebra I is an essential ideal in its multiplier algebra $M(I)$.

3.1.8. Theorem. *Let I be a closed ideal in a C*-algebra A. Then there is a unique *-homomorphism $\varphi: A \to M(I)$ extending the inclusion $I \to M(I)$. Moreover, φ is injective if I is essential in A.*

Proof. We have seen above that the inclusion map $I \to M(I)$ admits a *-homomorphic extension $\varphi: A \to M(I)$. Suppose that $\psi: A \to M(I)$ is another such extension. If $a \in A$ and $b \in I$, then $\varphi(a)b = \varphi(ab) = ab = \psi(ab) = \psi(a)b$. Hence, $(\varphi(a) - \psi(a))I = 0$, so $\varphi(a) = \psi(a)$, since I is essential in $M(I)$. Thus, $\varphi = \psi$.

Suppose now that I is essential in A and let $a \in \ker(\varphi)$. Then $aI = L_a(I) = 0$, so $a = 0$. Thus, φ is injective. $\qquad\qquad\square$

Theorem 3.1.8 tells us that the multiplier algebra $M(I)$ of I is the largest unital C*-algebra containing I as an essential closed ideal.

3.1.2. Example. If H is a Hilbert space, then $K(H)$ is an essential ideal in $B(H)$. For if u is an operator in $B(H)$ such that $uK(H) = 0$, then for all $x \in H$ we have $u(x) \otimes x = u(x \otimes x) = 0$, so $u(x) = 0$. By Theorem 3.1.8, the inclusion map $K(H) \to M(K(H))$ extends uniquely to an injective *-homomorphism $\varphi: B(H) \to M(K(H))$. We show that φ is surjective, that is, a *-isomorphism. Suppose that $(L, R) \in M(K(H))$, and fix a unit vector e in H. The linear map

$$u: H \to H, \quad x \mapsto (L(x \otimes e))(e),$$

is bounded, since $\|u(x)\| \leq \|L(x \otimes e)\| \leq \|L\|\|x \otimes e\| = \|L\|\|x\|$. If $x, y, z \in H$, then

$$
\begin{aligned}
(L_u(x \otimes y))(z) &= (u(x) \otimes y)(z) \\
&= \langle z, y \rangle (L(x \otimes e))(e) \\
&= (L(x \otimes e))(\langle z, y \rangle e) \\
&= (L(x \otimes e))(e \otimes y)(z).
\end{aligned}
$$

Hence, $L_u(x \otimes y) = L(x \otimes e)(e \otimes y) = L(x \otimes y)$ for all $x, y \in H$. Therefore, $(\varphi(u) - (L, R))K(H) = 0$, so $\varphi(u) = (L, R)$.

Thus, we may regard $B(H)$ as the multiplier algebra of $K(H)$. This example is the motivating one for the use of the multiplier algebra in K-theory.

3.1.3. Example. If Ω is a locally compact Hausdorff space, then it is easy to check that $C_0(\Omega)$ is an essential ideal in the C*-algebra $C_b(\Omega)$. Therefore, by Theorem 3.1.8 there is a unique injective *-homomorphism $\varphi: C_b(\Omega) \to M(C_0(\Omega))$ extending the inclusion $C_0(\Omega) \to M(C_0(\Omega))$. We show that φ is surjective, that is, a *-isomorphism. To see this, it suffices to show that if $g \in M(C_0(\Omega))$ is positive, then it is the range of φ. If $(u_\lambda)_{\lambda \in \Lambda}$ is an approximate unit for $C_0(\Omega)$, then for each $\omega \in \Omega$ the net of real numbers $(gu_\lambda(\omega))$ is increasing and bounded above by $\|g\|$, and therefore it converges to a number, $h(\omega)$ say. The function

$$h: \Omega \to \mathbf{C}, \quad \omega \mapsto h(\omega),$$

is bounded. Moreover, if $f \in C_0(\Omega)$, then $hf = gf$, since $f = \lim_\lambda fu_\lambda$. To see that h is continuous, let $(\omega_\mu)_{\mu \in M}$ be a net in Ω converging to a point ω. Let K be a compact neighbourhood of ω in Ω. To show that $h(\omega) = \lim_\mu h(\omega_\mu)$, we may suppose $\omega_\mu \in K$ for all indices μ (there exists μ_0 such that $\omega_\mu \in K$ for all indices $\mu \geq \mu_0$, so, if neccessary, replace the net $(\omega_\mu)_{\mu \in M}$ by the net $(\omega_\mu)_{\mu \geq \mu_0}$). Use Urysohn's lemma to choose a function $f \in C_0(\Omega)$ such that $f = 1$ on K. Since $fh \in C_0(\Omega)$,

$$h(\omega) = fh(\omega) = \lim_\mu fh(\omega_\mu) = \lim_\mu h(\omega_\mu).$$

Therefore, h is continuous, so $h \in C_b(\Omega)$. For f an arbitrary function in $C_0(\Omega)$ we have $\varphi(h)f = \varphi(hf) = hf = gf$, so $(\varphi(h) - g)C_0(\Omega) = 0$. Consequently, $g = \varphi(h)$.

3.2. Hereditary C*-Subalgebras

This section introduces a new class of C*-subalgebras, namely the hereditary ones. These are particularly well-behaved, especially with respect to extending positive linear functionals, an important topic to be taken up in the next section. We illustrate the nice behaviour of hereditary C*-subalgebras in connection with the concept of simplicity of an algebra.

A C*-subalgebra B of a C*-algebra A is said to be *hereditary* if for $a \in A^+$ and $b \in B^+$ the inequality $a \leq b$ implies $a \in B$. Obviously, 0 and A are hereditary C*-subalgebras of A, and any intersection of hereditary C*-subalgebras is one also. The hereditary C*-subalgebra *generated* by a subset S of A is the smallest hereditary C*-subalgebra of A containing S.

3.2.1. *Example*. If p is a projection in a C*-algebra A, the C*-subalgebra pAp is hereditary. For, assuming $0 \leq b \leq pap$, then $0 \leq (1-p)b(1-p) \leq (1-p)pap(1-p) = 0$, so $(1-p)b(1-p) = 0$. Hence, $\|b^{1/2}(1-p)\|^2 = \|(1-p)b(1-p)\| = 0$, so $b(1-p) = 0$. Therefore, $b = pbp \in pAp$.

The correspondence between hereditary C*-subalgebras and closed left ideals in the following theorem is very useful.

3.2.1. Theorem. *Let A be a C*-algebra.*

(1) *If L is a closed left ideal in A, then $L \cap L^*$ is a hereditary C*-subalgebra of A. The map $L \mapsto L \cap L^*$ is a bijection from the set of closed left ideals of A onto the set of hereditary C*-subalgebras of A.*

(2) *If L_1, L_2 are closed left ideals of A, then $L_1 \subseteq L_2$ if and only if $L_1 \cap L_1^* \subseteq L_2 \cap L_2^*$.*

(3) *If B is a hereditary C*-subalgebra of A, then the set*

$$L(B) = \{a \in A \mid a^*a \in B\}$$

is the unique closed left ideal of A corresponding to B.

Proof. If L is a closed left ideal of A, then clearly $B = L \cap L^*$ is a C*-subalgebra of A. Suppose that $a \in A^+$ and $b \in B^+$ and $a \leq b$. By Theorem 3.1.2 there is an increasing net $(u_\lambda)_{\lambda \in \Lambda}$ in the closed unit ball of L^+ such that $\lim_\lambda bu_\lambda = b$. Now $0 \leq (1-u_\lambda)a(1-u_\lambda) \leq (1-u_\lambda)b(1-u_\lambda)$, so $\|a^{1/2} - a^{1/2}u_\lambda\|^2 = \|(1-u_\lambda)a(1-u_\lambda)\| \leq \|(1-u_\lambda)b(1-u_\lambda)\| \leq \|b - bu_\lambda\|$. Hence, $a^{1/2} = \lim_\lambda a^{1/2}u_\lambda$, so $a^{1/2} \in L$, since $u_\lambda \in L$ ($\lambda \in \Lambda$). Therefore, $a \in B$, so B is hereditary in A.

Suppose now that L_1, L_2 are closed left ideals of A. It is evident that $L_1 \subseteq L_2 \Rightarrow L_1 \cap L_1^* \subseteq L_2 \cap L_2^*$. To show the reverse implication, suppose that $L_1 \cap L_1^* \subseteq L_2 \cap L_2^*$ and let $(u_\lambda)_{\lambda \in \Lambda}$ be an approximate unit for $L_1 \cap L_1^*$, and $a \in L_1$. Then $\lim_\lambda \|a - au_\lambda\|^2 = \lim_\lambda \|(1-u_\lambda)a^*a(1-u_\lambda)\| \leq \lim_\lambda \|a^*a(1-u_\lambda)\| = 0$, since $a^*a \in L_1 \cap L_1^*$. It follows that $\lim_\lambda au_\lambda = a$. Therefore, $a \in L_2$, since $u_\lambda \in L_1 \cap L_1^* \subseteq L_2$. This proves Condition (2).

Now let B be a hereditary C*-subalgebra of A and let $L = L(B)$. If $a, b \in L$, $(a+b)^*(a+b) \leq (a+b)^*(a+b) + (a-b)^*(a-b) = 2a^*a + 2b^*b \in B$, so $a + b \in L$. If $a \in A$ and $b \in L$, then $(ab)^*(ab) = b^*a^*ab \leq \|a\|^2b^*b \in B$, so $ab \in L$. Similarly, L is closed under scalar multiplication. Thus, L is a left ideal, and it is obviously closed, since B is closed. If $b \in B$, then $b^*b \in B$, so $b \in L$. Hence, $B \subseteq L \cap L^*$. If $0 \leq b \in L \cap L^*$, then $b^2 \in B$, so $b \in B$, and therefore $L \cap L^* \subseteq B$. Hence, $L \cap L^* = B$. This proves Condition (3), and Condition (1) follows directly. $\qquad\square$

3.2.2. Theorem. *Let B be a C*-subalgebra of a C*-algebra A. Then B is hereditary in A if and only if $bab' \in B$ for all $b, b' \in B$ and $a \in A$.*

Proof. If B is hereditary, then by Theorem 3.2.1 $B = L \cap L^*$ for some closed left ideal L of A. Hence, if $b, b' \in B$ and $a \in A$, we have $b(ab') \in L$ and $b'^*(a^*b^*) \in L$, so $bab' \in B$.

Conversely, suppose B has the property that $bab' \in B$ for all $b, b' \in B$ and $a \in A$. If $(u_\lambda)_{\lambda \in \Lambda}$ is an approximate unit for B and $a \in A^+$, $b \in B^+$, and $a \leq b$, then $0 \leq (1 - u_\lambda)a(1 - u_\lambda) \leq (1 - u_\lambda)b(1 - u_\lambda)$, and therefore $\|a^{1/2} - a^{1/2}u_\lambda\| \leq \|b^{1/2} - b^{1/2}u_\lambda\|$. Since $b^{1/2} = \lim_\lambda b^{1/2}u_\lambda$, therefore, $a^{1/2} = \lim_\lambda a^{1/2}u_\lambda$, so $a = \lim_\lambda u_\lambda a u_\lambda \in B$. Thus, B is hereditary. $\square$

The following corollary is obvious.

3.2.3. Corollary. *Every closed ideal of a C*-algebra is a hereditary C*-subalgebra.*

3.2.4. Corollary. *If A is a C*-algebra and $a \in A^+$, then $(aAa)^-$ is the hereditary C*-subalgebra of A generated by a.*

Proof. The only thing we show is that $a \in (aAa)^-$, because the rest is routine. If $(u_\lambda)_{\lambda \in \Lambda}$ is an approximate unit for A, then $a^2 = \lim_\lambda au_\lambda a$, so $a^2 \in (aAa)^-$. Since $(aAa)^-$ is a C*-algebra, $a = \sqrt{a^2} \in (aAa)^-$ also. $\square$

In the separable case, every hereditary C*-subalgebra is of the form in the preceding corollary:

3.2.5. Theorem. *Suppose that B is a separable hereditary C*-subalgebra of a C*-algebra A. Then there is a positive element $a \in B$ such that $B = (aAa)^-$.*

Proof. Since B is a separable C*-algebra, it admits a sequential approximate unit, $(u_n)_{n=1}^\infty$ say (*cf.* Remark 3.1.1). Set $a = \sum_{n=1}^\infty u_n/2^n$. Then $a \in B^+$, so B contains $(aAa)^-$. Since $u_n/2^n \leq a$, and $(aAa)^-$ is hereditary by Corollary 3.2.4, therefore $u_n \in (aAa)^-$. If $b \in B$, then $b = \lim_{n \to \infty} u_n b u_n$, and $u_n b u_n \in (aAa)^-$, so $b \in (aAa)^-$. This shows that $B = (aAa)^-$. $\square$

If the separability condition is dropped in Theorem 3.2.5, the result may fail. To see this let H be a Hilbert space, and suppose that u is a positive element of $B(H)$ such that $K(H) = (uB(H)u)^-$. If $x \in H$, then $x \otimes x = \lim_{n \to \infty} uv_n u$ for a sequence (v_n) in $B(H)$, and therefore x is in the closure of the range of u. This shows that $H = (u(H))^-$, and therefore H is separable, since the range of a compact operator is separable (*cf.* Remark 1.4.1). Thus, if H is a non-separable Hilbert space, then the hereditary C*-subalgebra $K(H)$ of $B(H)$ is not of the form $(uB(H)u)^-$ for any $u \in B(H)^+$.

3.2.6. Theorem. *Suppose that B is a hereditary C^*-subalgebra of a unital C^*-algebra A, and let $a \in A^+$. If for each $\varepsilon > 0$ there exists $b \in B^+$ such that $a \leq b + \varepsilon$, then $a \in B$.*

Proof. Let $\varepsilon > 0$. By the hypothesis there exists $b_\varepsilon \in B^+$ such that $a \leq b_\varepsilon^2 + \varepsilon^2$, so $a \leq (b_\varepsilon + \varepsilon)^2$. Hence, $(b_\varepsilon + \varepsilon)^{-1} a (b_\varepsilon + \varepsilon)^{-1} \leq 1$, and therefore $\|(b_\varepsilon + \varepsilon)^{-1} a (b_\varepsilon + \varepsilon)^{-1}\| \leq 1$. Using the fact that $1 - b_\varepsilon (b_\varepsilon + \varepsilon)^{-1} = \varepsilon (b_\varepsilon + \varepsilon)^{-1}$, we get

$$\|a^{1/2} - a^{1/2} b_\varepsilon (b_\varepsilon + \varepsilon)^{-1}\|^2 = \varepsilon^2 \|a^{1/2}(b_\varepsilon + \varepsilon)^{-1}\|^2$$
$$= \varepsilon^2 \|(b_\varepsilon + \varepsilon)^{-1} a (b_\varepsilon + \varepsilon)^{-1}\|$$
$$\leq \varepsilon^2.$$

Hence,

$$a^{1/2} = \lim_{\varepsilon \to 0} a^{1/2} b_\varepsilon (b_\varepsilon + \varepsilon)^{-1},$$

and therefore also

$$a^{1/2} = \lim_{\varepsilon \to 0} (b_\varepsilon + \varepsilon)^{-1} b_\varepsilon a^{1/2},$$

by taking adjoints. Thus,

$$a = \lim_{\varepsilon \to 0} (b_\varepsilon + \varepsilon)^{-1} b_\varepsilon a b_\varepsilon (b_\varepsilon + \varepsilon)^{-1}.$$

Now $b_\varepsilon (b_\varepsilon + \varepsilon)^{-1} \in B$, and therefore $(b_\varepsilon + \varepsilon)^{-1} b_\varepsilon a b_\varepsilon (b_\varepsilon + \varepsilon)^{-1} \in B$, since B is hereditary in A. It follows that $a \in B$. $\square$

We briefly indicate the connection between the ideal structure of a C^*-algebra and its hereditary C^*-subalgebras in the following results, but we shall defer to Chapter 5 a fuller consideration of this matter.

3.2.7. Theorem. *Let B be a hereditary C^*-subalgebra of a C^*-algebra A, and let J be a closed ideal of B. Then there exists a closed ideal I of A such that $J = B \cap I$.*

Proof. Let $I = AJA$. Then I is a closed ideal of A. Since J is a C^*-algebra, $J = J^3$, and since B is hereditary in A, we have $B \cap I = BIB$ (both of these assertions follow easily from the existence of approximate units). Therefore, $B \cap I = BIB = B(AJA)B = BAJ^3 AB \subseteq BJB$, because BAJ and JAB are contained in B by Theorem 3.2.2. Since $BJB = J$, because J is a closed ideal in B, we have $B \cap I \subseteq J$, and the reverse inclusion is obvious, so $B \cap I = J$. $\square$

A C^*-algebra A is said to be *simple* if 0 and A are its only closed ideals. These algebras are (loosely) thought of as the building blocks of the theory of C^*-algebras, and it is important to compile a large stock of examples. We shall be presenting some as we proceed, but for the present we content ourselves with one class of examples:

3.2.2. Example. If H is a Hilbert space, then the C*-algebra $K(H)$ is simple. For if I is a closed non-zero ideal of $K(H)$, it is also an ideal of $B(H)$ (*cf.* Remark 3.1.2), so I contains the ideal $F(H)$ by Theorem 2.4.7, and therefore $I = K(H)$.

It is not true that C*-subalgebras of simple C*-algebras are necessarily simple. For instance, if p, q are finite-rank non-zero projections on a Hilbert space H such that $pq = 0$, then $A = \mathbf{C}p + \mathbf{C}q$ is a non-simple C*-subalgebra of the simple C*-algebra $K(H)$ (the closed ideal $Ap = \mathbf{C}p$ of A is non-trivial).

3.2.8. Theorem. *Every hereditary C*-subalgebra of a simple C*-algebra is simple.*

Proof. Let B be a hereditary C*-subalgebra of a simple C*-algebra A. If J is a closed ideal of B, then $J = B \cap I$ for some closed ideal I of A by Theorem 3.2.7. Simplicity of A implies that $I = 0$ or A, and therefore $J = 0$ or B. $\qquad\square$

3.3. Positive Linear Functionals

For abelian C*-algebras we were able completely to determine the structure of the algebra in terms of the character space, that is, in terms of the one-dimensional representations. For the non-abelian case this is quite inadequate, and we have to look at representations of arbitrary dimension. There is a deep inter-relationship between the representations and the positive linear functionals of a C*-algebra. Representations will be defined and some aspects of this inter-relationship investigated in the next section. In this section we establish the basic properties of positive linear functionals.

If $\varphi \colon A \to B$ is a linear map between C*-algebras, it is said to be *positive* if $\varphi(A^+) \subseteq \varphi(B^+)$. In this case $\varphi(A_{sa}) \subseteq B_{sa}$, and the restriction map $\varphi \colon A_{sa} \to B_{sa}$ is increasing.

Every *-homomorphism is positive.

3.3.1. Example. Let $A = C(\mathbf{T})$ and let m be normalised arc length measure on $\mathbf{T}$. Then the linear functional

$$C(\mathbf{T}) \to \mathbf{C}, \quad f \mapsto \int f \, dm,$$

is positive (and not a homomorphism).

3.3.2. Example. Let $A = M_n(\mathbf{C})$. The linear functional

$$\mathrm{tr} \colon A \to \mathbf{C}, \quad (\lambda_{ij}) \mapsto \sum_{\iota=1}^{n} \lambda_{ii},$$

is positive. It is called the *trace*. Observe that there are no non-zero $*$-homomorphisms from $M_n(\mathbf{C})$ to $\mathbf{C}$ if $n > 1$.

Let A be a C*-algebra and τ a positive linear functional on A. Then the function

$$A^2 \to \mathbf{C}, \ (a, b) \mapsto \tau(b^*a),$$

is a positive sesquilinear form on A. Hence, $\tau(b^*a) = \tau(a^*b)^-$ and $|\tau(b^*a)| \leq \tau(a^*a)^{1/2}\tau(b^*b)^{1/2}$. Moreover, the function $a \mapsto \tau(a^*a)^{1/2}$ is a semi-norm on A.

Suppose now only that τ is a linear functional on A and that M is an element of $\mathbf{R}^+$ such that $|\tau(a)| \leq M$ for all positive elements of the closed unit ball of A. Then τ is bounded with norm $\|\tau\| \leq 4M$. We show this: First suppose that a is a hermitian element of A such that $\|a\| \leq 1$. Then a^+, a^- are positive elements of the closed unit ball of A, and therefore $|\tau(a)| = |\tau(a^+) - \tau(a^-)| \leq 2M$. Now suppose that a is an arbitrary element of the closed unit ball of A, so $a = b + ic$ where b, c are its real and imaginary parts, and $\|b\|, \|c\| \leq 1$. Then $|\tau(a)| = |\tau(b) + i\tau(c)| \leq 4M$.

3.3.1. Theorem. *If τ is a positive linear functional on a C*-algebra A, then it is bounded.*

Proof. If τ is not bounded, then by the preceding remarks $\sup_{a \in S} \tau(a) = +\infty$, where S is the set of all positive elements of A of norm not greater then 1. Hence, there is a sequence (a_n) in S such that $2^n \leq \tau(a_n)$ for all $n \in \mathbf{N}$. Set $a = \sum_{n=0}^{\infty} a_n/2^n$, so $a \in A^+$. Now $1 \leq \tau(a_n/2^n)$ and therefore $N \leq \sum_{n=0}^{N-1} \tau(a_n/2^n) = \tau(\sum_{n=0}^{N-1} a_n/2^n) \leq \tau(a)$. Hence, $\tau(a)$ is an upper bound for the set $\mathbf{N}$, which is impossible. This shows that τ is bounded. $\square$

3.3.2. Theorem. *If τ is a positive linear functional on a C*-algebra A, then $\tau(a^*) = \tau(a)^-$ and $|\tau(a)|^2 \leq \|\tau\|\tau(a^*a)$ for all $a \in A$.*

Proof. Let $(u_\lambda)_{\lambda \in \Lambda}$ be an approximate unit for A. Then

$$\tau(a^*) = \lim_\lambda \tau(a^*u_\lambda) = \lim_\lambda \tau(u_\lambda a)^- = \tau(a)^-.$$

Also, $|\tau(a)|^2 = \lim_\lambda |\tau(u_\lambda a)|^2 \leq \sup_\lambda \tau(u_\lambda^2)\tau(a^*a) \leq \|\tau\|\tau(a^*a)$. $\square$

3.3.3. Theorem. *Let τ be a bounded linear functional on a C*-algebra A. The following conditions are equivalent:*

(1) τ *is positive.*
(2) *For each approximate unit $(u_\lambda)_{\lambda \in \Lambda}$ of A, $\|\tau\| = \lim_\lambda \tau(u_\lambda)$.*
(3) *For some approximate unit $(u_\lambda)_{\lambda \in \Lambda}$ of A, $\|\tau\| = \lim_\lambda \tau(u_\lambda)$.*

Proof. We may suppose that $\|\tau\| = 1$. First we show the implication (1) $\Rightarrow$ (2) holds. Suppose that τ is positive, and let $(u_\lambda)_{\lambda \in \Lambda}$ be an approximate unit of A. Then $(\tau(u_\lambda))_{\lambda \in \Lambda}$ is an increasing net in $\mathbf{R}$, so it converges to its supremum, which is obviously not greater than 1. Thus, $\lim_\lambda \tau(u_\lambda) \leq 1$. Now suppose that $a \in A$ and $\|a\| \leq 1$. Then $|\tau(u_\lambda a)|^2 \leq \tau(u_\lambda^2)\tau(a^*a) \leq \tau(u_\lambda)\tau(a^*a) \leq \lim_\lambda \tau(u_\lambda)$, so $|\tau(a)|^2 \leq \lim_\lambda \tau(u_\lambda)$. Hence, $1 \leq \lim_\lambda \tau(u_\lambda)$. Therefore, $1 = \lim_\lambda \tau(u_\lambda)$, so (1) $\Rightarrow$ (2).

That (2) $\Rightarrow$ (3) is obvious.

Now we show that (3) $\Rightarrow$ (1). Suppose that $(u_\lambda)_{\lambda \in \Lambda}$ is an approximate unit such that $1 = \lim_\lambda \tau(u_\lambda)$. Let a be a self-adjoint element of A such that $\|a\| \leq 1$ and write $\tau(a) = \alpha + i\beta$ where α, β are real numbers. To show that $\tau(a) \in \mathbf{R}$, we may suppose that $\beta \leq 0$. If n is a positive integer, then

$$\|a - inu_\lambda\|^2 = \|(a + inu_\lambda)(a - inu_\lambda)\|$$
$$= \|a^2 + n^2 u_\lambda^2 - in(au_\lambda - u_\lambda a)\|$$
$$\leq 1 + n^2 + n\|au_\lambda - u_\lambda a\|,$$

so

$$|\tau(a - inu_\lambda)|^2 \leq 1 + n^2 + n\|au_\lambda - u_\lambda a\|.$$

However, $\lim_\lambda \tau(a - inu_\lambda) = \tau(a) - in$, and $\lim_\lambda au_\lambda - u_\lambda a = 0$, so in the limit as $\lambda \to \infty$ we get

$$|\alpha + i\beta - in|^2 \leq 1 + n^2.$$

The left-hand side of this inequality is $\alpha^2 + \beta^2 - 2n\beta + n^2$, so if we cancel and rearrange we get

$$-2n\beta \leq 1 - \beta^2 - \alpha^2.$$

Since β is not positive and this inequality holds for all positive integers n, β must be zero. Therefore, $\tau(a)$ is real if a is hermitian.

Now suppose that a is positive and $\|a\| \leq 1$. Then $u_\lambda - a$ is hermitian and $\|u_\lambda - a\| \leq 1$, so $\tau(u_\lambda - a) \leq 1$. But then $1 - \tau(a) = \lim_\lambda \tau(u_\lambda - a) \leq 1$, and therefore $\tau(a) \geq 0$. Thus, τ is positive and we have shown (3) $\Rightarrow$ (1). $\square$

3.3.4. Corollary. *If τ is a bounded linear functional on a unital C*-algebra, then τ is positive if and only if $\tau(1) = \|\tau\|$.*

Proof. The sequence which is constantly 1 is an approximate unit for the C*-algebra. Apply Theorem 3.3.3. $\square$

3.3.5. Corollary. *If τ, τ' are positive linear functionals on a C*-algebra, then $\|\tau + \tau'\| = \|\tau\| + \|\tau'\|$.*

Proof. If $(u_\lambda)_{\lambda \in \Lambda}$ is an approximate unit for the algebra, then $\|\tau + \tau'\| = \lim_\lambda (\tau + \tau')(u_\lambda) = \lim_\lambda \tau(u_\lambda) + \lim_\lambda \tau'(u_\lambda) = \|\tau\| + \|\tau'\|$. $\square$

A *state* on a C*-algebra A is a positive linear functional on A of norm one. We denote by $S(A)$ the set of states of A.

3.3.6. Theorem. *If a is a normal element of a non-zero C*-algebra A, then there is a state τ of A such that $\|a\| = |\tau(a)|$.*

Proof. We may assume that $a \neq 0$. Let B be the C*-algebra generated by 1 and a in $\tilde{A}$. Since B is abelian and $\hat{a}$ is continuous on the compact space $\Omega(B)$, there is a character τ_2 on B such that $\|a\| = \|\hat{a}\|_\infty = |\tau_2(a)|$. By the Hahn–Banach theorem, there is a bounded linear functional τ_1 on $\tilde{A}$ extending τ_2 and preserving the norm, so $\|\tau_1\| = 1$. Since $\tau_1(1) = \tau_2(1) = 1$, τ_1 is positive by Corollary 3.3.4. If τ denotes the restriction of τ_1 to A, then τ is a positive linear functional on A such that $\|a\| = |\tau(a)|$. Hence, $\|\tau\|\|a\| \geq |\tau(a)| = \|a\|$, so $\|\tau\| \geq 1$, and the reverse inequality is obvious. Therefore, τ is a state of A. $\quad\square$

3.3.7. Theorem. *Suppose that τ is a positive linear functional on a C*-algebra A.*

(1) *For each $a \in A$, $\tau(a^*a) = 0$ if and only if $\tau(ba) = 0$ for all $b \in A$.*
(2) *The inequality*

$$\tau(b^*a^*ab) \leq \|a^*a\|\tau(b^*b)$$

holds for all $a, b \in A$.

Proof. Condition (1) follows from the Cauchy–Schwarz inequality.

To show Condition (2), we may suppose, using Condition (1), that $\tau(b^*b) > 0$. The function

$$\rho\colon A \to \mathbf{C}, \quad c \mapsto \tau(b^*cb)/\tau(b^*b),$$

is positive and linear, so if $(u_\lambda)_{\lambda \in \Lambda}$ is any approximate unit for A, then

$$\|\rho\| = \lim_\lambda \rho(u_\lambda) = \lim_\lambda \tau(b^*u_\lambda b)/\tau(b^*b) = \tau(b^*b)/\tau(b^*b) = 1.$$

Hence, $\rho(a^*a) \leq \|a^*a\|$, and therefore $\tau(b^*a^*ab) \leq \|a^*a\|\tau(b^*b)$. $\quad\square$

We turn now to the problem of extending positive linear functionals.

3.3.8. Theorem. *Let B be a C*-subalgebra of a C*-algebra A, and suppose that τ is a positive linear functional on B. Then there is a positive linear functional τ' on A extending τ such that $\|\tau'\| = \|\tau\|$.*

Proof. Suppose first that $A = \tilde{B}$. Define a linear functional τ' on A by setting $\tau'(b + \lambda) = \tau(b) + \lambda\|\tau\|$ ($b \in B$, $\lambda \in \mathbf{C}$). Let $(u_\lambda)_{\lambda \in \Lambda}$ be an approximate unit for B. By Theorem 3.3.3, $\|\tau\| = \lim_\lambda \tau(u_\lambda)$. Now suppose that $b \in B$ and $\mu \in \mathbf{C}$. Then $|\tau'(b + \mu)| = |\lim_\lambda \tau(bu_\lambda) + \mu\lim_\lambda \tau(u_\lambda)| = |\lim_\lambda \tau((b + \mu)(u_\lambda))| \leq \sup_\lambda \|\tau\|\|(b + \mu)u_\lambda\| \leq \|\tau\|\|b + \mu\|$, since $\|u_\lambda\| \leq 1$. Hence, $\|\tau'\| \leq \|\tau\|$, and the reverse inequality is obvious. Thus, $\|\tau'\| = \|\tau\| = \tau'(1)$, so τ' is positive by Corollary 3.3.4. This proves the theorem in the case $A = \tilde{B}$.

Now suppose that A is an arbitrary C*-algebra containing B as a C*-subalgebra. Replacing B and A by $\tilde{B}$ and $\tilde{A}$ if necessary, we may suppose that A has a unit 1 which lies in B. By the Hahn–Banach theorem, there is a functional $\tau' \in A^*$ extending τ and of the same norm. Since $\tau'(1) = \tau(1) = \|\tau\| = \|\tau'\|$, it follows as before from Corollary 3.3.4 that τ' is positive.   □

In the case of hereditary C*-subalgebras, we can strengthen the above result—we can even write down an "expression" for τ':

3.3.9. Theorem. *Let B be a hereditary C*-subalgebra of a C*-algebra A. If τ is a positive linear functional on B, then there is a unique positive linear functional τ' on A extending τ and preserving the norm. Moreover, if $(u_\lambda)_{\lambda \in \Lambda}$ is an approximate unit for B, then*

$$\tau'(a) = \lim_\lambda \tau(u_\lambda a u_\lambda) \qquad (a \in A).$$

Proof. Of course we already have existence, so we only prove uniqueness. Let τ' be a positive linear functional on A extending τ and preserving the norm. We may in turn extend τ' in a norm-preserving fashion to a positive functional (also denoted τ') on $\tilde{A}$. Let $(u_\lambda)_{\lambda \in \Lambda}$ be an approximate unit for B. Then $\lim_\lambda \tau(u_\lambda) = \|\tau\| = \|\tau'\| = \tau'(1)$, so $\lim_\lambda \tau'(1 - u_\lambda) = 0$. Thus, for any element $a \in A$,

$$
\begin{aligned}
|\tau'(a) - \tau(u_\lambda a u_\lambda)| &\leq |\tau'(a - u_\lambda a)| + |\tau'(u_\lambda a - u_\lambda a u_\lambda)| \\
&\leq \tau'((1 - u_\lambda)^2)^{1/2} \tau'(a^* a)^{1/2} \\
&\qquad + \tau'(a^* u_\lambda^2 a)^{1/2} \tau'((1 - u_\lambda)^2)^{1/2} \\
&\leq (\tau'(1 - u_\lambda))^{1/2} \tau'(a^* a)^{1/2} + \tau'(a^* a)^{1/2} (\tau'(1 - u_\lambda))^{1/2}.
\end{aligned}
$$

Since $\lim_\lambda \tau'(1 - u_\lambda) = 0$, these inequalities imply $\lim_\lambda \tau(u_\lambda a u_\lambda) = \tau'(a)$.□

Let Ω be a compact Hausdorff space and denote by $C(\Omega, \mathbf{R})$ the real Banach space of all real-valued continuous functions on Ω. The operations on $C(\Omega, \mathbf{R})$ are the pointwise-defined ones and the norm is the sup-norm. The Riesz–Kakutani theorem asserts that if $\tau : C(\Omega, \mathbf{R}) \to \mathbf{R}$ is a bounded real-linear functional, then there is a unique real measure $\mu \in M(\Omega)$ such that $\tau(f) = \int f \, d\mu$ for all $f \in C(\Omega, \mathbf{R})$. Moreover, $\|\mu\| = \|\tau\|$, and μ is positive if and only if τ is positive; that is, $\tau(f) \geq 0$ for all $f \in C(\Omega, \mathbf{R})$ such that $f \geq 0$. The Jordan decomposition for a real measure $\mu \in M(\Omega)$ asserts that there are positive measures $\mu^+, \mu^- \in M(\Omega)$ such that $\mu = \mu^+ - \mu^-$ and $\|\mu\| = \|\mu^+\| + \|\mu^-\|$. We translate this via the Riesz–Kakutani theorem into a statement about linear functionals: If $\tau : C(\Omega, \mathbf{R}) \to \mathbf{R}$ is a bounded real-linear functional, then there exist positive bounded real-linear functionals $\tau_+, \tau_- : C(\Omega, \mathbf{R}) \to \mathbf{R}$ such that $\tau = \tau_+ - \tau_-$ and $\|\tau\| = \|\tau_+\| + \|\tau_-\|$. We are now going to prove an analogue of this result for C*-algebras.

Let A be a C*-algebra. If τ is a bounded linear functional on A, then

$$\|\tau\| = \sup_{\|a\| \le 1} |Re(\tau(a))|. \tag{1}$$

For if $a \in A$ and $\|a\| \le 1$, then there is a number $\lambda \in \mathbf{T}$ such that $\lambda\tau(a) \in \mathbf{R}$, so $|\tau(a)| = |Re(\tau(\lambda a))| \le \|\tau\|$, which implies Eq. (1).

If $\tau \in A^*$, we define $\tau^* \in A^*$ by setting $\tau^*(a) = \tau(a^*)^-$ for all $a \in A$. Note that $\tau^{**} = \tau$, $\|\tau^*\| = \|\tau\|$, and the map $\tau \mapsto \tau^*$ is conjugate-linear.

We say a functional $\tau \in A^*$ is *self-adjoint* if $\tau = \tau^*$. For any bounded linear functional τ on A, there are unique self-adjoint bounded linear functionals τ_1 and τ_2 on A such that $\tau = \tau_1 + i\tau_2$ (take $\tau_1 = (\tau + \tau^*)/2$ and $\tau_2 = (\tau - \tau^*)/2i$).

The condition $\tau = \tau^*$ is equivalent to $\tau(A_{sa}) \subseteq \mathbf{R}$, and therefore if τ is self-adjoint, the restriction $\tau': A_{sa} \to \mathbf{R}$ of τ is a bounded real-linear functional. Moreover, $\|\tau\| = \|\tau'\|$; that is,

$$\|\tau\| = \sup_{\substack{a \in A_{sa} \\ \|a\| \le 1}} |\tau(a)|.$$

For if $a \in A$, we have $Re(\tau(a)) = \tau(Re(a))$, so

$$\|\tau\| = \sup_{\|a\| \le 1} |Re(\tau(a))| \le \sup_{\substack{b \in A_{sa} \\ \|b\| \le 1}} |\tau(b)| \le \|\tau\|.$$

We denote by A^*_{sa} the set of self-adjoint functionals in A^*, and by A^*_+ the set of positive functionals in A^*.

We adopt some temporary notation for the proof of the next theorem: If X is a real-linear Banach space, we denote its dual (over $\mathbf{R}$) by $X^\natural$.

The space A_{sa} is a real-linear Banach space and it is an easy exercise to verify that A^*_{sa} is a real-linear vector subspace of A^* and that the map $A^*_{sa} \to A^\natural_{sa}$, $\tau \mapsto \tau'$, is an isometric real-linear isomorphism. We shall use these observations in the proof of the following result.

3.3.10. Theorem (Jordan Decomposition). *Let τ be a self-adjoint bounded linear functional on a C*-algebra A. Then there exist positive linear functionals τ_+, τ_- on A such that $\tau = \tau_+ - \tau_-$ and $\|\tau\| = \|\tau_+\| + \|\tau_-\|$.*

Proof. Let Ω denote the set of all $\tau \in A^*_+$ such that $\|\tau\| \le 1$. Then Ω is weak* closed in the unit ball of A^*, so by the Banach–Alaoglu theorem Ω is a (Hausdorff) weak* compact space. If $a \in A_{sa}$, define $\theta(a) \in C(\Omega, \mathbf{R})$ by setting $\theta(a)(\tau) = \tau(a)$. The map

$$\theta: A_{sa} \to C(\Omega, \mathbf{R}), \quad a \mapsto \theta(a),$$

is clearly real-linear, and also order-preserving; that is, if a is a positive element of A, then $\theta(a) \geq 0$ on Ω. Moreover, θ is isometric by Theorem 3.3.6.

If $\tau \in A_{sa}^*$, then $\tau' \in A_{sa}^\natural$. By the Hahn–Banach theorem, there exists a real-linear functional $\rho \in C(\Omega, \mathbf{R})^\natural$ such that $\rho\theta = \tau'$ and $\|\rho\| = \|\tau'\|$. By the remarks preceding this theorem, there exist positive functionals $\rho_+, \rho_- \in C(\Omega, \mathbf{R})^\natural$ such that $\rho = \rho_+ - \rho_-$ and $\|\rho\| = \|\rho_+\| + \|\rho_-\|$. Set $\tau'_+ = \rho_+ \circ \theta$ and $\tau'_- = \rho_- \circ \theta$. Clearly, $\tau'_+, \tau'_- \in A_{sa}^\natural$. We denote the corresponding self-adjoint functionals in A_{sa}^* by τ_+ and τ_-. Then $\tau = \tau_+ - \tau_-$, and since $\|\tau\| = \|\tau'\| = \|\rho\| = \|\rho_+\| + \|\rho_-\| \geq \|\tau'_+\| + \|\tau'_-\| = \|\tau_+\| + \|\tau_-\| \geq \|\tau\|$, we have $\|\tau\| = \|\tau_+\| + \|\tau_-\|$. Clearly, $\tau_+, \tau_- \in A_+^*$. $\qquad\square$

One can show that the functionals τ_+ and τ_- in the preceding theorem are unique ([Ped, Theorem 3.2.5]), but we shall have no need of this.

3.4. The Gelfand–Naimark Representation

In this section we introduce the important GNS construction and prove that every C*-algebra can be regarded as a C*-subalgebra of $B(H)$ for some Hilbert space H. It is partly due to this concrete realisation of the C*-algebras that their theory is so accessible in comparison with more general Banach algebras.

A *representation* of a C*-algebra A is a pair (H, φ) where H is a Hilbert space and $\varphi: A \to B(H)$ is a *-homomorphism. We say (H, φ) is *faithful* if φ is injective.

If $(H_\lambda, \varphi_\lambda)_{\lambda \in \Lambda}$ is a family of representations of A, their *direct sum* is the representation (H, φ) got by setting $H = \oplus_\lambda H_\lambda$, and $\varphi(a)((x_\lambda)_\lambda) = (\varphi_\lambda(a)(x_\lambda))_\lambda$ for all $a \in A$ and all $(x_\lambda)_\lambda \in H$. It is readily verified that (H, φ) is indeed a representation of A. If for each non-zero element $a \in A$ there is an index λ such that $\varphi_\lambda(a) \neq 0$, then (H, φ) is faithful.

Recall now that if H is an inner product space (that is, a pre-Hilbert space), then there is a unique inner product on the Banach space completion $\hat{H}$ of H extending the inner product of H and having as its associated norm the norm of $\hat{H}$. We call $\hat{H}$ endowed with this inner product the *Hilbert space completion* of H.

With each positive linear functional, there is associated a representation. Suppose that τ is a positive linear functional on a C*-algebra A. Setting

$$N_\tau = \{a \in A \mid \tau(a^*a) = 0\},$$

it is easy to check (using Theorem 3.3.7) that N_τ is a closed left ideal of A and that the map

$$(A/N_\tau)^2 \to \mathbf{C}, \quad (a + N_\tau, b + N_\tau) \mapsto \tau(b^*a),$$

is a well-defined inner product on A/N_τ. We denote by H_τ the Hilbert completion of A/N_τ.

If $a \in A$, define an operator $\varphi(a) \in B(A/N_\tau)$ by setting

$$\varphi(a)(b + N_\tau) = ab + N_\tau.$$

The inequality $\|\varphi(a)\| \leq \|a\|$ holds since we have $\|\varphi(a)(b + N_\tau)\|^2 = \tau(b^*a^*ab) \leq \|a\|^2\tau(b^*b) = \|a\|^2\|b + N_\tau\|^2$ (the latter inequality is given by Theorem 3.3.7). The operator $\varphi(a)$ has a unique extension to a bounded operator $\varphi_\tau(a)$ on H_τ. The map

$$\varphi_\tau \colon A \to B(H_\tau), \ a \mapsto \varphi_\tau(a),$$

is a *-homomorphism (this is an easy exercise).

The representation (H_τ, φ_τ) of A is the *Gelfand–Naimark–Segal* representation (or *GNS representation*) associated to τ.

If A is non-zero, we define its *universal* representation to be the direct sum of all the representations (H_τ, φ_τ), where τ ranges over $S(A)$.

3.4.1. Theorem (Gelfand–Naimark). *If A is a C^*-algebra, then it has a faithful representation. Specifically, its universal representation is faithful.*

Proof. Let (H, φ) be the universal representation of A and suppose that a is an element of A such that $\varphi(a) = 0$. By Theorem 3.3.6 there is a state τ on A such that $\|a^*a\| = \tau(a^*a)$. Hence, if $b = (a^*a)^{1/4}$, then $\|a\|^2 = \tau(a^*a) = \tau(b^4) = \|\varphi_\tau(b)(b + N_\tau)\|^2 = 0$ (since $\varphi_\tau(b^4) = \varphi_\tau(a^*a) = 0$, so $\varphi_\tau(b) = 0$). Hence, $a = 0$, and φ is injective. $\square$

The Gelfand–Naimark theorem is one of those results that are used all of the time. For the present we give just two applications.

The first application is to matrix algebras. If A is an algebra, $M_n(A)$ denotes the algebra of all $n \times n$ matrices with entries in A. (The operations are defined just as for scalar matrices.) If A is a *-algebra, so is $M_n(A)$, where the involution is given by $(a_{ij})^*_{i,j} = (a^*_{ji})_{i,j}$.

If $\varphi \colon A \to B$ is a *-homomorphism between *-algebras, its *inflation* is the *-homomorphism (also denoted φ)

$$\varphi \colon M_n(A) \to M_n(B), \ (a_{ij}) \mapsto (\varphi(a_{ij})).$$

If H is a Hilbert space, we write $H^{(n)}$ for the orthogonal sum of n copies of H. If $u \in M_n(B(H))$, we define $\varphi(u) \in B(H^{(n)})$ by setting

$$\varphi(u)(x_1, \ldots, x_n) = \left(\sum_{j=1}^n u_{1j}(x_j), \ldots, \sum_{j=1}^n u_{nj}(x_j)\right),$$

for all $(x_1, \ldots, x_n) \in H^{(n)}$. It is readily verified that the map

$$\varphi \colon M_n(B(H)) \to B(H^{(n)}), \quad u \mapsto \varphi(u),$$

is a $*$-isomorphism. We call φ the *canonical* $*$-isomorphism of $M_n(B(H))$ onto $B(H^{(n)})$, and use it to identify these two algebras. If v is an operator in $B(H^{(n)})$ such that $v = \varphi(u)$ where $u \in M_n(B(H))$, we call u the *operator matrix* of v. We define a norm on $M_n(B(H))$ making it a C*-algebra by setting $\|u\| = \|\varphi(u)\|$. The following inequalities for $u \in M_n(B(H))$ are easy to verify and are often useful:

$$\|u_{ij}\| \leq \|u\| \leq \sum_{k,l=1}^{n} \|u_{kl}\| \qquad (i, j = 1, \ldots, n).$$

3.4.2. Theorem. *If A is a C*-algebra, then there is a unique norm on $M_n(A)$ making it a C*-algebra.*

Proof. Let the pair (H, φ) be the universal representation of A, so the $*$-homomorphism $\varphi \colon M_n(A) \to M_n(B(H))$ is injective. We define a norm on $M_n(A)$ making it a C*-algebra by setting $\|a\| = \|\varphi(a)\|$ for $a \in M_n(A)$ (completeness can be easily checked using the inequalities preceding this theorem). Uniqueness is given by Corollary 2.1.2. $\qquad \square$

3.4.1. Remark. If A is a C*-algebra and $a \in M_n(A)$, then

$$\|a_{ij}\| \leq \|a\| \leq \sum_{k,l=1}^{n} \|a_{kl}\| \qquad (i, j = 1, \ldots, n).$$

These inequalities follow from the corresponding inequalities in $M_n(B(H))$.

Matrix algebras play a fundamental role in the K-theory of C*-algebras. The idea is to study not just the algebra A but simultaneously all of the matrix algebras $M_n(A)$ over A also.

Whereas it seems that the only way known of showing that matrix algebras over general C*-algebras are themselves normable as C*-algebras is to use the Gelfand–Naimark representation, for our second application of this representation alternative proofs exist, but the proof given here has the virtue of being very "natural."

3.4.3. Theorem. *Let a be a self-adjoint element of a C*-algebra A. Then $a \in A^+$ if and only if $\tau(a) \geq 0$ for all positive linear functionals τ on A.*

Proof. The forward implication is plain. Suppose conversely that $\tau(a) \geq 0$ for all positive linear functionals τ on A. Let (H, φ) be the universal representation of A, and let $x \in H$. Then the linear functional

$$\tau \colon A \to \mathbf{C}, \quad b \mapsto \langle \varphi(b)(x), x \rangle,$$

is positive, so $\tau(a) \geq 0$; that is, $\langle \varphi(a)(x), x \rangle \geq 0$. Since this is true for all $x \in H$, and since $\varphi(a)$ is self-adjoint, therefore $\varphi(a)$ is a positive operator on H. Hence, $\varphi(a) \in \varphi(A)^+$, so $a \in A^+$, because the map $\varphi \colon A \to \varphi(A)$ is a *-isomorphism. $\square$

3.5. Toeplitz Operators

In this section we apply some of the theory we have developed so far. Our objective is to develop some aspects of the theory of Toeplitz operators. The literature on these operators is vast, and their theory is deep. We shall for the most part confine ourselves to Toeplitz operators with continuous symbol, since their theory is directly accessible to C*-algebraic methods. Apart from their great intrinsic interest, we are concerned with these operators for another reason—the C*-algebra that they generate (called the Toeplitz algebra) will play an indispensible role in the proof of Bott Periodicity in K-theory that we present in Chapter 7. With this application in mind, we shall develop a number of the properties of this algebra.

Endow the circle group $\mathbf{T}$ with its normalised arc length measure (= Haar measure), denoted by $d\lambda$, and write $L^p(\mathbf{T})$ for $L^p(\mathbf{T}, d\lambda)$. Thus, if $f \in L^1(\mathbf{T})$, then $\int f(\lambda) \, d\lambda = \frac{1}{2\pi} \int_0^{2\pi} f(e^{it}) \, dt$.

For each integer n, the function $\varepsilon_n \colon \mathbf{T} \to \mathbf{T}$, $\lambda \mapsto \lambda^n$, is of course continuous. We denote by Γ the linear span of the ε_n ($n \in \mathbf{Z}$). The elements of Γ are called *trigonometric polynomials*. The set Γ is a *-subalgebra of $C(\mathbf{T})$, and as we have observed already (in Example 2.4.2) it follows from the Stone–Weierstrass theorem that Γ is norm-dense in $C(\mathbf{T})$. Since $C(\mathbf{T})$ is L^p-norm dense in $L^p(\mathbf{T})$ for $1 \leq p < +\infty$, therefore Γ is also L^p-norm dense in $L^p(\mathbf{T})$. Hence, $(\varepsilon_n)_{n \in \mathbf{Z}}$ is an orthonormal basis of the Hilbert space $L^2(\mathbf{T})$.

If $f \in L^1(\mathbf{T})$, recall that the nth *Fourier coefficient* of f is defined to be

$$\hat{f}(n) = \int f(\lambda) \bar{\lambda}^n \, d\lambda,$$

and that the function

$$\hat{f} \colon \mathbf{Z} \to \mathbf{C}, \quad n \mapsto \hat{f}(n),$$

is the *Fourier transform* of f.

If $\hat{f} = 0$, then $f = 0$ a.e. For in this case, $\int g(\lambda) f(\lambda) \, d\lambda = 0$ for all $g \in \Gamma$, and therefore for all $g \in C(\mathbf{T})$ by sup-norm density of Γ in $C(\mathbf{T})$. Hence, the measure $f d\lambda$ is zero, so $f = 0$ a.e.

For $p \in [1, +\infty]$ set

$$H^p = \{ f \in L^p(\mathbf{T}) \mid \hat{f}(n) = 0 \ (n < 0) \}.$$

This is an L^p-norm closed vector subspace of $L^p(\mathbf{T})$, called a *Hardy space*.

We write Γ_+ for the linear span of the functions ε_n $(n \in \mathbf{N})$, and call the elements of Γ_+ the *analytic* trigonometric polynomials. The set Γ_+ is L^2-norm dense in H^2 and $(\varepsilon_n)_{n \in \mathbf{N}}$ is an orthonormal basis for H^2.

It is an easy exercise to verify that for $\varphi \in L^\infty(\mathbf{T})$ the inclusion $\varphi H^2 \subseteq H^2$ holds if and only if $\varphi \in H^\infty$, and to show from this that H^∞ is a subalgebra of $L^\infty(\mathbf{T})$.

The Hardy spaces have an interpretation in terms of analytic functions on the open unit disc in the plane satisfying certain growth conditions approaching the boundary (see Exercise 3.10). This explains the "analytic-type" behaviour displayed in the following result.

3.5.1. Lemma. *If $f, \bar{f} \in H^1$, then there exists a scalar $\alpha \in \mathbf{C}$ such that $f = \alpha$ a.e.*

Proof. Suppose first that $f = \bar{f}$ a.e. Set $\alpha = \int f(\lambda) \, d\lambda$, and observe that $\bar{\alpha} = \int \overline{f(\lambda)} \, d\lambda = \int f(\lambda) \, d\lambda = \alpha$. If $n \leq 0$, then $(f - \alpha \varepsilon_0)\hat{\ }(n) = \int (f(\lambda) - \alpha) \varepsilon_n(\lambda) \, d\lambda = \int f(\lambda) \overline{\varepsilon_n(\lambda)} \, d\lambda - \alpha \int \overline{\varepsilon_n(\lambda)} \, d\lambda = 0$, and hence, also, $(f - \alpha \varepsilon_0)\hat{\ }(n) = 0$ if $n > 0$, so $f - \alpha \varepsilon_0$ has zero Fourier transform, and therefore $f = \alpha$ a.e.

If we now suppose only that $f, \bar{f} \in H^1$, then $\mathrm{Re}(f)$ and $\mathrm{Im}(f)$ are in H^1, so by what we have just shown, these functions are constant a.e., and therefore f is constant a.e. $\qquad\square$

Recall from Example 2.5.1 that if $\varphi \in L^\infty(\mathbf{T})$, then the multiplication operator M_φ with symbol φ is defined by $M_\varphi(f) = \varphi f$ $(f \in L^2(\mathbf{T}))$; that $M_\varphi \in B(L^2(\mathbf{T}))$; and that the map

$$L^\infty(\mathbf{T}) \to B(L^2(\mathbf{T})), \quad \varphi \mapsto M_\varphi,$$

is an isometric *-homomorphism.

In this context M_φ is called a *Laurent* operator.

Set $v = M_{\varepsilon_1}$ and observe that v is the bilateral shift on the basis $(\varepsilon_n)_{n \in \mathbf{Z}}$, since $v(\varepsilon_n) = \varepsilon_{n+1}$ for all $n \in \mathbf{Z}$. The restriction u of v to H^2 is the unilateral shift on the basis $(\varepsilon_n)_{n \in \mathbf{N}}$ of H^2.

We are now going to characterise the invariant subspaces of v, and for this we determine the commutant of v, that is, the set of operators commuting with v.

3.5.2. Theorem. *If w is a bounded operator on $L^2(\mathbf{T})$, then w commutes with v if and only if $w = M_\varphi$ for some $\varphi \in L^\infty(\mathbf{T})$.*

Proof. We show the forward implication only because the reverse is clear. Suppose then $wv = vw$. If $\psi \in \Gamma$, then M_ψ is in the linear span of all the powers v^n $(n \in \mathbf{Z})$, so M_ψ commutes with w. If now ψ is an arbitrary element of $L^\infty(\mathbf{T})$, then there is a sequence (ψ_n) in Γ converging in the L^2-norm to ψ. Hence, $\lim_{n \to \infty} \|w(\psi_n) - w(\psi)\|_2 = 0$, and, by going to subsequences if necessary, we may suppose that (ψ_n) converges to ψ a.e. and $(w(\psi_n))$ converges to $w(\psi)$ a.e. If $\varphi = w(\varepsilon_0)$, then $w(\psi_n) = wM_{\psi_n}(\varepsilon_0) = M_{\psi_n}w(\varepsilon_0) = \psi_n\varphi$ a.e. Hence, $w(\psi) = \psi\varphi$ a.e.

Let $E_n = \{\lambda \in \mathbf{T} \mid |\varphi(\lambda)| > \|w\| + 1/n\}$, so E_n is a measurable set, and since

$$\|w\|^2\|\chi_{E_n}\|_2^2 \geq \|w(\chi_{E_n})\|_2^2$$
$$= \int |\varphi(\lambda)|^2 \chi_{E_n}(\lambda)\, d\lambda$$
$$\geq \int (\|w\| + 1/n)^2 \chi_{E_n}(\lambda)\, d\lambda$$
$$= (\|w\| + 1/n)^2\|\chi_{E_n}\|_2^2,$$

E_n is of measure zero. Hence, the set of points $\lambda \in \mathbf{T}$ such that $|\varphi(\lambda)| > \|w\|$, which is just the union $\cup_{n=1}^{\infty}E_n$, is a set of measure zero. It follows that $|\varphi(\lambda)| \leq \|w\|$ a.e., and therefore $\varphi \in L^{\infty}(\mathbf{T})$. Because w is equal to M_φ on $L^{\infty}(\mathbf{T})$, and therefore on $L^2(\mathbf{T})$ by L^2-norm density of $L^{\infty}(\mathbf{T})$ in $L^2(\mathbf{T})$, the theorem is proved. $\square$

If E is a Borel set of $\mathbf{T}$, then M_{χ_E} is a projection on $L^2(\mathbf{T})$. We call its range K a *Wiener* vector subspace of $L^2(\mathbf{T})$. Note that $v(K) = K$. If φ is a unitary of $L^{\infty}(\mathbf{T})$, then φH^2 is a closed vector subspace of $L^2(\mathbf{T})$, called a *Beurling* vector subspace. Note that φH^2 is invariant for v also, but $v(\varphi H^2) \neq \varphi H^2$ (otherwise, $\varphi H^2 = v\varphi H^2 = \varphi v H^2$; therefore, $H^2 = u(H^2)$, so the unilateral shift is surjective and therefore invertible, which is false).

3.5.3. Theorem. *The closed vector subspaces of $L^2(\mathbf{T})$ invariant for the bilateral shift $v = M_{\varepsilon_1}$ are precisely the Wiener and Beurling spaces. If K is an invariant closed vector subspace for v, then*

(1) $v(K) = K$ *if and only if K is a Wiener space,*
and
(2) $v(K) \neq K$ *if and only if K is a Beurling space.*

Proof. Let K be a closed vector subspace of $L^2(\mathbf{T})$ invariant for v. Suppose first that $v(K) = K$, and let p be the projection of $L^2(\mathbf{T})$ onto K. Since K reduces v therefore $pv = vp$. Hence, by Theorem 3.5.2 there is an element $\varphi \in L^{\infty}(\mathbf{T})$ such that $p = M_\varphi$. Because p is a projection, so is φ, and therefore $\varphi = \chi_E$ a.e. where E is a measurable set. Hence, $p = M_{\chi_E}$, so K is a Wiener space.

Now suppose instead that $v(K) \neq K$. Then there is a unit vector $\varphi \in K$ such that φ is orthogonal to $v(K)$. Since $v^n(\varphi) \in v(K)$ for all $n > 0$, it follows that $0 = \langle v^n(\varphi), \varphi \rangle = \int \varepsilon_n(\lambda)|\varphi(\lambda)|^2\, d\lambda$. Therefore, for any non-zero integer n, we have $\int \varepsilon_n(\lambda)|\varphi(\lambda)|^2\, d\lambda = 0$, and so $|\varphi|^2 = \alpha$ a.e. for some scalar α. Since $\|\varphi\|_2 = 1$, therefore $\alpha = 1$. Thus, φ is a unitary in $L^{\infty}(\mathbf{T})$, and clearly, $\varphi H^2 \subseteq K$. Also, $(\varepsilon_n\varphi)_{n\in\mathbf{Z}}$ is an orthonormal basis for $L^2(\mathbf{T})$, and $\varepsilon_n\varphi \in K^{\perp}$ for $n < 0$ (because $\langle \varepsilon_n\varphi, \psi \rangle = \langle \varphi, \varepsilon_{-n}\psi \rangle = 0$ for all $\psi \in K$, since $\varepsilon_{-n}\psi \in v(K)$). It follows from these observations that $(\varepsilon_n\varphi)_{n\in\mathbf{N}}$ is an orthonormal basis for K, and therefore $K = \varphi H^2$. Thus, K is a Beurling space, and the theorem is proved. $\qquad\square$

We give an interesting application of Theorem 3.5.3 to derive an important result in function theory.

3.5.4. Theorem (F. and M. Riesz). *If f is a function in H^2 that does not vanish a.e., then the set of points of $\mathbf{T}$ where f vanishes is a set of measure zero.*

Proof. Let $E = f^{-1}\{0\}$, and let K be the L^2-norm closed vector subspace of H^2 consisting of all elements $g \in H^2$ such that $g\chi_E = 0$ a.e. Obviously, K is invariant for u, and therefore for v. Observe that $\cap_{n=0}^{\infty} u^n(K) \subseteq \cap_{n=0}^{\infty} u^n(H^2) = 0$. Hence, if $v(K) = K$, then $K = 0$, and therefore since $f \in K$, $f = 0$ a.e. This contradicts the hypothesis. Hence, $v(K) \neq K$, so by Theorem 3.5.3, $K = \varphi H^2$ for some unitary element $\varphi \in L^{\infty}(\mathbf{T})$. Consequently, $\varphi\chi_E = 0$ a.e., so $\chi_E = 0$ a.e. Therefore, E is of measure zero. $\qquad\square$

3.5.5. Theorem. *The only closed vector subspaces of H^2 reducing for the unilateral shift u are the trivially invariant spaces 0 and H^2.*

Proof. Suppose K is a non-trivial closed vector subspace of H^2 that reduces u. Since $\cap_{n=1}^{\infty} u^n(K) \subseteq \cap_{n=1}^{\infty} u^n(H^2) = 0$ and $K \neq 0$, therefore $u(K) \neq K$, so K is a Beurling space by Theorem 3.5.3. Similarly, $H^2 \ominus K$ is a Beurling space. Hence, there are unitaries $\varphi, \psi \in L^{\infty}(\mathbf{T})$ such that $K = \varphi H^2$ and $H^2 \ominus K = \psi H^2$. For all $n \geq 0$, we have $\varepsilon_n\varphi \in K$ and $\varepsilon_n\psi \in H \ominus K$, so $\langle \varepsilon_n\varphi, \psi \rangle = \langle \varphi, \varepsilon_n\psi \rangle = 0$. Hence, $\varphi\bar{\psi}$ has zero Fourier transform, so $\varphi\bar{\psi} = 0$ a.e., a contradiction (since φ, ψ are unitaries). Thus, the only reducing closed vector subspaces for u are the spaces 0 and H^2. $\qquad\square$

The irreducibility of u is important for the analysis of the Toeplitz algebra which we undertake below.

Denote by p the projection of $L^2(\mathbf{T})$ onto H^2. If $\varphi \in L^{\infty}(\mathbf{T})$, the operator

$$T_\varphi \colon H^2 \to H^2, \quad \varphi \mapsto p(\varphi f),$$

(that is, the compression of M_φ to H^2) has norm $\|T_\varphi\| \leq \|\varphi\|_{\infty}$. We call T_φ the *Toeplitz operator* with *symbol* φ. The map

$$L^{\infty}(\mathbf{T}) \to B(H^2), \quad \varphi \mapsto T_\varphi,$$

is linear and preserves adjoints; that is, $T_\varphi^* = T_{\bar{\varphi}}$. The latter is true because if $f, g \in H^2$, then $\langle T_\varphi^*(f), g \rangle = \langle f, T_\varphi(g) \rangle = \langle f, p(\varphi g) \rangle = \langle p(f), \varphi g \rangle = \langle \bar{\varphi}f, g \rangle = \langle \bar{\varphi}f, p(g) \rangle = \langle p(\bar{\varphi}f), g \rangle = \langle T_{\bar{\varphi}}(f), g \rangle$.

Therefore, if $\bar{\varphi} = \varphi$, then T_φ is self-adjoint.

If $\varphi \in L^\infty(\mathbf{T})$, then the matrix (λ_{ij}) of M_φ with respect to the basis $(\varepsilon_n)_{n \in \mathbf{Z}}$ is constant along diagonals; that is, $\lambda_{ij} = \lambda_{i+1,j+1}$ for all i,j. This follows from the fact that M_φ commutes with $v = M_{\varepsilon_1}$. Conversely, if w is a bounded operator on $L^2(\mathbf{T})$ whose matrix with respect to (ε_n) is constant along the diagonals, then it is easily verified that w commutes with v, and therefore w is a Laurent operator by Theorem 3.5.2. From these remarks it is clear that the matrix of a Toeplitz operator with respect to the basis $(\varepsilon_n)_{n \in \mathbf{N}}$ is also constant along its diagonals. One can show conversely (but we shall not) that a bounded operator on H^2 whose matrix with respect to (ε_n) is constant along the diagonals is a Toeplitz operator.

The F. and M. Riesz theorem has a bearing on the spectral theory of Toeplitz operators: If $\varphi \in H^\infty$ and φ is not a scalar a.e., then T_φ has no eigenvalues. For suppose that $f \in H^2$ and $\lambda \in \mathbf{C}$ and $(T_\varphi - \lambda)(f) = 0$. Then $(\varphi - \lambda)f = 0$ a.e. Since $\varphi - \lambda \in H^2$ and is not the zero element, the set of points where it vanishes is a null set by Theorem 3.5.4. Therefore, $f = 0$ a.e.

A complication that arises in the theory of Toeplitz operators and distinguishes it from the theory of Laurent operators is that although $M_\varphi M_\psi = M_{\varphi\psi}$ for arbitrary $\varphi, \psi \in L^\infty(\mathbf{T})$, the corresponding statement for Toeplitz operators is not in general true. For instance, if $\varphi = \varepsilon_1$ and $\psi = \varepsilon_{-1}$, then $T_\varphi = u$, the unilateral shift, and $T_\psi = u^*$. As $u^*(\varepsilon_0) = 0$, $uu^* \neq 1$, but $T_{\varphi\psi} = T_{\varepsilon_0} = 1$, so $T_\varphi T_\psi \neq T_{\varphi\psi}$.

We can get $T_\varphi T_\psi = T_{\varphi\psi}$ in certain important special cases, as for instance in the following result.

3.5.6. Theorem. Let $\varphi \in L^\infty$ and $\psi \in H^\infty$. Then

$$T_{\varphi\psi} = T_\varphi T_\psi \qquad \text{and} \qquad T_{\bar{\psi}\varphi} = T_{\bar{\psi}} T_\varphi.$$

Proof. Since $\psi \in H^\infty$, therefore $\psi H^2 \subseteq H^2$. If $f \in H^2$, then $T_\varphi T_\psi(f) = p(\varphi p(\psi f)) = p(\varphi \psi f) = T_{\varphi\psi}(f)$, so $T_\varphi T_\psi = T_{\varphi\psi}$.

To get the second equality in the statement of the theorem, observe that $T_{\bar{\varphi}} T_\psi = T_{\bar{\varphi}\psi}$, so by taking adjoints, $T_\psi^* T_{\bar{\varphi}}^* = T_{\bar{\varphi}\psi}^*$; that is, $T_{\bar{\psi}} T_\varphi = T_{\bar{\psi}\varphi}$. $\quad\square$

Now we examine some aspects of the elementary spectral theory of Toeplitz operators.

3.5.7. Theorem (Hartman–Wintner). Let $\varphi \in L^\infty(\mathbf{T})$ and let $\sigma(\varphi)$ denote the spectrum of φ in $L^\infty(\mathbf{T})$. Then

$$\sigma(\varphi) \subseteq \sigma(T_\varphi)$$

and

$$r(T_\varphi) = \|T_\varphi\| = \|\varphi\|_\infty.$$

Proof. Since $T_\varphi - \lambda = T_{\varphi-\lambda}$ if $\lambda \in \mathbf{C}$, in order to show that $\sigma(\varphi) \subseteq \sigma(T_\varphi)$, it suffices to show that if T_φ is invertible, then φ is invertible in $L^\infty(\mathbf{T})$. Assume then T_φ is invertible and denote by M the positive number $\|T_\varphi^{-1}\|$. For all $f \in H^2$, $\|T_\varphi^{-1}(f)\| \leq M\|f\|$, so replacing f by $T_\varphi(f)$ we get $\|f\| \leq M\|T_\varphi(f)\|$. If $n \in \mathbf{Z}$, then $\|M_\varphi(\varepsilon_n f)\| = \|\varphi \varepsilon_n f\| = \|\varphi f\| \geq \|T_\varphi(f)\| \geq \|f\|/M = \|\varepsilon_n f\|/M$. However, the functions $\varepsilon_n f$ are dense in $L^2(\mathbf{T})$ relative to the L^2-norm, since Γ is L^2-norm dense in $L^2(\mathbf{T})$. Hence, for all $g \in L^2(\mathbf{T})$ we have $\|M_\varphi(g)\| \geq \|g\|/M$, and therefore $\langle M_\varphi^* M_\varphi(g), g \rangle \geq \langle g, g \rangle / M^2$, so $M_\varphi^* M_\varphi \geq M^{-2} > 0$. It follows that $M_\varphi^* M_\varphi$ is invertible, so M_φ is invertible (by normality of M_φ). Since the map

$$L^\infty(\mathbf{T}) \to B(L^2(\mathbf{T})), \quad \varphi \mapsto M_\varphi,$$

is an isometric $*$-homomorphism, φ is invertible in $L^\infty(\mathbf{T})$.

Now suppose that φ is an arbitrary element of $L^\infty(\mathbf{T})$. Since $\sigma(\varphi) \subseteq \sigma(T_\varphi)$, we get $\|T_\varphi\| \leq \|\varphi\|_\infty = r(\varphi) \leq r(T_\varphi) \leq \|T_\varphi\|$, so we have $\|T_\varphi\| = r(T_\varphi) = \|\varphi\|_\infty$. $\qquad\square$

3.5.8. Theorem. If $\varphi \in L^\infty(\mathbf{T})$, then T_φ is compact if and only if $\varphi = 0$.

Proof. Let u denote the unilateral shift. Then

$$u^{*n}(\varepsilon_m) = \begin{cases} \varepsilon_{m-n}, & \text{if } m \geq n \\ 0, & \text{if } m < n. \end{cases}$$

Therefore, if $f \in H^2$, then

$$\|u^{*n}(f)\|^2 = \left\| \sum_{m=n}^\infty \langle f, \varepsilon_m \rangle \varepsilon_{m-n} \right\|^2 = \sum_{m=n}^\infty |\langle f, \varepsilon_m \rangle|^2,$$

so the sequence $(u^{*n}(f))$ converges to zero as $n \to \infty$. If $v \in B(H^2)$ is of finite rank, then by Theorem 2.4.6 there exists $f_1, \ldots, f_n$ and $g_1, \ldots, g_n$ in H^2 such that $v = \sum_{j=1}^n f_j \otimes g_j$. Therefore, for each positive integer m we have $u^{*m} v = \sum_{j=1}^n u^{*m}(f_j) \otimes g_j$, so $\lim_{m \to \infty} u^{*m} v = 0$. Hence, for all $v \in K(H^2)$, $\lim_{m \to \infty} u^{*m} v = 0$, because the finite-rank operators are norm-dense in $K(H^2)$. Observe that $u^* T_\varphi u = T_{\bar\varepsilon_1} T_\varphi T_{\varepsilon_1} = T_{\bar\varepsilon_1 \varphi \varepsilon_1} = T_\varphi$, so if T_φ is compact, then since $\|T_\varphi\| = \|u^{*m} T_\varphi u^m\| \leq \|u^{*m} T_\varphi\|$ and $\lim_{m \to \infty} u^{*m} T_\varphi = 0$, we have $T_\varphi = 0$, and therefore $\varphi = 0$. $\qquad\square$

3.5.9. Lemma. If $\varphi \in C(\mathbf{T})$ and $\psi \in L^\infty(\mathbf{T})$, then $T_\varphi T_\psi - T_{\varphi\psi}$ and $T_\psi T_\varphi - T_{\psi\varphi}$ are compact operators.

Proof. We show $T_\psi T_\varphi - T_{\psi\varphi} \in K(H^2)$, and this implies $T_\varphi T_\psi - T_{\varphi\psi} = (T_{\bar\psi} T_{\bar\varphi} - T_{\overline{\psi\varphi}})^* \in K(H^2)$. By density of the set Γ of trigonometric polynomials in $C(\mathbf{T})$, we may suppose that φ is a trigonometric polynomial, and by linearity of the map $\varphi \mapsto T_\varphi$, we may even suppose that $\varphi = \varepsilon_n$ for some integer n. If $n \geq 0$, then by Theorem 3.5.6 $T_\psi T_{\varepsilon_n} = T_{\psi \varepsilon_n}$. Therefore, we need only show $T_\psi T_{\varepsilon_{-k}} - T_{\psi \varepsilon_{-k}} \in K(H^2)$ for all k positive. We show this by induction on k.

If $f \in H^2$, then

$$
\begin{aligned}
T_\psi T_{\varepsilon_{-1}}(f) &= p(\psi p(\varepsilon_{-1} f)) \\
&= p(\psi(\varepsilon_{-1} f - \langle f, \varepsilon_0 \rangle \varepsilon_{-1})) \\
&= T_{\psi \varepsilon_{-1}}(f) - \langle f, \varepsilon_0 \rangle p(\psi \varepsilon_{-1}).
\end{aligned}
$$

Hence, $T_\psi T_{\varepsilon_{-1}} - T_{\psi \varepsilon_{-1}}$ is an operator of rank not greater than one.

Suppose now we have shown that $T_\psi T_{\varepsilon_{-k}} - T_{\psi \varepsilon_{-k}} \in K(H^2)$ for all $\psi \in L^\infty(\mathbf{T})$ and some k. Then

$$
T_\psi T_{\varepsilon_{-k-1}} - T_{\psi \varepsilon_{-k-1}} = (T_\psi T_{\varepsilon_{-k}} - T_{\psi \varepsilon_{-k}}) T_{\varepsilon_{-1}} + T_{\psi \varepsilon_{-k}} T_{\varepsilon_{-1}} - T_{(\psi \varepsilon_{-k}) \varepsilon_{-1}}
$$

is compact. This proves the result.   □

Let $\mathbf{A}$ denote the C*-algebra generated by all Toeplitz operators T_φ with continuous symbol φ, and call $\mathbf{A}$ the *Toeplitz algebra*. We are going to use $\mathbf{A}$ to analyse these operators. To do this we need to identify the commutator ideal of $\mathbf{A}$.

If A is a C*-algebra, then its *commutator* ideal I is the closed ideal generated by the commutators $[a, b] = ab - ba$ ($a, b \in A$). It is easily verified that the commutator ideal is the smallest closed ideal I in A such that A/I is abelian.

3.5.10. Theorem. *The commutator ideal of the Toeplitz algebra $\mathbf{A}$ is $K(H^2)$.*

Proof. If K is a closed vector subspace of H^2 invariant for $\mathbf{A}$, then K is reducing for the unilateral shift u, so by Theorem 3.5.5 $K = 0$ or $K = H^2$. Thus, $\mathbf{A}$ is an irreducible subalgebra of $B(H^2)$. Now $p = 1 - uu^*$ is a rank-one operator, so $p \in \mathbf{A} \cap K(H^2)$, and therefore $K(H^2) \subseteq \mathbf{A}$ by Theorem 2.4.9. The quotient algebra $\mathbf{A}/K(H^2)$ is abelian, since it is generated by the elements $T_\varphi + K(H^2)$ ($\varphi \in C(\mathbf{T})$), which are commuting and normal by Lemma 3.5.9. Hence, $K(H^2)$ contains the commutator ideal I of $\mathbf{A}$. Since I contains $p = [u^*, u]$, it is non-zero. Therefore, $I = K(H^2)$ because $K(H^2)$ is simple (*cf.* Example 3.2.2).   □

3.5.11. Theorem. *The map*

$$
\psi \colon C(\mathbf{T}) \to \mathbf{A}/K(H^2), \quad \varphi \mapsto T_\varphi + K(H^2),
$$

*is a *-isomorphism.*

Proof. That ψ is linear and preserves adjoints is clear, and by Lemma 3.5.9 it is multiplicative, so it is a $*$-homomorphism. Since the Toeplitz operators T_φ ($\varphi \in C(\mathbf{T})$) generate $\mathbf{A}$, the elements $T_\varphi + K(H^2)$ ($\varphi \in C(\mathbf{T})$) generate $\mathbf{A}/K(H^2)$, and therefore ψ is surjective. Injectivity of ψ is immediate from Theorem 3.5.8. $\qquad\qquad\square$

3.5.12. Corollary. *If $\varphi \in C(\mathbf{T})$, then T_φ is a Fredholm operator if and only if φ vanishes nowhere.*

Proof. By the Atkinson characterisation (Theorem 1.4.16), T_φ is Fredholm if and only if $T_\varphi + K(H^2)$ is invertible in the quotient $B(H^2)/K(H^2)$. Hence, T_φ is Fredholm if and only if $T_\varphi + K(H^2)$ is invertible in $\mathbf{A}/K(H^2)$. By Theorem 3.5.11, therefore, T_φ is Fredholm if and only if φ is invertible in $C(\mathbf{T})$. $\qquad\qquad\square$

3.5.13. Corollary. *If $\varphi \in C(\mathbf{T})$, then $\sigma_e(T_\varphi) = \varphi(\mathbf{T})$. Hence, a Toeplitz operator with continuous symbol has connected essential spectrum.*

Proof. From the Atkinson characterisation of Fredholm operators and from Theorem 3.5.11, $\sigma_e(T_\varphi) = \sigma(T_\varphi + K(H^2)) = \sigma(\varphi) = \varphi(\mathbf{T})$. $\qquad\square$

A case of particular interest is the unilateral shift $u = T_{\varepsilon_1}$. It follows from the corollary that $\sigma_e(u) = \mathbf{T}$.

If $n \in \mathbf{Z}$, then the Fredholm index of T_{ε_n} is $-n$. To see this, one may suppose that $n > 0$, and then observe that in this case $\mathrm{ind}(T_{\varepsilon_n}) = \mathrm{ind}(u^n) = n\,\mathrm{ind}(u) = n(-1)$ (u is the unilateral shift). We shall be generalising this remark shortly and shall need the following elementary result.

3.5.14. Lemma. *If φ is an invertible function in $C(\mathbf{T})$, then there exists a unique integer $n \in \mathbf{Z}$ such that $\varphi = \varepsilon_n e^\psi$ for some $\psi \in C(\mathbf{T})$.*

Proof. First let us remark that if $\|1 - \varphi\| < 1$, then $\varphi = e^\psi$ for some $\psi \in C(\mathbf{T})$. In fact we can take $\psi = ln \circ \varphi$, where $ln\colon \mathbf{C} \setminus (-\infty, 0] \to \mathbf{C}$ is the principal branch of the logarithmic function (the hypothesis $\|1 - \varphi\| < 1$ implies that the range of φ lies in the domain of ln). Hence, if φ, φ' are invertible elements of $C(\mathbf{T})$ such that $\|\varphi - \varphi'\| < \|\varphi^{-1}\|^{-1}$, then $\varphi' = \varphi e^\psi$ for some $\psi \in C(\mathbf{T})$.

Suppose then φ is an invertible element of $C(\mathbf{T})$, and we shall show that $\varphi = \varepsilon_n e^\psi$ for some $n \in \mathbf{Z}$ and some $\psi \in C(\mathbf{T})$. Since Γ is dense in $C(\mathbf{T})$, we may suppose that $\varphi \in \Gamma$, by the observations in the first paragraph of this proof. Hence, we may write $\varphi = \sum_{|n| \leq N} \lambda_n \varepsilon_n$, for some $N > 0$ and some $\lambda_n \in \mathbf{C}$. Therefore, $\varphi = \varepsilon_{-N} \varphi'$ for some $\varphi' \in \Gamma_+$, so we may suppose that $\varphi \in \Gamma_+$. In this case φ is a polynomial in $z = \varepsilon_1$, and therefore a product of a constant and factors of the form $z - \lambda$, where $\lambda \notin \mathbf{T}$. Thus, we may further reduce and suppose that $\varphi = z - \lambda$ with $|\lambda| \neq 1$. If $|\lambda| < 1$, then $\|\varphi - z\| = |\lambda| < 1 = \|z^{-1}\|^{-1}$, so φ is of the form ze^ψ for some

$\psi \in C(\mathbf{T})$. Likewise, if $|\lambda| > 1$, then $\|(1 - \lambda^{-1}z) - 1\| < 1$, so $1 - \lambda^{-1}z$ is of the form e^ψ for some $\psi \in C(\mathbf{T})$, and therefore $\varphi = -\lambda e^\psi = e^{\psi'}$ for some $\psi' \in C(\mathbf{T})$. Thus, we have shown that if φ is invertible in $C(\mathbf{T})$, then $\varphi = \varepsilon_n e^\psi$ for some $n \in \mathbf{Z}$ and $\psi \in C(\mathbf{T})$.

To show uniqueness of n, we need only show that if ε_n is of the form e^ψ for some $\psi \in C(\mathbf{T})$, then $n = 0$. Suppose then that $\varepsilon_n = e^\psi$, where $n \in \mathbf{Z}$ and $\psi \in C(\mathbf{T})$. The map

$$\alpha \colon [0,1] \to \mathbf{Z}, \quad t \mapsto \operatorname{ind}(T_{e^{t\psi}}),$$

is continuous and has discrete range and connected domain, so it is necessarily constant. Hence, $-n = \operatorname{ind}(T_{e^\psi}) = \alpha(1) = \alpha(0) = \operatorname{ind}(T_1) = \operatorname{ind}(1) = 0$. This completes the proof. $\square$

The integer n in Lemma 3.5.14 is called the *winding number* of φ (with respect to the origin). We denote it by $\operatorname{wn}(\varphi)$.

3.5.15. Theorem. Let φ be an invertible element in $C(\mathbf{T})$. Then the Fredholm index of T_φ is minus the winding number of φ, that is,

$$\operatorname{ind}(T_\varphi) = -\operatorname{wn}(\varphi).$$

Moreover, T_φ is invertible if and only if it is Fredholm of index zero, if and only if $\varphi = e^\psi$ for some $\psi \in C(\mathbf{T})$.

Proof. If ψ is a trigonometric polynomial, say $\psi = \sum_{|n|\leq N} \lambda_n \varepsilon_n$ for some $N > 0$ and $\lambda_n \in \mathbf{C}$, write $\psi' = \sum_{n=0}^N \lambda_n \varepsilon_n$ and $\psi'' = \sum_{n=1}^N \lambda_{-n}\varepsilon_{-n}$. Then $\psi = \psi' + \psi''$ and $\psi', \bar{\psi}'' \in H^\infty$. Since H^∞ is a closed subalgebra of $L^\infty(\mathbf{T})$, it follows that $e^{\psi'}, e^{\bar{\psi}''} \in H^\infty$. Hence, $T_{e^{-\psi'}}T_{e^{\psi'}} = T_{e^{-\psi'}e^{\psi'}}$ (by Theorem 3.5.6), so $T_{e^{-\psi'}}T_{e^{\psi'}} = T_1 = 1$. Likewise, $T_{e^{\psi'}}T_{e^{-\psi'}} = T_1 = 1$. Thus, $T_{e^{\psi'}}$ is invertible. By a similar argument $T_{e^{\psi''}}$ is invertible, with inverse $T_{e^{-\psi''}}$. If we suppose now that φ is an arbitrary element of $C(\mathbf{T})$, then using the density of Γ in $C(\mathbf{T})$ we may choose a trigonometric polynomial ψ as above such that $\|1 - e^{\varphi - \psi}\| < 1$. Then $T_{e^\varphi} = T_{e^{\psi''}e^{\varphi - \psi}e^{\psi'}} = T_{e^{\psi''}}T_{e^{\varphi - \psi}}T_{e^{\psi'}}$ (by Theorem 3.5.6). Since $\|1 - T_{e^{\varphi - \psi}}\| = \|1 - e^{\varphi - \psi}\| < 1$, the operator $T_{e^{\varphi - \psi}}$ is invertible, and we have already seen that $T_{e^{\psi''}}$ and $T_{e^{\psi'}}$ are invertible. Hence, T_{e^φ} is a product of invertible operators and is therefore invertible.

Now suppose that φ is an arbitrary invertible element of $C(\mathbf{T})$ with winding number n. We show $\operatorname{ind}(T_\varphi) = -n$, and to do this we may suppose $n \geq 0$ (replace φ by $\bar{\varphi}$ if necessary). Now $\varphi = e^\psi \varepsilon_n$ for some $\psi \in C(\mathbf{T})$, and $T_\varphi = T_{e^\psi}T_{\varepsilon_n}$ by Theorem 3.5.6. Hence,

$$\operatorname{ind}(T_\varphi) = \operatorname{ind}(T_{e^\psi}) + \operatorname{ind}(T_{\varepsilon_n}) = -n,$$

since T_{e^ψ} is of index zero because it is invertible. The theorem follows. $\square$

3.5.16. Theorem. *The spectrum of a Toeplitz operator with continuous symbol is connected.*

Proof. If $\varphi \in C(\mathbf{T})$, then by Theorem 3.5.15

$$\sigma(T_\varphi) = \varphi(\mathbf{T}) \cup \{\lambda \in \mathbf{C} \mid T_\varphi - \lambda \text{ is Fredholm of non-zero index}\}.$$

Therefore, $\sigma(T_\varphi)$ is a compact set consisting of the connected compact set $\varphi(\mathbf{T})$ and some of its holes, and therefore by elementary plane topology $\sigma(T_\varphi)$ is connected.   □

The preceding theorem is a simple special case of a deep theorem of Widom which asserts that all Toeplitz operators have connected spectra [Dou 1, Corollary 7.46].

If $(H_\lambda)_{\lambda \in \Lambda}$ is a family of Hilbert spaces, $u_\lambda \in B(H_\lambda)$ for all $\lambda \in \Lambda$, and $M = \sup_\lambda \|u_\lambda\| < \infty$, we define $u \in B(\oplus_\lambda H_\lambda)$ by

$$u((x_\lambda)_\lambda) = (u_\lambda(x_\lambda))_\lambda \qquad ((x_\lambda)_\lambda \in \oplus_\lambda H_\lambda).$$

It is easily checked that $\|u\| = M$. We call u the *direct sum* of the family $(u_\lambda)_{\lambda \in \Lambda}$, and denote it by $\oplus_{\lambda \in \Lambda} u_\lambda$. It is straightforward to verify that the map

$$\oplus_\lambda B(H_\lambda) \to B(\oplus_\lambda H_\lambda), \quad (u_\lambda)_\lambda \mapsto \oplus_\lambda u_\lambda,$$

is an isometric $*$-homomorphism of C*-algebras.

If $u = T_{\varepsilon_1}$, then it is easily seen that u generates the C*-algebra **A** using the fact that ε_1 generates $C(\mathbf{T})$. As we mentioned in the introduction to this section, the algebra **A** plays a role in K-theory. What makes it useful is that it is the "universal" C*-algebra generated by a non-unitary isometry. This is made precise in Theorem 3.5.18. To establish that theorem we shall need the following result, which is an important structure theorem for isometries.

3.5.17. Theorem (Wold–von Neumann). *If v is an isometry on a Hilbert space H, then v is a unitary, or a direct sum of copies of the unilateral shift, or a direct sum of a unitary and copies of the unilateral shift.*

Proof. We may suppose that v is neither a unitary nor a sum of copies of the unilateral shift. Set $K = \cap_{n=0}^\infty v^n(H)$. Then $v(K) = K$, and therefore K reduces v. Let w be the compression of v to K and w' the compression of v to $K^\perp$. Since w is an isometry, and the equation $v(K) = K$ implies that w is surjective, therefore w is a unitary.

Now set $L = (vH)^\perp$. For all $n > 0$, $v^n(L) \subseteq v(H) = L^\perp$, so if $m, n \in \mathbf{N}$ and $m \neq n$, then $v^n(L)$ is orthogonal to $v^m(L)$. We claim that the internal orthogonal sum $\oplus_{n=0}^{\infty} v^n(L)$ is equal to $K^\perp$. To see this, first observe that $v^n(L)$ is orthogonal to $v^n(L^\perp) = v^{n+1}(H)$, so $v^n(L) \subseteq K^\perp$, since $K \subseteq v^{n+1}(H)$. Consequently, $\oplus_{n=0}^{\infty} v^n(L) \subseteq K^\perp$. To show the opposite inclusion, it suffices to show that if $x \in H$ is orthogonal to $v^n(L)$ for all $n \in \mathbf{N}$, then $x \in K$. Suppose then $x \in \cap_{n=0}^{\infty}(v^n(L))^\perp$. We show by induction that $x \in v^n(H)$ for all n. This is trivially true for $n = 0$. If $x \in v^n(H)$, then $x = v^n(y)$ for some $y \in H$, and since $v^n(y) \perp v^n(L)$, then $y \in L^\perp = v(H)$, and therefore $x \in v^{n+1}(H)$. This proves our claim.

If E is an orthonormal basis for L, then $\cup_{n=0}^{\infty} v^n(E)$ is an orthonormal basis for $K^\perp$. For each $e \in E$, let L_e be the Hilbert subspace of $K^\perp$ having $(v^n(e))_{n=0}^{\infty}$ as orthonormal basis. Then $K^\perp$ is the internal orthogonal sum $\oplus_{e \in E} L_e$, each L_e is invariant for v, the compression v_e of v to L_e is the unilateral shift, and $w' = \oplus_{e \in E} v_e$. $\square$

An interesting consequence of the Wold–von Neumann decomposition is that every non-unitary isometry has spectrum the closed unit disc. This is the case since the unilateral shift has such a spectrum (*cf.* Example 2.3.2), and is a direct summand of every non-unitary isometry.

3.5.1. Remark. If v is a unitary in a unital C*-algebra B, and $z \colon \mathbf{T} \to \mathbf{C}$ is the inclusion function, then there is a unique unital $*$-homomorphism $\varphi \colon C(\mathbf{T}) \to B$ such that $\varphi(z) = v$. To construct φ, first observe that $\sigma(v) \subseteq \mathbf{T}$ and that the "restriction" map

$$C(\mathbf{T}) \to C(\sigma(v)), \quad f \mapsto f_{\sigma(v)},$$

is a unital $*$-homomorphism. One gets φ by composing this map with the functional calculus $C(\sigma(v)) \to B$ at v. Since z generates $C(\mathbf{T})$, φ is unique.

3.5.18. Theorem (Coburn). *Suppose that v is an isometry in a unital C*-algebra B, and let $u = T_{e_1} \in \mathbf{A}$. Then there exists a unique unital $*$-homomorphism $\varphi \colon \mathbf{A} \to B$ such that $\varphi(u) = v$. Moreover, if $vv^* \neq 1$, then φ is isometric.*

Proof. Since u generates $\mathbf{A}$, therefore φ is unique. We use the universal representation of B to reduce to the case where B is a C*-subalgebra of $B(H)$ for some Hilbert space H, and $\mathrm{id}_H \in B$. By Theorem 3.5.17, we can write $H = \oplus_{\lambda \in \Lambda} H_\lambda$ and $v = \oplus_{\lambda \in \Lambda} v_\lambda$, where H_λ are Hilbert spaces and each $v_\lambda \in B(H_\lambda)$ is a unitary or a unitlateral shift.

If v_λ is a unitary, then combining Theorem 3.5.11 and Remark 3.5.1, there is a unital $*$-homomorphism $\varphi_\lambda \colon \mathbf{A} \to B(H_\lambda)$ such that $\varphi_\lambda(u) = v_\lambda$.

If v_λ is a unilateral shift, then there exists a unitary $w_\lambda \colon H^2 \to H_\lambda$ such that $v_\lambda = w_\lambda u w_\lambda^*$ (*cf.* Example 2.3.2). Hence, the map

$$\varphi_\lambda \colon \mathbf{A} \to B(H_\lambda), \quad a \mapsto w_\lambda a w_\lambda^*,$$

is an isometric unital $*$-homomorphism such that $\varphi_\lambda(u) = v_\lambda$.

Let (H, φ) be the direct sum of the family of representations $(H_\lambda, \varphi_\lambda)_\lambda$ of $\mathbf{A}$. Then $\varphi \colon \mathbf{A} \to B(H)$ is a unital $*$-homomorphism such that $\varphi(u) = \oplus_\lambda v_\lambda = v$. Moreover, since $\varphi(u) \in B$ and u generates $\mathbf{A}$, therefore $\varphi(\mathbf{A})$ is contained in B.

Now suppose that $vv^* \neq 1$. Then some v_{λ_0} is a unilateral shift. Hence, the representation $(H_{\lambda_0}, \varphi_{\lambda_0})$ is faithful, so (H, φ) is faithful. Therefore, φ is isometric. □

3. Exercises

In Exercises 1 to 7, A denotes an arbitrary C*-algebra.

1. Let a, b be normal elements of a C*-algebra A, and c an element of A such that $ac = cb$. Show that $a^*c = cb^*$, using Fuglede's theorem (Exercise 2.8) and the fact that the element

$$d = \begin{pmatrix} a & 0 \\ 0 & b \end{pmatrix}$$

is normal in $M_2(A)$ and commutes with

$$d' = \begin{pmatrix} 0 & c \\ 0 & 0 \end{pmatrix}.$$

This more general result is called the Putnam–Fuglede theorem.

2. Let τ be a positive linear functional on A.
(a) If I is a closed ideal in A, show that $I \subseteq \ker(\tau)$ if and only if $I \subseteq \ker(\varphi_\tau)$.
(b) We say τ is *faithful* if $\tau(a) = 0 \Rightarrow a = 0$ for all $a \in A^+$. Show that if τ is faithful, then the GNS representation (H_τ, φ_τ) is faithful.
(c) Suppose that α is an automorphism of A such that $\tau(\alpha(a)) = \tau(a)$ for all $a \in A$. Define a unitary on H_τ by setting $u(a + N_\tau) = \alpha(a) + N_\tau$ $(a \in A)$. Show that $\varphi_\tau(\alpha(a)) = u\varphi(a)u^*$ $(a \in A)$.

3. If $\varphi \colon A \to B$ is a positive linear map between C*-algebras, show that φ is necessarily bounded.

4. Suppose that A is unital. Let α be an automorphism of A such that $\alpha^2 = \mathrm{id}_A$. Define B to be the set of all matrices

$$c = \begin{pmatrix} a & b \\ \alpha(b) & \alpha(a) \end{pmatrix},$$

where $a, b \in A$. Show that B is a C*-subalgebra of $M_2(A)$. Define a map $\varphi: A \to B$ by setting

$$\varphi(a) = \begin{pmatrix} a & 0 \\ 0 & \alpha(a) \end{pmatrix}.$$

Show that φ is an injective *-homomorphism. We can thus identify A as a C*-subalgebra of B. If we set $u = \begin{pmatrix} 0 & 1 \\ 1 & 0 \end{pmatrix}$, then u is a self-adjoint unitary and $B = A + Au$. If C is any unital C*-algebra with a self-adjoint unitary element v, and $\psi: A \to C$ is a *-homomorphism such that

$$\psi(\alpha(a)) = v\psi(a)v^* \qquad (a \in A),$$

show that there is a unique *-homomorphism $\psi': B \to C$ extending ψ (that is, $\psi' \circ \varphi = \psi$) such that $\psi'(u) = v$.

(This establishes that B is a (very easy) example of a crossed product, namely $B = A \times_\alpha Z_2$, the crossed product of A by the two-element group Z_2 under the action α. The theory of crossed products is a vast area of the modern theory of C*-algebras. One of its primary uses is to generate new examples of simple C*-algebras. For an account of this theory, see [Ped].)

5. An element a of A^+ is *strictly positive* if the hereditary C*-subalgebra of A generated by a is A itself, that is, if $(aAa)^- = A$.

(a) Show that if A is unital, then $a \in A^+$ is strictly positive if and only if a is invertible.

(b) If H is a Hilbert space, show that a positive compact operator on H is strictly positive in $K(H)$ if and only if it has dense range.

(c) Show that if a is strictly positive in A, then $\tau(a) > 0$ for all non-zero positive linear functionals τ on A.

6. We say that A is *σ-unital* if it admits a sequence $(u_n)_{n=1}^\infty$ which is an approximate unit for A. It follows from Remark 3.1.1 that every separable C*-algebra is σ-unital.

(a) Let a be a strictly positive element of A, and set $u_n = a(a + 1/n)^{-1}$ for each positive integer n. Show that (u_n) is an approximate unit for A. (Hint: Define $g_n: \sigma(a) \to \mathbf{R}$ by $g_n(t) = t^2/(t + 1/n)$. Show that the sequence (g_n) is pointwise-increasing and pointwise-convergent to the inclusion $z: \sigma(a) \to \mathbf{R}$, and use Dini's theorem to deduce that (g_n) converges uniformly to z. Hence, $a = \lim_{n \to \infty} au_n$.)

(b) If $(u_n)_{n=1}^\infty$ is an approximate unit for A, show that $a = \sum_{n=1}^\infty u_n/2^n$ is a strictly positive element of A.

Thus, A is σ-unital if and only if it admits a strictly positive element.

7. Let Ω be a locally compact Hausdorff space. Show that $C_0(\Omega)$ admits an approximate unit $(p_n)_{n=1}^{\infty}$, where all the p_n are projections, if and only if Ω is the union of a sequence of compact open sets. Deduce that if a C*-algebra A admits a strictly positive element a such that $\sigma(a) \setminus \{0\}$ is discrete, then A admits an approximate unit $(p_n)_{n=1}^{\infty}$ consisting of projections. (Show that $C^*(a)$ is *-isomorphic to $C_0(\sigma(a) \setminus \{0\})$.)

8. Let $z \colon \mathbf{T} \to \mathbf{C}$ be the inclusion map. Let $\theta \in [0,1]$. Show that there is a unique automorphism α of $C(\mathbf{T})$ such that $\alpha(z) = e^{i2\pi\theta}z$. Define the faithful positive linear functional $\tau \colon C(\mathbf{T}) \to \mathbf{C}$ by setting $\tau(f) = \int f \, dm$ where m is normalised arc length on $\mathbf{T}$. Show that $\tau(\alpha(f)) = \tau(f)$ for all $f \in C(\mathbf{T})$. Deduce from Exercise 3.2 that there is a unitary v on the Hilbert space H_τ such that $\varphi_\tau(\alpha(f)) = v\varphi_\tau(f)v^*$ for all $f \in C(\mathbf{T})$. Let u be the unitary $\varphi_\tau(z)$. Show that $vu = e^{i2\pi\theta}uv$. If θ is irrational, the C*-algebra A_θ generated by u and v is called an *irrational rotation* algebra, and A_θ can be shown to be simple. See [Rie] for more details concerning A_θ. These algebras form a very important class of examples in C*-algebra theory. They are motivating examples in Connes' development of "non-commutative differential geometry," a subject of great future promise [Con 2].

9. Let m be normalised Haar measure on $\mathbf{T}$. If $\lambda \in \mathbf{C}$, $|\lambda| < 1$, define $\tau_\lambda \colon H^1 \to \mathbf{C}$ by setting

$$\tau_\lambda(f) = \int \frac{f(w)}{1 - \lambda\bar{w}} \, dmw \qquad (f \in H^1).$$

Show that $\tau_\lambda \in (H^1)^*$. By expanding $(1 - \lambda\bar{w})^{-1}$ in a power series, show that $\tau_\lambda(f) = \sum_{n=0}^{\infty} \hat{f}(n)\lambda^n$. Deduce that the function

$$\tilde{f} \colon \text{int } \mathbf{D} \to \mathbf{C}, \quad \lambda \mapsto \tau_\lambda(f),$$

is analytic, where $\text{int } \mathbf{D} = \{\lambda \in \mathbf{C} \mid |\lambda| < 1\}$. If $f, g \in H^2$, show that $fg \in H^1$ and $\tau_\lambda(fg) = \tau_\lambda(f)\tau_\lambda(g)$. (Hint: There exist sequences (φ_n) and (ψ_n) in Γ_+ converging to f and g, respectively, in the L^2-norm. Show that the sequence $(\varphi_n\psi_n)$ converges to fg in the L^1-norm, and deduce the result by first showing it for functions in Γ_+.)

10. If $f \colon \text{int } \mathbf{D} \to \mathbf{C}$ is an analytic function and $0 < r < 1$, define $f_r \in C(\mathbf{T})$ by setting $f_r(\lambda) = f(r\lambda)$. Set $\|f\|_2 = \sup_{0 < r < 1} \|f_r\|_2$, and let $H^2(\mathbf{D})$ denote the set of all analytic functions $f \colon \text{int } \mathbf{D} \to \mathbf{C}$ such that $\|f\|_2 < \infty$. If $f \in H^2(\mathbf{D})$, show that $\|f\|_2 = \sqrt{\sum_{n=0}^{\infty} |\lambda_n|^2}$, where $f(\lambda) = \sum_{n=0}^{\infty} \lambda_n\lambda^n$ is the Taylor series expansion of f. Show that $H^2(\mathbf{D})$ is a Hilbert space with inner product $\langle f, g \rangle = \sum_{n=0}^{\infty} \lambda_n\bar{\mu}_n$, where $\lambda_n = f^{(n)}(0)/n!$ and $\mu_n =$

$g^{(n)}(0)/n!$ (the operations are pointwise-defined), and show also that the map

$$H^2 \to H^2(\mathbf{D}), \quad f \mapsto \tilde{f},$$

is a unitary operator. (Thus, the elements of H^2 can be interpreted as analytic functions on int $\mathbf{D}$ satisfying a growth condition approaching the boundary. A similar interpretation can be given for the other H^p-spaces.)

11. Show that if φ is a function in $L^\infty(\mathbf{T})$ not almost everywhere zero, then either T_φ or T_φ^* is injective (Coburn). (Hint: If $f \in \ker(T_\varphi)$ and $g \in \ker(T_\varphi^*)$, show that $\varphi f \bar{g}$ and $\bar{\varphi} \bar{f} g \in H^1$. Deduce that $\varphi f \bar{g} = 0$ a.e. and apply Theorem 3.5.4 to show that f or $g = 0$ a.e.) Deduce that T_φ is invertible if and only if it is a Fredholm operator of index zero.

3. Addenda

An *ordered group* is a pair $(G, \leq)$ consisting of an abelian (discrete) group G and a partial order $\leq$ on G such that for all $x, y, z \in G$ the implication $x \leq y \Rightarrow x + z \leq y + z$ holds, and either $x \leq y$ or $y \leq x$. If G is an arbitrary abelian group, then there exists an order $\leq$ on G such that $(G, \leq)$ is an ordered group if and only if G is torsion-free, if and only if the Pontryagin dual group $\hat{G}$ is connected.

Let $(G, \leq)$ be an ordered group, and set $G^+ = \{x \in G \mid 0 \leq x\}$. If $f \in L^1(\hat{G})$, denote by $\hat{f}: G \to \mathbf{C}$ its Fourier transform,

$$\hat{f}(x) = \int_{\hat{G}} f(\gamma) \overline{\gamma(x)} \, dm\gamma \qquad (x \in G).$$

Here m is the unique Haar measure on $\hat{G}$ such that $m(\hat{G}) = 1$, and $L^p(\hat{G}) = L^p(\hat{G}, m)$. The generalised Hardy space $H^p = H^p(G, \leq)$ is defined to be

$$H^p = \{f \in L^p(\hat{G}) \mid \hat{f}(x) = 0 \ (x \in G, \ x < 0)\}.$$

This is an L^p-closed vector subspace of $L^p(\hat{G})$ for all $p \in [1, \infty]$.

Denote by q the projection of the Hilbert space $L^2(\hat{G})$ onto H^2. If $\varphi \in L^\infty(\hat{G})$, define $T_\varphi \in B(H^2)$ by setting $T_\varphi(f) = q(\varphi f)$, and call T_φ a (generalised) Toeplitz operator on H^2. Much of the classical theory of Toeplitz operators carries over to this situation.

Denote by $T(G)$ the C*-subalgebra of $B(H^2)$ generated by all T_φ where $\varphi \in C(\hat{G})$, and let $KT(G)$ be the commutator ideal of $T(G)$. Then $T(G)$ acts irreducibly on H^2. Let $V_x = T_{\varepsilon_x}$, where $\varepsilon_x: \hat{G} \to \mathbf{T}$, $\gamma \mapsto \gamma(x)$. If $W: G^+ \to B$ is a map to a unital C*-algebra B such that all W_x are isometries and $W_{x+y} = W_x W_y$ $(x, y \in G^+)$, then there is a unique *-homomorphism $\varphi: T(G) \to B$ such that $\varphi(V_x) = W_x$ $(x \in G^+)$.

If G is an ordered subgroup of $\mathbf{R}$, that is, G is a subgroup of $\mathbf{R}$ with the order induced from that of $\mathbf{R}$, then $KT(G)$ is a simple C*-algebra (it is *-isomorphic to some $K(H)$ for a Hilbert space H if and only if G is order isomorphic to 0 or $\mathbf{Z}$). Conversely, if $KT(G)$ is simple, then G is order isomorphic to an ordered subgroup of $\mathbf{R}$.

References: [Dou 2], [Mur].

CHAPTER 4

Von Neumann Algebras

A useful way of thinking of the theory of C*-algebras is as "non-commutative topology." This is justified by the correspondence between abelian C*-algebras and locally compact Hausdorff spaces given by the Gelfand representation. The algebras studied in this chapter, von Neumann algebras, are a class of C*-algebras whose study can be thought of as "non-commutative measure theory." The reason for the analogy in this case is that the abelian von Neumann algebras are (up to isomorphism) of the form $L^\infty(\Omega, \mu)$, where (Ω, μ) is a measure space.

The theory of von Neumann algebras is a vast and very well-developed area of the theory of operator algebras. We shall be able only to cover some of the basics. The main results of this chapter are the von Neumann double commutant theorem and the Kaplansky density theorem.

4.1. The Double Commutant Theorem

There are a number of topologies on $B(H)$ (H a Hilbert space), apart from the norm topology, that play a crucial role, and each has valuable properties that the others lack. The two most important are the strong (operator) topology and the weak (operator) topology. This section is concerned with the former (we shall introduce the weak topology in the next section). One of the reasons for the usefulness of the strong topology is the "order completeness" property asserted in Vigier's theorem (Theorem 4.1.1) which is analogous to the order completeness property of the real numbers **R**.

Henceforth, we shall be using some results concerning locally convex spaces. The relevant definitions and the required results are given in the appendix.

Let H be a Hilbert space, and $x \in H$. Then the function

$$p_x \colon B(H) \to \mathbf{R}^+, \quad u \mapsto \|u(x)\|,$$

is a semi-norm on $B(H)$. The locally convex topology on $B(H)$ generated by the separating family $(p_x)_{x\in H}$ is called the *strong* topology on $B(H)$. Thus, a net $(u_\lambda)_{\lambda\in\Lambda}$ converges strongly to an operator u on H if and only if $u(x) = \lim_\lambda u_\lambda(x)$ for all $x \in H$. It follows that the strong topology is weaker than the norm topology on $B(H)$.

With respect to the strong topology, $B(H)$ is a topological vector space, so the operations of addition and scalar multiplication are strongly continuous. This is not the case in general for the multiplication and involution operations.

4.1.1. Example. Let H be an infinite-dimensional Hilbert space with an orthonormal basis $(e_n)_{n=1}^\infty$. Set $u_n = e_1 \otimes e_n$. If $x \in H$, then $u_n(x) = \langle x, e_n\rangle e_1$, so $\lim_{n\to\infty}\|u_n(x)\| = \lim_{n\to\infty}|\langle x, e_n\rangle| = 0$. Thus, the sequence (u_n) is strongly convergent to zero in $B(H)$. Now $u_n^* = e_n \otimes e_1$, so $\|u_n^*(x)\| = |\langle x, e_1\rangle|$. Therefore, $\lim_{n\to\infty}\|u_n^*(x)\| = 1$ for $x = e_1$, so the sequence (u_n^*) does not converge strongly to zero. This shows that the operation $u \mapsto u^*$ on $B(H)$ is not strongly continuous, and therefore the strong and the norm topologies on $B(H)$ do not coincide.

Observe also that $\|u_n\| = \|e_1\|\|e_n\| = 1$, so the sequence $(\|u_n\|)$ is not convergent to zero, and therefore the norm $\|.\|: B(H) \to \mathbf{R}^+$ is not strongly continuous.

The operation of multiplication $B(H) \times B(H) \to B(H)$, $(u, v) \mapsto uv$, is not strongly continuous either (*cf.* Exercise 4.3).

The preceding example shows that the strong topology behaves badly in some respects, but it also has some very good qualities, as we shall prove in the next theorem.

Let H be an arbitrary Hilbert space and suppose that $(u_\lambda)_{\lambda\in\Lambda}$ is an increasing net in $B(H)_{sa}$ that converges strongly to u (so u also belongs to $B(H)_{sa}$). Then $u = \sup_\lambda u_\lambda$ and $\langle u(x), x\rangle = \sup_\lambda\langle u_\lambda(x), x\rangle$ ($x \in H$). The corresponding statement for decreasing nets in $B(H)_{sa}$ is also true. Both of these observations follow from the fact that if a net $(u_\lambda)_{\lambda\in\Lambda}$ converges strongly to an operator u, then $\langle u(x), y\rangle = \lim_\lambda\langle u_\lambda(x), y\rangle$ ($x, y \in H$).

4.1.1. Theorem (Vigier). *Let $(u_\lambda)_{\lambda\in\Lambda}$ be a net of hermitian operators on a Hilbert space H. Then $(u_\lambda)_{\lambda\in\Lambda}$ is strongly convergent if it is increasing and bounded above, or if it is decreasing and bounded below.*

Proof. We prove only the case where $(u_\lambda)_{\lambda\in\Lambda}$ is increasing, since the decreasing case can be got from this by multiplying by minus one.

Suppose then that (u_λ) is increasing and bounded above. By truncating the net (that is, by choosing a point $\lambda_0 \in \Lambda$ and considering the truncated net $(u_\lambda)_{\lambda\geq\lambda_0}$) we may suppose that (u_λ) is also bounded below, by v say. We may further suppose that all u_λ are positive (by considering the net $(u_\lambda - v)$ if necessary). Hence, there is a positive number M such that $\|u_\lambda\| \le M$ for indices λ. It follows that the increasing net $(\langle u_\lambda(x), x\rangle)$ is bounded above (by $M\|x\|^2$), so this net is convergent. Using the polarisation identity

$$\langle u_\lambda(x), y\rangle = \tfrac{1}{4} \sum_{k=0}^{3} i^k \langle u_\lambda(x + i^k y), x + i^k y\rangle,$$

we see that $(\langle u_\lambda(x), y\rangle)$ is a convergent net for all $x, y \in H$. Letting $\sigma(x, y)$ denote its limit, it is easy to check that the function

$$\sigma: H^2 \to \mathbf{C}, \ (x, y) \mapsto \sigma(x, y),$$

is a sesquilinear form on H. Moreover, $|\sigma(x, y)| = \lim_\lambda |\langle u_\lambda(x), y\rangle| \le M\|x\|\|y\|$, so σ is bounded. Hence, there is an operator u on H such that $\langle u(x), y\rangle = \sigma(x, y)$ for all x, y. Clearly, $\|u\| \le M$, u is hermitian, and $u_\lambda \le u$ for all $\lambda \in \Lambda$. Also,

$$\begin{aligned}
\|u(x) - u_\lambda(x)\|^2 &= \|(u - u_\lambda)^{1/2}(u - u_\lambda)^{1/2}(x)\|^2 \\
&\le \|u - u_\lambda\|\|(u - u_\lambda)^{1/2}(x)\|^2 \\
&\le 2M\langle(u - u_\lambda)(x), x\rangle,
\end{aligned}$$

and $\lim_\lambda\langle(u - u_\lambda)(x), x\rangle = 0$, so $u(x) = \lim_\lambda u_\lambda(x)$. Thus, (u_λ) converges strongly to u.   □

4.1.1. Remark. If (p_λ) is a net of projections on a Hilbert space strongly convergent to an operator u, then u is a projection. For u is self-adjoint and $\langle u(x), y\rangle = \lim_\lambda\langle p_\lambda(x), y\rangle = \lim_\lambda\langle p_\lambda(x), p_\lambda(y)\rangle = \langle u(x), u(y)\rangle = \langle u^2(x), y\rangle$, so $u = u^2$.

4.1.2. Theorem. *Suppose that $(p_\lambda)_{\lambda \in \Lambda}$ is a net of projections on a Hilbert space H.*

(1) *If (p_λ) is increasing, then it is strongly convergent to the projection of H onto the closed vector subspace $(\cup_\lambda p_\lambda(H))^-$.*

(2) *If (p_λ) is decreasing, then it is strongly convergent to the projection of H onto $\cap_\lambda p_\lambda(H)$.*

Proof. An easy exercise.   □

Just as for normed vector spaces, we say a family $(x_\lambda)_{\lambda \in \Lambda}$ of elements of a locally convex space is *summable* to a point x if the net $(\sum_{\lambda \in F} x_\lambda)_F$ (where F runs over all non-empty finite subsets of Λ) is convergent to x, and in this case we write $x = \sum_{\lambda \in \Lambda} x_\lambda$.

4.1.3. Theorem. *Let* $(p_\lambda)_{\lambda \in \Lambda}$ *be a family of projections on a Hilbert space H that are pairwise orthogonal (that is, $p_\lambda p_{\lambda'} = 0$ if λ, λ' are distinct indices in Λ). Then (p_λ) is summable in the strong topology on $B(H)$ to a projection, p say, such that*

$$\|p(x)\| = (\sum_{\lambda \in \Lambda} \|p_\lambda(x)\|^2)^{1/2} \qquad (x \in H).$$

If $p = 1$, then the map

$$H \to \bigoplus_\lambda p_\lambda(H), \quad x \mapsto (p_\lambda(x)),$$

is a unitary.

Proof. If F is a finite non-empty subset of Λ, then $p_F = \sum_{\lambda \in F} p_\lambda$ is a projection. Therefore, $(p_F)_F$ is an increasing net of projections, hence strongly convergent to a projection p; that is, the family (p_λ) is strongly summable to p. Moreover,

$$\|p(x)\|^2 = \lim_F \|p_F(x)\|^2 = \lim_F \sum_{\lambda \in F} \|p_\lambda(x)\|^2 = \sum_{\lambda \in \Lambda} \|p_\lambda(x)\|^2.$$

The observation concerning the case where $p = 1$ is clear. $\qquad\qquad\square$

If C is a subset of an algebra A, we define its *commutant* C' to be the set of all elements of A that commute with all the elements of C. Observe that C' is a subalgebra of A. The *double commutant* C'' of C is $(C')'$. Similarly, $C''' = (C'')'$. Always $C \subseteq C''$ and $C' = C'''$. If A is a normed algebra, then C' is closed. If A is a $*$-algebra and C is self-adjoint, then C' is a $*$-subalgebra of A. All of these facts are elementary with easy proofs.

4.1.4. Lemma. *Let H be a Hilbert space and A a $*$-subalgebra of $B(H)$ containing id_H. Then A is strongly dense in A''.*

Proof. Let $u \in A''$, $x \in H$, and $K = cl\{v(x) \mid v \in A\}$. Then K is a closed vector subspace of H which is invariant, and therefore reducing, for all $v \in A$, since A is self-adjoint. Thus, if p is the projection of H onto K, then $p \in A'$, so $pu = up$. Hence, $u(x) \in K$, and therefore there is a sequence (v_n) in A such that $u(x) = \lim_{n \to \infty} v_n(x)$.

For each positive integer n the map

$$\varphi : B(H) \to B(H^{(n)}), \quad v \mapsto (\delta_{ij}v),$$

is a unital $*$-homomorphism, so $\varphi(A)$ is a $*$-subalgebra of $B(H^{(n)})$ containing $\mathrm{id}_{H^{(n)}}$. Moreover, $\varphi(u) \in (\varphi(A))''$, for if $w \in (\varphi(A))'$ and $v \in A$ then $\varphi(v)w = w\varphi(v) \Rightarrow vw_{ij} = w_{ij}v$. Hence, $w_{ij} \in A'$, so $uw_{ij} = w_{ij}u$.

Therefore, $\varphi(u)w = w\varphi(u)$. Suppose now that $x = (x_1, \ldots, x_n) \in H^{(n)}$. Then by the first paragraph of this proof there is a sequence $(v_m)_m$ in A such that $\varphi(u)(x) = \lim_{m\to\infty} \varphi(v_m)(x)$. Hence, $u(x_j) = \lim_{m\to\infty} v_m(x_j)$ for $j = 1, \ldots, n$.

We show that this implies that u is in the strong closure of A. If W is a strong neighbourhood of u, we must show that $W \cap A$ is non-empty, and to do this we may suppose that $W - u$ is a basic neighbourhood of 0. Therefore, there are elements $x_1, \ldots, x_n \in H$ and a positive number ε such that

$$W - u = \{v \in B(H) \mid \|v(x_j)\| < \varepsilon \ (j = 1, \ldots, n)\}.$$

Hence, there is a sequence $(v_m)_m$ in A such that

$$u(x_j) = \lim_{m\to\infty} v_m(x_j) \qquad (j = 1, \ldots, n).$$

Consequently, for some N the operator $v_N \in W$, so $W \cap A \neq \emptyset$. $\qquad\square$

Let H be a Hilbert space. If A is a strongly closed $*$-subalgebra of $B(H)$, we call A a *von Neumann* algebra on H. Since the strong topology is weaker than the norm topology, a strongly closed set is also norm-closed. Hence, a von Neumann algebra is a C*-algebra.

Obviously, $B(H)$ is a von Neumann algebra on H, as is $\mathbf{C}1$ where 1 is the identity map on H. If $(H_\lambda)_{\lambda\in\Lambda}$ is a family of Hilbert spaces and A_λ is a von Neumann algebra on H_λ for each index λ, then it is an easy exercise to show that the direct sum $\oplus_\lambda A_\lambda$ is a von Neumann algebra on $\oplus_\lambda H_\lambda$.

If A is a $*$-algebra on a Hilbert space H, then its commutant A' is a von Neumann algebra on H (it is straightforward to verify that A' is strongly closed).

If A is a von Neumann algebra on H and p is a projection in A, then pAp is a von Neumann algebra on H. Also, $M_n(A)$ is a von Neumann algebra on $H^{(n)}$.

If H is infinite-dimensional, then $K(H)$ is not a von Neumann algebra on H. To see this, let E be an orthonormal basis for H, and for each finite non-empty subset F of E let $p_F = \sum_{e\in F} e\otimes e$. Then p_F is a finite-rank projection and the net $(p_F)_F$ (where F runs over all finite non-empty subsets of E) converges strongly to 1 on H. If $K(H)$ were a von Neumann algebra, this would imply that it contains 1, and so $\dim(H) < \infty$, contradicting our assumption on H.

A fundamental result concerning von Neumann algebras is the following, known as the *double commutant* theorem.

4.1.5. Theorem (von Neumann). *Let A be a $*$-algebra on a Hilbert space H and suppose that $\mathrm{id}_H \in A$. Then A is a von Neumann algebra on H if and only if $A = A''$.*

Proof. Immediate from Lemma 4.1.4.   □

The intersection of a family of von Neumann algebras on a Hilbert space H is also a von Neumann algebra. Thus, for any set $C \subseteq B(H)$ there is a smallest von Neumann algebra A containing C. We call A the von Neumann algebra *generated* by C. If C is self-adjoint and contains id_H, then $A = C''$. If in addition C consists of commuting elements, then A is abelian (for in this case $C \subseteq C' \Rightarrow A = C'' \subseteq C' \Rightarrow A \subseteq A'$). This implies that there are non-trivial examples of abelian von Neumann algebras. We give an explicit example:

4.1.2. Example. Let Ω be a compact Hausdorff space, and suppose that μ is a finite positive regular Borel measure on Ω. We saw in Example 2.5.1 that the map

$$L^\infty(\Omega, \mu) \to B(L^2(\Omega, \mu)), \quad \varphi \mapsto M_\varphi,$$

is an isometric *-homomorphism. Its range A is a C*-subalgebra of $B(H)$. Denote by B the C*-subalgebra of A of all multiplication operators on $L^2(\Omega, \mu)$ with continuous symbol. The commutant of B is A (to see this, mimic the proof of Theorem 3.5.2 using the L^2-norm density of $C(\Omega)$ in $L^2(\Omega, \mu)$). Hence, A is a von Neumann algebra on the Hilbert space $L^2(\Omega, \mu)$. Since $A \subseteq A'$ (because A is abelian) and $A' \subseteq B' = A$, therefore $A = A'$. Consequently, $A = B''$, so B is strongly dense in A by Lemma 4.1.4.

Let K be a closed vector subspace of a Hilbert space H and let p be the projection of H onto K. If $u \in B(H)$, let $u_p = u_K$ be the compression of u to K. It is easy to verify that the map

$$pB(H)p \to B(K), \quad u \to u_K,$$

is a *-isomorphism.

If A is a *-algebra on H and p is in A', set $A_p = \{u_p \mid u \in A\}$.

4.1.6. Lemma. *Let A be a *-algebra on a Hilbert space H, and p a projection in A'. Then pAp and A_p are *-algebras on H and $p(H)$, respectively, and the map*

$$pAp \to A_p, \quad u \mapsto u_p,$$

*is a *-isomorphism. Moreover, if also $p \in A''$, then $(A')_p = (A_p)'$.*

Proof. We show only that $p \in A'' \Rightarrow (A')_p = (A_p)'$, because the rest is a routine exercise.

Suppose that $u \in (A')_p$ and $v \in A_p$. Then there exist $\tilde{u}$ and $\tilde{v}$ in A' and A, respectively, such that $u = \tilde{u}_p$ and $v = \tilde{v}_p$. Hence, for any $x \in p(H)$ we have $uv(x) = p\tilde{u}p\tilde{v}(x) = p\tilde{v}p\tilde{u}(x) = vu(x)$. Therefore, $u \in (A_p)'$, so $(A')_p \subseteq (A_p)'$.

Conversely, suppose now that $u \in (A_p)'$, and write $u = \tilde{u}_p$ for some $\tilde{u} \in pB(H)p$. If $v \in A$, then $v_p u = u v_p$, so $(pvp)_p \tilde{u}_p = \tilde{u}_p (pvp)_p$. Hence, $pvp\tilde{u} = \tilde{u}pvp$, so $v\tilde{u} = \tilde{u}v$. Consequently, $\tilde{u} \in A'$, and therefore $u \in (A')_p$. This shows that the inclusion $(A_p)' \subseteq (A')_p$ holds.   □

The reader should be aware that some authors define a von Neumann algebra on a Hilbert space H to be a *-algebra A on H such that $A = A''$. This automatically ensures that $\mathrm{id}_H \in A$. However, proofs appear to run more smoothly if von Neumann algebras are defined as we have done. Moreover, we can frequently reduce to the case where $A = A''$, by the trick explained in Remark 4.1.2. Using our definition von Neumann algebras are still unital, but the unit may not be the identity map of the underlying Hilbert space:

4.1.7. Theorem. *If A is a non-zero von Neumann algebra, then it is unital.*

Proof. Suppose that A acts on the Hilbert space H, and let $(u_\lambda)_{\lambda \in \Lambda}$ be an approximate unit for A. By Theorem 4.1.1, (u_λ) converges strongly to a self-adjoint operator, p say, and obviously $p \in A$, since A is strongly closed. If $x \in H$ and $u \in A$, then $pu(x) = \lim_\lambda u_\lambda u(x) = u(x)$, so $pu = u$. Hence, p is a unit for A.   □

If p is a projection in a von Neumann algebra A, then pAp is a strongly closed hereditary C*-subalgebra of A, and if also $p \in A'$, then Ap is a strongly closed ideal of A. We now prove the converse of these statements.

4.1.8. Theorem. *Let A be a von Neumann algebra.*

(1) *If B is a strongly closed hereditary C*-subalgebra of A, then there is a unique projection $p \in B$ such that $B = pAp$.*

(2) *If I is a strongly closed ideal in A, then there is a unique projection q in I such that $I = Aq$. Moreover, $q \in A'$.*

Proof. The existence of p and q follows from the observation that B and I are von Neumann algebras and therefore unital by Theorem 4.1.7. Uniqueness is clear in each case.   □

4.1.2. Remark. Let A be a von Neumann algebra on a Hilbert space H and let p be the unit of A. Of course, p is a projection in A'. The map

$$A \to A_p, \quad u \mapsto u_p,$$

is a *-isomorphism (by Lemma 4.1.6), and A_p is a von Neumann algebra on $p(H)$ containing $\mathrm{id}_{p(H)}$, so $A_p = (A_p)''$. This device will be frequently used to reduce to the case where the von Neumann algebra is its own double commutant.

If u is an operator on a Hilbert space H, then its *range projection* $[u]$ is the projection of H on $(u(H))^-$. We have $[u] = [(uu^*)^{1/2}]$, since $\overline{u(H)}^\perp = \ker(u^*) = \ker(uu^*)^{1/2}$ (by the polar decomposition of u^*) $= \overline{(uu^*)^{1/2}(H)}^\perp$.

4.1.9. Theorem. *If A is a von Neumann algebra, then it contains the range projections of all of its elements.*

Proof. Let A act on H, and let $u \in A$. Since $(uu^*)^{1/2} \in A$, to show that the range projection of u is in A, we may suppose that $u \geq 0$. Obviously, we may also assume that $u \leq 1$. Let $u_n = u^{1/2^n}$ for $n \in \mathbf{N}$. Then (u_n) is an increasing sequence of positive elements in the closed unit ball of A, so by Theorem 4.1.1 (u_n) is strongly convergent to a positive operator, p say. If $x \in H$, then

$$\|(p^2 - u_n^2)(x)\| \leq \|(p^2 - u_np)(x)\| + \|(u_np - u_n^2)(x)\|$$
$$\leq \|(p - u_n)p(x)\| + \|(p - u_n)(x)\|.$$

Therefore, (u_n^2) converges to p^2 strongly. But $u_n^2 = u_{n-1}$ for all $n > 0$, so $p = p^2$.

The sequence (u_n) is in $C^*(u)$, so $p(H) \subseteq (u(H))^-$. The continuous functions

$$\sigma(u) \to \mathbf{R}, \ t \mapsto t^{1+1/2^n},$$

form an increasing sequence and converge pointwise to the identity function $t \mapsto t$, so by Dini's theorem, they converge uniformly. Therefore, by the functional calculus, $u = \lim_{n \to \infty} u^{1+1/2^n}$; that is, $u = \lim_{n \to \infty} uu_n$. Hence, $u = up = pu$, so $(u(H))^- \subseteq p(H)$. Therefore, $[u] = p \in A$.   □

4.1.10. Theorem. *Let A be a von Neumann algebra on a Hilbert space H and v an element of A with polar decomposition $v = u|v|$. Then $u \in A$.*

Proof. Let w be a unitary in the unital C*-algebra A'. Then $\tilde{w} = w^*uw$ is a partial isometry on H such that $v = \tilde{w}|v|$ and $\ker(\tilde{w}) = \ker(v)$. It follows, therefore, from the uniqueness of the polar decomposition that $\tilde{w} = u$, so u and w commute. But A' is the linear span of its unitaries, so u must commute with all elements of A', and therefore $u \in A'' = (A + \mathbf{C}1)''$. By Lemma 4.1.4 there is a net $(u_\lambda)_{\lambda \in \Lambda}$ in A and a net $(\alpha_\lambda)_{\lambda \in \Lambda}$ in $\mathbf{C}$ such that the net $(u_\lambda + \alpha_\lambda 1)$ converges strongly to u on H. If $p = [|v|]$, then by Theorem 4.1.9 $p \in A$. Since $(1 - p)(H) = \overline{|v|(H)}^\perp = \ker(|v|) = \ker(v) = \ker(u)$, we have $u(1 - p) = 0$; that is, $u = up$. Therefore, u is the strong limit of the net $(u_\lambda p + \alpha_\lambda p)$ which lies in A, so $u \in A$.   □

4.1.11. Theorem. *Suppose that A is a von Neumann algebra on a Hilbert space H.*

(1) *A is the closed linear span of its projections.*

(2) If $\mathrm{id}_H \in A$ and u is a normal element of A, then $E(S) \in A$ for every Borel set S of $\sigma(u)$, where E is the spectral resolution of the identity for u.

(3) If $\mathrm{id}_H \in A$ and $v \in B(H)$, then $v \in A$ if and only if v commutes with all the projections of A'.

Proof. We may suppose in all cases that $\mathrm{id}_H \in A$. We prove Condition (2) first. Let u be a normal element of A with spectral resolution of the identity for u denoted by E. If $v \in A'$, then $vu = uv$ and $vu^* = u^*v$, so $vf(u) = f(u)v$ for every $f \in B_\infty(\sigma(u))$. In particular, $vE(S) = E(S)v$ for every Borel set S of $\sigma(u)$. Therefore, $E(S) \in A'' = A$.

Condition (1) follows directly from Condition (2), using the fact that the closed linear span of the characteristic functions χ_S (S a Borel set of $\sigma(u)$) is $B_\infty(\sigma(u))$ for each normal element u of A.

Condition (3) follows immediately from Condition (1), since A' is a von Neumann algebra, and therefore it is the closed linear span of its projections.   □

We give an immediate and important application of this result in the next theorem. First we make some observations.

4.1.3. Remark. If H is a Hilbert space, then $B(H)' = \mathbf{C}1$. For it is obvious that $\mathbf{C}' = B(H)$, and since $\mathbf{C}$ is a von Neumann algebra containing id_H, Theorem 4.1.5 implies that $\mathbf{C} = \mathbf{C}''$, so $\mathbf{C} = B(H)'$.

4.1.4. Remark. If A is a C*-algebra acting on a Hilbert space H and $S \subseteq H$, denote by AS the linear span of the set $\{u(x) \mid u \in A,\ x \in S\}$, and denote by $[AS]$ the closure of AS.

We say A acts *non-degenerately* on H if $[AH] = H$. Equivalently, for each non-zero element $x \in H$ there exists $u \in A$ such that $u(x) \neq 0$.

If A acts non-degenerately on H and $(u_\lambda)_{\lambda \in \Lambda}$ is an approximate unit for A, then $(u_\lambda)_{\lambda \in \Lambda}$ converges to $1 = \mathrm{id}_H$ strongly on H. (We have to show that $\lim_\lambda u_\lambda(x) = x$, for all $x \in H$. This is clear if $x = u(y)$ for some $u \in A$ and $y \in H$. By taking linear combinations, one gets $\lim_\lambda u_\lambda(x) = x$ for all x in AH. Using density of AH in H, it follows that $\lim_\lambda u_\lambda(x) = x$ for arbitrary x.)

If A acts irreducibly on H and $A \neq 0$, then it acts non-degenerately, since $[AH]$ is a non-zero closed vector subspace of H invariant for A, and therefore equals H.

4.1.12. Theorem. Let A be a non-zero C*-algebra acting on a Hilbert space H. The following conditions are equivalent:

(1) A acts irreducibly on H.

(2) $A' = \mathbf{C}1$.

(3) A is strongly dense in $B(H)$.

Proof. If p is a projection in $B(H)$, then $p \in A'$ if and only if the closed vector subspace $p(H)$ of H is invariant for A. Since A' is a von Neumann algebra, it is the closed linear span of its projections by Theorem 4.1.11, so if A acts irreducibly, then A' has no projections except the trivial ones, and therefore $A' = \mathbf{C}1$. Therefore, $(1) \Rightarrow (2)$. The reverse implication $(2) \Rightarrow (1)$ is clear.

If we suppose now that $A' = \mathbf{C}1$, then $(A + \mathbf{C}1)' = \mathbf{C}1$, and therefore $(A + \mathbf{C}1)'' = B(H)$. By Lemma 4.1.4 the C*-algebra $A + \mathbf{C}1$ is strongly dense in $B(H)$. However, since A acts irreducibly on H, it acts non-degenerately, and therefore if $(u_\lambda)_{\lambda \in \Lambda}$ is an approximate unit for A, it is strongly convergent to 1. Therefore, $1 \in A^-$ and $A^- = (A + \mathbf{C}1)^- = B(H)$, where the symbol $^-$ denotes strong closure. Hence, $(2) \Rightarrow (3)$.

Finally, if A is strongly dense in $B(H)$, then $A' = B(H)' = \mathbf{C}$, so $(3) \Rightarrow (2)$. $\square$

If $(\varepsilon_n)_{n=0}^\infty$ is the usual orthonormal basis for the Hardy space H^2 and u is the unilateral shift on this basis, then the C*-algebra $\mathbf{A}$ generated by u acts irreducibly on H^2, by Theorem 3.5.5. Hence, $\mathbf{A}$ is strongly dense in $B(H^2)$ by the preceding theorem, and therefore the von Neumann algebra generated by u is $B(H^2)$.

Von Neumann algebras have a plentiful supply of projections, as we saw in Theorem 4.1.11, and this is again illustrated by the following "binary expansion."

4.1.13. Theorem. *If A is a hereditary C*-subalgebra of a von Neumann algebra and a is a positive element of A such that $\|a\| \leq 1$, then there is a sequence of projections $(p_n)_{n=1}^\infty$ in A such that $a = \sum_{n=1}^\infty p_n/2^n$.*

Proof. First suppose that A itself is a von Neumann algebra. We may suppose that A acts on a Hilbert space H such that $\mathrm{id}_H \in A$. We construct by induction a sequence of projections (p_n) in A such that $0 \leq a - \sum_{j=1}^n p_j/2^j \leq 1/2^n$, for $n \geq 1$, and this will prove the result.

Let χ be the Borel function from $\sigma(a)$ to $\mathbf{C}$ which is defined by

$$\chi(t) = \begin{cases} 1, & \text{if } t \geq 1/2 \\ 0, & \text{if } t < 1/2. \end{cases}$$

If z is the inclusion map of $\sigma(a)$ in $\mathbf{C}$, then $0 \leq z - \frac{1}{2}\chi \leq \frac{1}{2}$, so $0 \leq a - \frac{1}{2}p_1 \leq \frac{1}{2}$, where $p_1 = \chi(a)$. We are, of course, using the Borel functional calculus at a here. Note that p_1 is a projection and lies in A, since A is a von Neumann algebra containing id_H (use Theorem 4.1.11). Thus, we have started the inductive construction.

Suppose then that $p_1, \ldots, p_n$ have been constructed with the required properties. Then $b = a - \sum_{j=1}^n p_j/2^j$ is positive and $\sigma(b) \subseteq [0, 1/2^n]$. Define the Borel function $\chi\colon \sigma(b) \to \mathbf{C}$ by

$$\chi(t) = \begin{cases} 1, & \text{if } 1/2^{n+1} \leq t \leq 1/2^n \\ 0, & \text{otherwise.} \end{cases}$$

Then if z is the inclusion function of $\sigma(b)$ in $\mathbf{C}$, we have $0 \leq z - \chi/2^{n+1} \leq 1/2^{n+1}$, so again by the Borel functional calculus the element $p_{n+1} = \chi(b)$ is a projection in A, and $0 \leq b - p_{n+1}/2^{n+1} \leq 1/2^{n+1}$. Therefore, $0 \leq a - \sum_{j=1}^{n+1} p_j/2^j \leq 1/2^{n+1}$. This completes the induction.

Now let us suppose only that A is a hereditary C*-subalgebra of a von Neumann algebra, B say. As before, if a is in the closed unit ball of A^+, then $a = \sum_{n=1}^{\infty} p_n/2^n$ for a sequence (p_n) of projections in B. Since $a \geq p_n/2^n$ and A is hereditary in B, therefore $p_n \in A$. This proves the theorem. $\square$

It follows from Theorem 4.1.13 that the closed unit ball of A^+ is the closed convex hull of the projections of A. This of course is not true for all C*-algebras—consider $C[0,1]$, for instance.

4.1.14. Corollary. *If A is a hereditary C*-subalgebra of a von Neumann algebra, then it is the closed linear span of its projections.*

Proof. The algebra A is the linear span of A^+, and A^+ is contained in the closed linear span of the projections by Theorem 4.1.13. $\square$

If p, q are projections in a C*-algebra A, we say they are *(Murray-von Neumann) equivalent*, and we write $p \sim q$, if there exists $u \in A$ such that $p = u^*u$ and $q = uu^*$. It is a straightforward exercise to show that this is indeed an equivalence relation on the projections of A. The relation $\sim$ is of fundamental importance in the classification theory of von Neumann algebras (see the Addenda section of this chapter), and in K-theory for C*-algebras (Chapter 7). However, for the moment we need only one small result concerning $\sim$:

4.1.5. Remark. If p, q are infinite-rank projections on a separable Hilbert space H, then $p \sim q$. To see this, choose orthonormal bases $(e_n)_{n=1}^{\infty}$ and $(f_n)_{n=1}^{\infty}$ for $p(H)$ and $q(H)$, respectively. Let $v\colon p(H) \to q(H)$ be the unitary such that $v(e_n) = f_n$ for all n. Define $u \in B(H)$ by setting $u = v$ on $p(H)$ and $u = 0$ on $(1-p)(H)$. It is easily verified that $p = u^*u$ and $q = uu^*$.

4.1.15. Theorem. *If H is a separable infinite-dimensional Hilbert space, then $K(H)$ is the unique non-trivial closed ideal of $B(H)$.*

Proof. Let I be a non-zero closed ideal of $B(H)$. By Theorems 2.4.5 and 2.4.7, $K(H) \subseteq I$. Now if $I \not\subseteq K(H)$, then by Corollary 4.1.14 $K(H)$ does not contain all of the projections of I. Hence, I has an infinite-rank projection, p say. If q is any other infinite-rank projection on H, then as we saw in Remark 4.1.5 there is an element $u \in B(H)$ such that $p = u^*u$ and $q = uu^*$. Therefore, $q = q^2 = upu^*$, so q also belongs to I. Hence, I contains all the projections of $B(H)$, whether their rank is finite or infinite, and therefore $I = B(H)$. Thus, we have shown that the only closed ideals of $B(H)$ are $0, K(H)$, and $B(H)$. $\square$

4.1.6. Remark. Let H be a separable infinite-dimensional Hilbert space. If $(x_n)_{n=1}^\infty$ is a dense sequence in H, it is easy to check that $K(H)$ is the closed linear span of the operators $x_n \otimes x_m$ $(n, m \geq 1)$ using the fact that $K(H)$ is the closed linear span of the rank-one operators (Theorems 2.4.5 and 2.4.6). Hence, $K(H)$ is separable. However, $B(H)$ is non-separable (and therefore $B(H)/K(H)$ is non-separable). To see this, choose an ortho-normal basis $(e_n)_{n=1}^\infty$ for H. For each set S of positive integers, let p_S be the projection in $B(H)$ defined by setting $p_S(e_n) = e_n$ if $n \in S$, and $p_S(e_n) = 0$ otherwise. Clearly, $\|p_S - p_{S'}\| = 1$ if $S \neq S'$. Hence, the family of operators $(p_S)_S$ cannot be in the closure of the range of any sequence in $B(H)$.

4.1.16. Theorem. *If H is an infinite-dimensional separable Hilbert space, then the Calkin algebra $B(H)/K(H)$ is a simple C*-algebra.*

Proof. Let C denote the Calkin algebra and $\pi: B(H) \to C$ the quotient map. To see that C is simple, let I be a closed ideal in C. Then $\pi^{-1}(I)$ is a closed ideal in $B(H)$, and therefore by Theorem 4.1.15, $\pi^{-1}(I) = 0, K(H)$, or $B(H)$. Hence, $I = 0$ or C. $\square$

4.1.7. Remark. Let u, v be operators on a Hilbert space H such that $uu^* \leq vv^*$. Then there is an operator $w \in B(H)$ such that $u = vw$. To see this, observe that $\|u^*(x)\|^2 = \langle uu^*(x), x\rangle \leq \langle vv^*(x), x\rangle = \|v^*(x)\|^2$. Hence, we get a well-defined norm-decreasing linear map $w_0: v^*(H) \to H$ by setting $w_0 v^*(x) = u^*(x)$. Clearly, we can extend w_0 to a bounded linear w_1 on H. Setting $w = w_1^*$, we get the required result, $u = vw$.

We give an application of Theorem 4.1.13 to single operator theory.

4.1.17. Theorem. *Let H be a Hilbert space and u a bounded operator on H. Then u is compact if and only if its range contains no infinite-dimensional closed vector subspace.*

Proof. Suppose that u is compact. Let K be a closed vector subspace of $u(H)$ and p the projection of H on K. The linear map

$$v: \ker(pu)^\perp \to K, \quad x \mapsto pu(x),$$

is compact, and bijective, therefore invertible by the open mapping theorem. Hence, $\mathrm{id}_K = vv^{-1}$ is compact, so K is finite-dimensional.

Now suppose conversely that $u(H)$ contains no infinite-dimensional closed vector subspaces. Observe that the range of u is the same as the range of $|u^*|$ by the polar decomposition of u^*, and that u is compact if and only if $|u^*|$ is compact. Thus, to show that u is compact we may suppose that it is positive, and by rescaling if necessary we may also suppose that $u \leq 1$. Hence, $0 \leq u^2 \leq 1$. It follows from Theorem 4.1.13 that there is a sequence of projections $(p_n)_{n=1}^\infty$ on H such that $u^2 = \sum_{n=1}^\infty p_n/2^n$. Now $p_n/2^n \leq u^2$, so by Remark 4.1.7 there exists $w_n \in B(H)$ such that $p_n = uw_n$. Hence, $p_n(H)$ is a closed vector subspace of $u(H)$ and is therefore finite-dimensional by assumption. Consequently, all of the projections p_n are compact operators, and therefore u is compact also.   □

4.2. The Weak and Ultraweak Topologies

Preparatory to our introduction of the weak and ultraweak topologies, we show now that $L^1(H)$ is the dual of $K(H)$, and $B(H)$ is the dual of $L^1(H)$.

Let H be a Hilbert space, and suppose that $u \in L^1(H)$. It follows from Theorem 2.4.16 that the function

$$\mathrm{tr}(u\cdot)\colon K(H) \to \mathbf{C}, \ v \mapsto \mathrm{tr}(uv),$$

is linear and bounded, and $\|\mathrm{tr}(u\cdot)\| \leq \|u\|_1$. We therefore have a map

$$L^1(H) \to K(H)^*, \ u \mapsto \mathrm{tr}(u\cdot),$$

which is clearly linear and norm-decreasing. We call this map the *canonical* map from $L^1(H)$ to $K(H)^*$.

4.2.1. Theorem. *If H is a Hilbert space, then the canonical map from $L^1(H)$ to $K(H)^*$ is an isometric linear isomorphism.*

Proof. Let θ denote this map. If $\theta(u) = 0$, then $\mathrm{tr}(u^*u) = 0$, so $u = 0$. Thus, θ is injective.

Now suppose that $\tau \in K(H)^*$. Then the function

$$\sigma\colon H^2 \to \mathbf{C}, \ (x,y) \mapsto \tau(x \otimes y),$$

is a sesquilinear form on H and $\|\sigma\| \leq \|\tau\|$. Hence, there is a unique operator u on H such that $\langle u(x), y \rangle = \sigma(x,y) = \tau(x \otimes y) \ (x,y \in H)$. Also, $\|u\| = \|\sigma\|$. Let E be an orthonormal basis for H and let $u = v|u|$ be the polar decomposition of u. If F is a finite subset of E, set $p_F = \sum_{e \in F} e \otimes e$ (so p_F is a projection). Then

$$
\sum_{e \in F} \langle |u|(e), e \rangle = \sum_{x \in F} \langle u(e), v(e) \rangle
$$

$$
= \sum_{x \in F} \sigma(e, v(e))
$$

$$
= \sum_{x \in F} \tau(e \otimes v(e))
$$

$$
= \tau\big(\big(\sum_{x \in F} e \otimes e \big) v^* \big)
$$

$$
= \tau(p_F v^*)
$$

$$
\leq \|\tau\|.
$$

Hence, $\|u\|_1 \leq \|\tau\|$, so $u \in L^1(H)$. If $x, y \in H$, then $\mathrm{tr}(u(x \otimes y)) = \langle u(x), y \rangle = \tau(x \otimes y)$, so $\mathrm{tr}(u \cdot)$ equals τ on $F(H)$ and therefore on $K(H)$; that is, $\theta(u) = \tau$. Therefore, θ is an isometric linear isomorphism. $\square$

4.2.2. Corollary. $L^1(H)$ is a Banach $*$-algebra under the trace-class norm.

Proof.   It is a dual space, and is therefore complete. $\square$

Suppose again that H is a Hilbert space and suppose that $v \in B(H)$. Then the function

$$
\mathrm{tr}(\cdot v) \colon L^1(H) \to \mathbf{C}, \quad u \mapsto \mathrm{tr}(uv),
$$

is linear and bounded, and $\| \mathrm{tr}(\cdot v) \| \leq \|v\|$, by Theorem 2.4.16. We call the norm-decreasing linear map

$$
B(H) \to L^1(H)^*, \quad v \mapsto \mathrm{tr}(\cdot v),
$$

the *canonical* map from $B(H)$ to $L^1(H)^*$.

4.2.3. Theorem. If H is a Hilbert space, then the canonical map from $B(H)$ to $L^1(H)^*$ is an isometric linear isomorphism.

Proof.   Let θ denote this map. If $\theta(v) = 0$, then $\theta(v)(x \otimes y) = 0$, that is, $\mathrm{tr}(x \otimes v^*(y)) = 0$, so $\langle x, v^*(y) \rangle = 0$, for all $x, y \in H$. Hence, $v = 0$, and therefore θ is injective.

Now let $\tau \in L^1(H)^*$. The map

$$
\sigma \colon H^2 \to \mathbf{C}, \quad (x, y) \mapsto \tau(x \otimes y),
$$

is a sesquilinear form on H, and $\|\sigma\| \leq \|\tau\|$. Hence, there is a unique operator $v \in B(H)$ such that $\langle v(x), y \rangle = \tau(x \otimes y)$ for all $x, y \in H$. Also,

$\|v\| = \|\sigma\| \leq \|\tau\|$. Now $\theta(v)(x \otimes y) = \mathrm{tr}(x \otimes v^*(y)) = \langle v(x), y \rangle = \tau(x \otimes y)$, so $\theta(v)$ equals τ on $F(H)$, and therefore on $L^1(H)$ (by Theorem 2.4.17, $F(H)$ is dense in $L^1(H)$ with respect to the trace-class norm). Therefore, θ is an isometric linear isomorphism. $\square$

If H is a Hilbert space, the Hausdorff locally convex topology on $B(H)$ generated by the separating family of semi-norms

$$B(H) \to \mathbf{R}^+, \quad u \mapsto |\langle u(x), y \rangle|, \qquad (x, y \in H)$$

is called the *weak (operator)* topology on $B(H)$. If $(u_\lambda)_{\lambda \in \Lambda}$ is a net in $B(H)$, then (u_λ) converges weakly to an operator u if and only if $\langle u(x), y \rangle = \lim_\lambda \langle u_\lambda(x), y \rangle$ $(x, y \in H)$.

The *ultraweak* or σ-*weak* topology on $B(H)$ is the Hausdorff locally convex topology on $B(H)$ generated by the semi-norms

$$B(H) \to \mathbf{R}^+, \quad u \mapsto |\mathrm{tr}(uv)|, \qquad (v \in L^1(H)).$$

The weak topology is weaker than the ultraweak topology. For if $(u_\lambda)_{\lambda \in \Lambda}$ is a net converging ultraweakly to an operator u, then for each $x, y \in H$ the net $(\langle u_\lambda(x) - u(x), y \rangle)_\lambda = (\mathrm{tr}((u_\lambda - u)(x \otimes y)))_\lambda$ converges to 0, so (u_λ) converges to u weakly. Clearly, the weak topology is also weaker than the strong topology.

The operations of addition and scalar multiplication are of course weakly and ultraweakly continuous, and it is easy to see that the involution operation $u \mapsto u^*$ is also continuous for these topologies. We showed in Example 4.1.1 that the involution is not strongly continuous in general, so the weak and strong topologies do not coincide in general.

Continuity of multiplication in the weak topology does not hold in general: Let H be a Hilbert space with an orthonormal basis $(e_n)_{n=1}^\infty$ and let u be the unilateral shift on this basis. Then the sequences (u^{*n}) and (u^n) both converge weakly to zero, but the product sequence $(u^{*n} u^n)$ is the constant 1.

We have seen that for an arbitrary Hilbert space H the Banach space $B(H)$ is the dual of $L^1(H)$. It is clear from this that the ultraweak topology is just the weak* topology on $B(H)$. Hence, the closed unit ball of $B(H)$ is ultraweakly compact, by the Banach–Alaoglu theorem.

4.2.4. Theorem. *If H is a Hilbert space, then the relative weak and ultraweak topologies on the closed unit ball of $B(H)$ coincide, and hence the ball is weakly compact.*

Proof. The identity map from the ball with the ultraweak topology to the ball with the weak topology is a continuous bijection from a compact space to a Hausdorff space and is therefore a homeomorphism. $\square$

4.2.1. *Remark.* The closed unit ball of $B(H)$ is not strongly compact in general. For if we suppose the contrary, then the identity map of the ball with the relative strong topology to the ball with the relative weak topology is a continuous bijection from a compact space to a Hausdorff one, and therefore a homeomorphism. This means the relative strong and weak topologies on the ball coincide. The involution operation $u \mapsto u^*$ is weakly continuous, but it follows from Example 4.1.1 that in general this operation is not strongly continuous when restricted to the ball. We therefore have a contradiction.

If C is a set of operators on a Hilbert space H, then the weak closure of C is contained in C''. For if u is in this weak closure, there is a net $(u_\lambda)_{\lambda \in \Lambda}$ in C converging weakly to u, and therefore if $v \in C'$ we have $\langle uv(x), y \rangle = \lim_\lambda \langle u_\lambda v(x), y \rangle = \lim_\lambda \langle vu_\lambda(x), y \rangle = \lim_\lambda \langle u_\lambda(x), v^*(y) \rangle = \langle u(x), v^*(y) \rangle = \langle vu(x), y \rangle$, so u and v commute, which implies that $u \in C''$.

4.2.5. Theorem. *Suppose that A is a $*$-algebra on a Hilbert space H containing id_H.*

(1) *The weak closure of A is A''.*

(2) *A is a von Neumann algebra if and only if it is weakly closed.*

Proof. This is immediate from the preceding remark and Theorem 4.1.5. In fact, this result is basically just the completion of that theorem.   □

4.2.6. Theorem. *Let H be a Hilbert space and τ a linear functional on $B(H)$. The following conditions are equivalent:*

(1) *τ is weakly continuous.*

(2) *τ is strongly continuous.*

(3) *There are vectors $x_1, \ldots, x_n$ and $y_1, \ldots, y_n$ in H such that*

$$\tau(u) = \sum_{j=1}^{n} \langle u(x_j), y_j \rangle \qquad (u \in B(H)).$$

Proof. The implications $(3) \Rightarrow (1) \Rightarrow (2)$ are clear. To show that $(2) \Rightarrow (3)$, suppose that τ is strongly continuous. Then by Theorem A.1 there is a positive number M and there exist $x_1, \ldots, x_n \in H$ such that $|\tau(u)| \le M \max_{1 \le j \le n} \|u(x_j)\|$ for all $u \in B(H)$. We may obviously suppose that $M = 1$. Then

$$|\tau(u)| \le \left(\sum_{j=1}^{n} \|u(x_j)\|^2 \right)^{1/2} \qquad (u \in B(H)).$$

Let K_0 be the vector subspace of $H^{(n)}$ consisting of all $(u(x_1), \ldots, u(x_n))$ $(u \in B(H))$, and let K be its norm closure. The function

$$\sigma \colon K_0 \to \mathbf{C}, \quad (u(x_1), \ldots, u(x_n)) \mapsto \tau(u),$$

is well-defined, linear, and bounded, with $\|\sigma\| \leq 1$, so it extends to a linear
norm-decreasing functional on K which we also denote by σ. By the Riesz
representation theorem for linear functionals on Hilbert spaces, there is a
unique element $y = (y_1, \ldots, y_n) \in K$ such that $\sigma(z) = \langle z, y \rangle$ $(z \in K)$.
Hence,

$$\tau(u) = \sigma(u(x_1), \ldots, u(x_n)) = \sum_{j=1}^{n} \langle u(x_j), y_j \rangle \qquad (u \in B(H)).$$

This shows that $(2) \Rightarrow (3)$. $\square$

4.2.7. Theorem. *Let H be a Hilbert space and C a convex subset of
$B(H)$. Then C is strongly closed if and only if it is weakly closed.*

Proof. Since the weak topology is weaker than the strong topology, a
weakly closed set is strongly closed. Suppose, therefore, that C is strongly
closed, and let u be a point in its weak closure. Then there is a net of
operators $(u_\lambda)_{\lambda \in \Lambda}$ in C converging to u weakly, and hence, for every weakly
continuous linear functional τ on $B(H)$, we have $\tau(u) = \lim_\lambda \tau(u_\lambda)$. By
Theorem 4.2.6, the weakly continuous linear functionals on $B(H)$ are the
same as the strongly continuous, so by Corollary A.8, u is in the strong
closure of C; that is, $u \in C$. Hence, C is weakly closed. $\square$

4.2.8. Corollary. *If A is a *-algebra on H, then A is a von Neumann
algebra if and only if it is weakly closed.*

Proof. Immediate, since A is convex. $\square$

We show now that a von Neumann algebra is the dual space of a
Banach space. This is not true for arbitrary C*-algebras. In fact, by a
theorem of Sakai every C*-algebra that is a dual space is isomorphic to a
von Neumann algebra ([Sak, Theorem 1.16.7]).

Suppose that A is a von Neumann algebra on a Hilbert space H. We
set

$$A^\perp = \{v \in L^1(H) \mid \mathrm{tr}(uv) = 0 \ (u \in A)\}.$$

This is a vector subspace of $L^1(H)$, closed with respect to the trace-class
norm. Set $A_* = L^1(H)/A^\perp$. Then A_* is a Banach space when endowed
with the quotient norm (corresponding to the trace-class norm). If $u \in A$,
we have a well-defined bounded linear functional

$$\theta(u): A_* \to \mathbf{C}, \quad v + A^\perp \mapsto \mathrm{tr}(uv).$$

The map

$$\theta: A \to (A_*)^*, \quad u \mapsto \theta(u),$$

is clearly norm-decreasing and linear. We call it the *canonical* map from A
to $(A_*)^*$.

4.2.9. Theorem. *Let A be a von Neumann algebra on a Hilbert space H. Then the canonical map from A to $(A_*)^*$ is an isometric linear isomorphism.*

Proof. Let $\theta: B(H) \to L^1(H)^*$ and $\theta': A \to (A_*)^*$ be the canonical maps. By Theorem 4.2.3, θ is an isometric linear isomorphism. If $\theta'(u) = 0$, then $\theta(u) = 0$, so $u = 0$. Thus, θ' is injective. If $\tau \in (A_*)^*$ and π is the quotient map from $L^1(H)$ to $L^1(H)/A^\perp$, then $\tau\pi \in L^1(H)^*$, so $\tau\pi = \theta(u)$ for some $u \in B(H)$. To show that $u \in A$, we need only show that $\mathrm{tr}(uw) = 0$ for all $w \in A^\perp$, using the fact that A is strongly closed, the characterisation of strongly continuous linear functionals on $B(H)$ given in Theorem 4.2.6, and Corollary A.9. But $\mathrm{tr}(uw) = \theta(u)(w) = \tau\pi(w) = \tau(0) = 0$. Therefore, $u \in A$. If $v \in L^1(H)$, then $\theta'(u)(\pi(v)) = \mathrm{tr}(uv) = \theta(u)(v) = \tau(\pi(v))$, so $\theta'(u) = \tau$. Therefore, θ' is a bijection. Observe also that if $\varepsilon > 0$, then there exists $v \in L^1(H)$ such that $\|v\|_1 \le 1$ and $|\theta(u)(v)| > \|\theta(u)\| - \varepsilon$, so $\|\tau\| \ge |\tau(\pi(v))| > \|u\| - \varepsilon$. Since ε was arbitrary, this shows that $\|\tau\| \ge \|u\|$. It follows that θ' is isometric. $\square$

It is easy to check that the weak* topology on A is just the relative σ-weak topology on A.

4.2.10. Theorem. *Let A be a von Neumann algebra on a Hilbert space H, and let $\tau: A \to \mathbf{C}$ be a linear functional. Then τ is σ-weakly continuous if and only if there exists $u \in L^1(H)$ such that $\tau(v) = \mathrm{tr}(uv)$ for all $v \in A$.*

Proof. This follows from the identification $A = (A_*)^*$, the remark preceding this theorem, and Theorem A.2. $\square$

4.2.2. Remark. Let A be a von Neumann algebra on a Hilbert space H containing id_H. If u is a normal element of A, then for each $f \in B_\infty(\sigma(u))$ the element $f(u)$ is in A. This is so because f is in the closed linear span of the characteristic functions in $B_\infty(\sigma(u))$, and if S is a Borel set of $\sigma(u)$, then the spectral projection $\chi_S(u) = E(S)$ belongs to A by Theorem 4.1.11. From this and the proof of Theorem 2.5.8, it follows that if $u \in A$ is a unitary, then $u = e^{iv}$ for some hermitian element $v \in A$.

4.3. The Kaplansky Density Theorem

We prepare the way for the density theorem with some useful results on strong convergence.

4.3.1. Theorem. *If H is a Hilbert space, the involution $u \mapsto u^*$ is strongly continuous when restricted to the set of normal operators of $B(H)$.*

Proof. Let $x \in H$ and suppose that u, v are normal operators in $B(H)$. Then

$$
\begin{aligned}
\|(v^* - u^*)(x)\|^2 &= \langle v^*(x) - u^*(x), v^*(x) - u^*(x) \rangle \\
&= \|v(x)\|^2 - \|u(x)\|^2 + \langle uu^*(x), x \rangle - \langle vu^*(x), x \rangle \\
&\quad + \langle uu^*(x), x \rangle - \langle uv^*(x), x \rangle \\
&= \|v(x)\|^2 - \|u(x)\|^2 + \langle (u - v)u^*(x), x \rangle + \langle x, (u - v)u^*(x) \rangle \\
&\leq \|v(x)\|^2 - \|u(x)\|^2 + 2\|(u - v)u^*(x)\|\|x\|.
\end{aligned}
$$

If $(v_\lambda)_{\lambda \in \Lambda}$ is a net of normal operators strongly convergent to a normal operator u, then the net $(\|v_\lambda(x)\|^2)$ is convergent to $\|u(x)\|^2$ and the net $((u - v_\lambda)u^*(x))$ is convergent to 0, so $(v_\lambda^*(x) - u^*(x))$ is convergent to 0. Therefore, (v_λ^*) is strongly convergent to u^*. $\quad\square$

4.3.1. Remark. If S is a bounded subset of $B(H)$, where H is a Hilbert space, then the map

$$
S \times B(H) \to B(H), \quad (v, u) \mapsto vu,
$$

is strongly continuous. The proof of this is the inequality

$$
\|vu(x) - v_1 u_1(x)\| \leq \|v\|\|u(x) - u_1(x)\| + \|(v - v_1)u_1(x)\|.
$$

We say that a continuous function f from $\mathbf{R}$ to $\mathbf{C}$ is *strongly continuous* if for every Hilbert space H and each net $(u_\lambda)_{\lambda \in \Lambda}$ of hermitian operators on H converging strongly to a hermitian operator u, we have $(f(u_\lambda))$ converges strongly to $f(u)$.

4.3.2. Theorem. *If $f \colon \mathbf{R} \to \mathbf{C}$ is a continuous bounded function, then f is strongly continuous.*

Proof. Let A denote the set of strongly continuous functions. This is clearly a vector space (for the pointwise-defined operations), and it follows from Remark 4.3.1 that if f, g belong to A and one of them is bounded, then $fg \in A$.

We show first that $C_0(\mathbf{R}) \subseteq A$: Let $A_0 = A \cap C_0(\mathbf{R})$. It is easy to verify that A_0 is a closed subalgebra of $C_0(\mathbf{R})$, and by Theorem 4.3.1 it is self-adjoint. If $z \colon \mathbf{R} \to \mathbf{C}$ is the inclusion function, then $f = 1/(1 + z^2)$ and $g = zf$ belong to $C_0(\mathbf{R})$ and $\|f\|_\infty, \|g\|_\infty \leq 1$. We show that $f, g \in A_0$. Let H be a Hilbert space and suppose that v, u are hermitian operators on H. Then

$$
\begin{aligned}
g(u) - g(v) &= u(1 + u^2)^{-1} - v(1 + v^2)^{-1} \\
&= (1 + u^2)^{-1}[u(1 + v^2) - (1 + u^2)v](1 + v^2)^{-1} \\
&= (1 + u^2)^{-1}[u - v + u(v - u)v](1 + v^2)^{-1}.
\end{aligned}
$$

Therefore, if $x \in H$,

$$
\begin{aligned}
\|g(u)(x) - g(v)(x)\| &\leq \|(1 + u^2)^{-1}(u - v)(1 + v^2)^{-1}(x)\| \\
&\quad + \|(1 + u^2)^{-1}(u(v - u)v)(1 + v^2)^{-1}(x)\| \\
&\leq \|(u - v)(1 + v^2)^{-1}(x)\| + \|(v - u)v(1 + v^2)^{-1}(x)\|,
\end{aligned}
$$

since $\|(1 + u^2)^{-1}\|$ and $\|(1 + u^2)^{-1}u\| \leq 1$. Hence, g is strongly continuous, and therefore $g \in A_0$. Since $z \in A$, we have $zg \in A$, so $f = 1 - zg \in A$, and therefore $f \in A_0$. The set $\{f, g\}$ separates the points of $\mathbf{R}$, and $f(t) > 0$ for all t, so by the Stone–Weierstrass theorem the C*-subalgebra generated by f and g is $C_0(\mathbf{R})$. Hence, $A_0 = C_0(\mathbf{R})$.

Now suppose that $h \in C_b(\mathbf{R})$. Then $hf, hg \in C_0(\mathbf{R})$, so $hf, hg \in A$, and therefore $zhg \in A$ also. Consequently, $h = hf + zhg \in A$. $\square$

If C is a convex set of operators on a Hilbert space H, then its strong and its weak closures coincide, by Theorem 4.2.7. If A is a *-subalgebra of $B(H)$, then its weak closure is a von Neumann algebra. These observations are used in the proof of the following theorem, which is known as the *density theorem*.

4.3.3. Theorem (Kaplansky).   *Let H be a Hilbert space and A a C*-subalgebra of $B(H)$ with strong closure B.*

(1) *A_{sa} is strongly dense in B_{sa}.*
(2) *The closed unit ball of A_{sa} is strongly dense in the closed unit ball of B_{sa}.*
(3) *The closed unit ball of A is strongly dense in the closed unit ball of B.*
(4) *If A contains id_H, then the unitaries of A are strongly dense in the unitaries of B.*

Proof.   If $u \in B_{sa}$, then there is a net $(u_\lambda)_{\lambda \in \Lambda}$ in A strongly convergent to u, so (u_λ^*) converges weakly to u^*, and therefore $(\mathrm{Re}(u_\lambda))$ is weakly convergent to u. Hence, u is in the weak closure of A_{sa}, and therefore in the strong closure of this set, since its weak and strong closures coincide (because it is convex). This proves Condition (1).

Suppose now that u is in the closed unit ball of B_{sa}. Then by Condition (1) there is a net $(u_\lambda)_{\lambda \in \Lambda}$ in A_{sa} strongly convergent to u. The function $f \colon \mathbf{R} \to \mathbf{C}$ defined by setting $f(t) = t$ for $t \in [-1, 1]$ and $f(t) = 1/t$ elsewhere belongs to $C_0(\mathbf{R})$, and therefore by Theorem 4.3.2, it is strongly continuous. Hence, $(f(u_\lambda))$ is strongly convergent to $f(u)$. But clearly, $f(u) = u$, since $\sigma(u) \subseteq [-1, 1]$. Moreover, $f(u_\lambda)$ is in the closed unit ball of A_{sa} for all indices λ, since $\bar{f} = f$ and $\|f\|_\infty \leq 1$. This proves Condition (2).

The algebra $M_2(A)$ is a C*-subalgebra of $M_2(B(H)) = B(H^{(2)})$, and strongly dense in the von Neumann algebra $M_2(B)$. If u is in the closed unit ball of B, then $v = \begin{pmatrix} 0 & u \\ u^* & 0 \end{pmatrix}$ is a hermitian operator on $H^{(2)}$ lying in the strong closure of $M_2(A)$, and since $\|v\| \leq 1$, it follows from Condition (2) that there is a net $(v_\lambda)_{\lambda \in \Lambda}$ in the closed unit ball of $M_2(A)_{sa}$ that strongly converges to v. Hence, $((v_\lambda)_{12})$ is strongly convergent on H to u, and $(v_\lambda)_{12}$ is in the closed unit ball of A for all indices λ. Thus, Condition (3) is proved.

Suppose now that A contains id_H and let $U(A)$ and $U(B)$ denote the sets of unitaries in A and B, respectively. If $u \in U(B)$, then by Remark 4.2.2 there is a hermitian element v of B such that $u = e^{iv}$. By Condition (1) there is a net $(v_\lambda)_{\lambda \in \Lambda}$ in A_{sa} strongly convergent to v. However, the function

$$f: \mathbf{R} \to \mathbf{C}, \quad t \mapsto e^{it},$$

is strongly continuous by Theorem 4.3.2, so $(f(v_\lambda))$ converges strongly to $f(v)$. Since $f(v_\lambda) = e^{iv_\lambda} \in U(A)$ and $f(v) = u$, Condition (4) is proved. $\square$

4.3.4. Theorem. *Let H_1 and H_2 be Hilbert spaces, A a von Neumann algebra on H_1, and $\varphi: A \to B(H_2)$ a weakly continuous $*$-homomorphism. Then $\varphi(A)$ is a von Neumann algebra on H_2.*

Proof. Observe first that $\varphi(A)$ is a C*-algebra on H_2. By applying Remark 4.1.2, we may suppose that A contains id_{H_1}.

Let $v \in \varphi(A)$ and suppose that $\|v\| < 1$, so there is a number α such that $\|v\| < \alpha < 1$. Write $v = \varphi(u)$ for some element $u \in A$ and let $u = w|u|$ be the polar decomposition of u. By Theorem 4.1.10 $w \in A$. Let E be the spectral resolution of the identity for $|u|$ and G the set of all points of the spectrum of $|u|$ not less than α. Then $E(G) \in A$, $\alpha E(G) \leq |u|E(G)$, and $|u|(1 - E(G)) \leq \alpha(1 - E(G))$. Hence, $0 \leq \alpha \varphi(E(G)) \leq \varphi(|u|)\varphi(E(G))$, so $\alpha\|\varphi(E(G))\| \leq \|\varphi(|u|)\|\|\varphi(E(G))\| \leq \|\varphi(|u|)\| = \|\varphi(w^*w|u|)\| \leq \|\varphi(u)\| = \|v\| < \alpha$. Therefore, $\|\varphi(E(G))\| < 1$, so $\varphi(E(G)) = 0$, since $\varphi(E(G))$ is a projection. Consequently, $v = \varphi(u(1 - E(G)))$. Also,

$$\|u(1 - E(G))\| \leq \||u|(1 - E(G))\| \leq \alpha\|1 - E(G)\| \leq \alpha < 1.$$

Set

$$R = \{u_1 \in A \mid \|u_1\| < 1\}.$$

We have shown that $\varphi(R) = \{v \in \varphi(A) \mid \|v\| < 1\}$. The closed unit ball S of A is weakly compact by Theorem 4.2.4, so $\varphi(S)$ is weakly compact. Observe that S is the weak closure of R. We claim that $\varphi(S)$ is the closed unit ball of $\varphi(A)$: Let v be an element of the closed unit ball of $\varphi(A)$ and take a sequence $\varepsilon_n \in (0, 1)$ converging to 1. Now $v = \varphi(u)$ for some $u \in A$, so $\varepsilon_n v \in \varphi(A)$ and $\|\varepsilon_n v\| < 1$. Hence, $\varepsilon_n v = \varphi(u_n)$ for some $u_n \in R$. Thus, $\varepsilon_n v$ is a sequence in $\varphi(S)$ converging in norm to v, so $v \in \varphi(S)$. This shows that the closed unit ball of $\varphi(A)$ is contained in $\varphi(S)$, and the reverse inclusion is obvious.

Now let u be a non-zero element of the weak closure of $\varphi(A)$ in $B(H_2)$. By the Kaplansky density theorem, $u/\|u\|$ is in the weak closure of the closed unit ball of $\varphi(A)$, and therefore $u/\|u\| \in \varphi(S)$, so $u \in \varphi(A)$. This shows that the weak closure of $\varphi(A)$ is equal to $\varphi(A)$, so $\varphi(A)$ is a von Neumann algebra. $\square$

4.4. Abelian Von Neumann Algebras

In this section we represent abelian von Neumann algebras acting on separable Hilbert spaces in terms of "L^∞-algebras." We begin by making some observations concerning separability.

4.4.1. Remark. Let A be a C*-algebra. If A is separable, then of course it is countably generated as a C*-algebra. The converse is also true (the proof is an easy exercise).

If A is abelian and separable, then its character space $\Omega = \Omega(A)$ is second countable. For if $(a_n)_{n=1}^\infty$ is a dense sequence in A, let τ be the smallest topology on Ω making all $\hat{a}_n$ continuous. The set B of all elements $a \in A$ such that $\hat{a}$ is continuous with respect to τ is a C*-subalgebra of A containing all the a_n, so $B = A$. Since the weak* topology on Ω is the smallest one making continuous all of the functions $\hat{a}$ ($a \in A$), therefore it is equal to τ. If we now choose a countable base E for the topology of $\mathbf{C}$, it is easily checked that the finite intersections of the sets $\hat{a}_n^{-1}(U)$ ($n \geq 1$, $U \in E$) form a countable base for the topology of Ω.

4.4.2. Remark. If H is a separable Hilbert space, then the closed unit ball S of $B(H)$ is separable in the strong topology. We show this: By Remark 4.1.6, $K(H)$ is separable for the norm topology. If $u \in K(H)'$, then $u(x \otimes x) = (x \otimes x)u$, that is, $u(x) \otimes x = x \otimes u^*(x)$, for all $x \in H$. Since $(x \otimes y)(H) = \mathbf{C}x$ if $x, y \neq 0$, therefore $u(x) = \tau(x)x$ for some scalar $\tau(x) \in \mathbf{C}$. If x, y are linearly independent in H, then the equations

$$u(x + y) = \tau(x + y)(x + y)$$

and

$$u(x + y) = \tau(x)x + \tau(y)y$$

imply that $\tau(x + y) = \tau(x) = \tau(y)$. From this it is immediately seen that $u \in \mathbf{C}1$. Therefore, $K(H)' = \mathbf{C}1$, so $K(H)$ is strongly dense in $B(H)$ by Theorem 4.1.12. It follows from Theorem 4.3.3 that the closed unit ball of $K(H)$ is strongly dense in S, and since the ball of $K(H)$ is separable for the norm topology, S is therefore separable for the strong topology as claimed.

The ball S has another useful property in the case that H is separable: It is metrisable for the relative strong topology. For if (x_n) is a dense sequence in the ball of H, then the equation

$$d(u, v) = \sum_{n=1}^{\infty} \frac{\|(u - v)(x_n)\|}{2^n}$$

defines a metric on S inducing the strong topology. The proof is a straight-forward exercise.

Let A be a C*-algebra acting on an arbitrary Hilbert space H and let $x \in H$. If $[Ax] = H$ we call x a *cyclic* vector for A. We say that $y \in H$ is a *separating* vector for A if for all $u \in A$, $u(y) = 0 \Rightarrow u = 0$. If x is cyclic for A, then it is separating for A'. For suppose $v \in A'$ and $v(x) = 0$. If $u \in A$, then $vu(x) = uv(x) = 0$. Hence, $v(H) = v[Ax] = 0$, and therefore $v = 0$. If A acts non-degenerately on H and if x is a separating vector for A', then it is cyclic for A. To see this, let p denote the projection of H on $[Ax]$. Then $p \in A'$, since $[Ax]$ is invariant for A. Note that $x \in [Ax]$, for if $(u_\lambda)_{\lambda \in \Lambda}$ is an approximate unit for A, then $x = \lim_\lambda u_\lambda(x)$, and $u_\lambda(x) \in [Ax]$. Hence, $(1 - p)(x) = 0$. By the separating property, $1 - p = 0$, so $[Ax] = H$; that is, x is cyclic for A, as claimed.

An abelian von Neumann algebra A on a Hilbert space H is *maximal* if it is not contained in any other abelian von Neumann algebra on H. It is easily verified that A is maximal if and only if $A' = A$. A simple application of Zorn's lemma shows that every abelian von Neumann algebra is contained in a maximal abelian von Neumann algebra.

4.4.1. *Example.* Let μ be a finite positive regular Borel measure on a compact Hausdorff space Ω. The vector $1 \in L^2(\Omega, \mu)$ is cyclic for the C*-algebra A on $L^2(\Omega, \mu)$ consisting of all multiplication operators M_φ with continuous symbol φ (because $C(\Omega)$ is L^2-norm dense in $L^2(\Omega, \mu)$). Recall that A' is the algebra of all multiplication operators on $L^2(\Omega, \mu)$ (by Example 4.1.2). The vector 1 is separating (and cyclic) for A'. The von Neumann algebra A' is maximal abelian, since $A'' = A'$.

4.4.1. Lemma. *If A is an abelian von Neumann algebra acting non-degenerately on a separable Hilbert space H, then A has a separating vector.*

Proof. Let E be a maximal set in H of unit vectors such that the spaces $[Ax]$ ($x \in E$) are pairwise orthogonal (E exists by Zorn's lemma). If $y \in H$ is a unit vector orthogonal to all $[Ax]$ ($x \in E$), then $[Ay]$ is orthogonal to all $[Ax]$ also, contradicting the maximality of E. Hence, H is the (inner) orthogonal sum of the spaces $[Ax]$ ($x \in E$). Since H is separable, the set E is necessarily countable, so we may write $E = \{x_n \mid n \geq 1\}$, where (x_n) is a sequence of unit vectors in H. Set $x = \sum_{n=1}^{\infty} x_n/2^n$. If $u \in A$ and

$u(x) = 0$, then $u(x_n) = 0$ for all n, because the sequence $(u(x_n))$ consists of pairwise orthogonal elements. Hence, if $v \in A$, then $uv(x_n) = vu(x_n) = 0$, so $u[Ax_n] = 0$, for all n. It follows that $u = 0$, so x is a separating vector for A.   □

4.4.2. Theorem. *If A is a maximal abelian von Neumann algebra on a separable Hilbert space, then A has a cyclic vector.*

Proof. By Lemma 4.4.1 $A = A'$ has a separating vector, x say. Hence, x is cyclic for A.   □

4.4.3. Theorem. *Let A be an abelian von Neumann algebra acting on a separable Hilbert space H, which contains id_H and has a cyclic vector. Then there exists a second countable compact Hausdorff space Ω, a positive measure $\mu \in M(\Omega)$, and a unitary $u: H \to L^2(\Omega, \mu)$, such that uAu^* is the von Neumann algebra of all multiplication operators M_φ on $L^2(\Omega, \mu)$.*

Proof. Let x be a cyclic vector for A. The closed unit ball of $B(H)$ is metrisable and separable for the strong topology by Remark 4.4.2, so the same is true for the ball of A. It follows that there is a separable C*-subalgebra B of A which is strongly dense in A. We may assume $1 = \mathrm{id}_H \in B$. Let $\varphi: B \to C(\Omega)$ be the Gelfand representation and note that the compact Hausdorff space Ω is second countable by Remark 4.4.1. We define a positive linear functional τ on $C(\Omega)$ by setting $\tau(f) = \langle \varphi^{-1}(f)(x), x \rangle$. By the Riesz–Kakutani theorem, there exists a positive measure $\mu \in M(\Omega)$ such that $\tau(f) = \int f \, d\mu$ for all $f \in C(\Omega)$. The map

$$\pi: B \to B(L^2(\Omega, \mu)), \quad v \mapsto M_{\varphi(v)},$$

is an injective *-homomorphism.

If $v \in B$, then

$$\int |\varphi(v)|^2 \, d\mu = \tau(|\varphi(v)|^2) = \langle \varphi^{-1}\varphi(v^*v)(x), x \rangle = \|v(x)\|^2.$$

Hence, the map $u: v(x) \mapsto \varphi(v)$ from $Bx = \{v(x) \mid v \in B\}$ to the L^2-norm dense subset $C(\Omega)$ of $L^2(\Omega, \mu)$ is well-defined and isometric, and it is clearly linear. Since $[Ax] = H$ and B is strongly dense in A, therefore $[Bx] = H$. We may therefore extend u to a unitary (also denoted by u) from H onto $L^2(\Omega, \mu)$. If $v, w \in B$, then $\pi(v)uw(x) = \varphi(vw) = uvw(x)$. Hence, $\pi(v)u = uv$ for all $v \in B$. Therefore, the *-isomorphism

$$\mathrm{Ad}\, u: B(H) \to B(L^2(\Omega, \mu)), \quad v \mapsto uvu^*,$$

is equal to π on B. Denote by C the von Neumann algebra of all multiplication operators on $L^2(\Omega, \mu)$. Since B is strongly dense in A, and the algebra of multiplication operators with continuous symbol is strongly dense in C (by Example 4.1.2), therefore $uAu^* = C$, because $\mathrm{Ad}\, u$ is a homeomorphism for the strong topologies.   □

4.4.4. Theorem. *Let A be an abelian von Neumann algebra on a separable Hilbert space H. Then there exists a second countable compact Hausdorff space Ω, and a positive measure $\mu \in M(\Omega)$ such that A is $*$-isomorphic to the C^*-algebra $L^\infty(\Omega, \mu)$.*

Proof. We may assume that $\mathrm{id}_H \in A$. By Lemma 4.4.1 there exists a separating vector x for A. If p is the projection of H onto $[Ax]$, then $p \in A'$. The map

$$\varphi: A \to A_p, \quad u \mapsto u_p,$$

is a $*$-homomorphism onto A_p, and since x is separating for A, this map is injective and therefore a $*$-isomorphism. Clearly, φ is weakly continuous, so by Theorem 4.3.4, $\varphi(A) = A_p$ is a von Neumann algebra on $p(H)$. Obviously, x is cyclic for A_p. Note also that $\mathrm{id}_{p(H)} \in A_p$.

Thus, to prove the theorem we have shown we may reduce to the case where A contains id_H and has some cyclic vector x. The result now follows from Theorem 4.4.3. $\qquad\square$

Suppose that v is a normal operator on a separable Hilbert space H. The von Neumann algebra generated by v is abelian, so there is a maximal abelian von Neumann algebra A containing v. By Theorems 4.4.2 and 4.4.3, there is a second countable compact Hausdorff space Ω and a positive measure $\mu \in M(\Omega)$, and a unitary $u: H \to L^2(\Omega, \mu)$, such that uAu^* is the von Neumann algebra of all multiplication operators on $L^2(\Omega, \mu)$. In particular, v is unitarily equivalent to a multiplication operator.

4. Exercises

1. Let H be a separable Hilbert space with an orthonormal basis $(e_n)_{n=1}^\infty$. Prove that the relative weak topology on the closed unit ball S of $B(H)$ is metrisable by showing that the equation

$$d(u, v) = \sum_{n,m=1}^{\infty} \frac{|\langle (u - v)(e_n), e_m \rangle|}{2^{n+m}}$$

defines a metric on S inducing the weak topology.

2. Let H be a Hilbert space.
(a) Show that a weakly convergent sequence of operators on H is necessarily norm-bounded.
(b) Show that if (u_n) and (v_n) are sequences of operators on H converging strongly to the operators u and v, respectively, then $(u_n v_n)$ converges strongly to uv.
(c) Show that if (u_n) is a sequence of operators on H converging strongly to u, and if $v \in K(H)$, then $(u_n v)$ converges in norm to uv. Show that $(v u_n)$ may not converge to vu in norm.

3. Let H be a Hilbert space with an orthonormal basis $(e_n)_{n=1}^{\infty}$.

(a) Denote by Λ the set of all pairs (n, U) where n is a positive integer, and U is a neighbourhood of 0 in the strong topology of $B(H)$. For (n, U) and (n', U') in Λ, write $(n, U) \leq (n', U')$ if $n \leq n'$ and $U' \subseteq U$. Show that Λ is a poset under the relation $\leq$, and that it is upwards-directed.

(b) Let u denote the unilateral shift on (e_n), and note that (u^{n*}) is strongly convergent to zero. If $\lambda = (n_\lambda, U_\lambda) \in \Lambda$, then $\lim_{n\to\infty}(n_\lambda u^{n*}) = 0$ in the strong topology, so for some n we have $n_\lambda u^{n*} \in U_\lambda$. Set $u_\lambda = n_\lambda u^{n*}$ and $v_\lambda = \frac{1}{n_\lambda}u^n$. Show that $\lim_\lambda u_\lambda = 0$ in the strong topology and $\lim_\lambda v_\lambda = 0$ in the norm topology. (Since $u_\lambda v_\lambda = 1$, this shows that the operation of multiplication

$$B(H) \times B(H) \to B(H), \quad (u, v) \mapsto uv,$$

is not jointly continuous in either the weak or the strong topologies.)

(c) Show that neither the weak nor the strong topologies on $B(H)$ are metrisable, using Exercise 4.2 and the nets (u_λ) and (v_λ) from part (b) of this exercise.

4. Let A be a von Neumann algebra on a Hilbert space H, and suppose that τ is a bounded linear functional on A. We say that τ is *normal* if, whenever an increasing net $(u_\lambda)_{\lambda \in \Lambda}$ in A_{sa} converges strongly to an operator $u \in A_{sa}$, we have $\lim_\lambda \tau(u_\lambda) = \tau(u)$. Show that every σ-weakly continuous functional $\tau \in A^*$ is normal (use Theorem 4.2.10 and show that if $(v_\lambda)_\lambda$ is a bounded net strongly convergent to v, and if $u \in L^1(H)$, then $\lim_\lambda \|v_\lambda u - vu\|_1 = 0$.)

5. The existence and characterisation of extreme points is very important in many contexts (for example, we shall be concerned with this in the next chapter in connection with pure states). See the Appendix for the definition of extreme points.

Let H be a non-zero Hilbert space.

(a) Show that the extreme points of the closed unit ball of H are precisely the unit vectors.

(b) Deduce that the isometries and co-isometries of $B(H)$ are extreme points of the closed unit ball of $B(H)$. (It can be shown that these are all of the extreme points. This follows from [Tak, Theorem I.10.2].)

6. Let A be a C*-algebra.

(a) Show that if A is unital, then its unit is an extreme point of its closed unit ball.

(b) If p is a projection of A, show that it is an extreme point of the closed unit ball of A^+ (use the unital algebra pAp and part (a)). The converse of this result is also true, but more difficult. It follows from [Tak, Lemma I.10.1].

(c) Show that if H is an infinite-dimensional Hilbert space, then the closed unit ball of $B(H)^+$ is not the convex hull of the projections of $B(H)$.

7. Let A be a C*-algebra. Show that if p, q are equivalent projections in A, and r is a projection orthogonal to both (that is, $rp = rq = 0$), then the projections $r + p$ and $r + q$ are equivalent.

If H is a separable Hilbert space and p is a projection not of finite rank, set $\operatorname{rank}(p) = \infty$. If p has finite rank, set $\operatorname{rank}(p) = \dim p(H)$. Show that $p \sim q$ in $B(H)$ if and only if $\operatorname{rank}(p) = \operatorname{rank}(q)$.

Thus, the equivalence class of a projection in a C*-algebra can be thought of as its "generalised rank."

We say a projection p in a C*-algebra A is *finite* if for any projection q such that $q \sim p$ and $q \leq p$ we necessarily have $q = p$. Otherwise, the projection is said to be *infinite*. Show that if p, q are projections such that $q \leq p$ and p is finite, then q is finite.

A projection p in a von Neumann algebra A is *abelian* if the algebra pAp is abelian. Show that abelian projections are finite.

A von Neumann algebra is said to be *finite* or *infinite* according as its unit is a finite or infinite projection. If H is a Hilbert space, show that the von Neumann algebra $B(H)$ is finite or infinite according as H is finite- or infinite-dimensional.

4. Addenda

Let τ be a bounded linear functional on a von Neumann algebra A. The following are equivalent conditions:
(i) τ is normal.
(ii) The restriction of τ to the closed unit ball of A is weakly continuous.
(iii) τ is σ-weakly continuous.

Reference: [Ped, Theorem 3.6.4].

A projection in a von Neumann algebra A is *central* if it commutes with every element of A.

We say that A is *Type* I if every non-zero central projection in A majorises a non-zero abelian projection in A. Thus, abelian von Neumann algebras are trivially Type I. Just as easy, $B(H)$ is Type I for every Hilbert space H.

We say that A is *Type* II if it has no non-zero abelian projections and every non-zero central projection majorises a non-zero finite projection.

We say that A is *Type* III if it contains no non-zero finite projections.

We say that A is *properly infinite* if it has no non-zero finite central projection.

If A is Type II and properly infinite, it is said to be *Type* II$_\infty$, and if it is Type II and finite, it is said to be *Type* II$_1$.

Each von Neumann algebra can be decomposed into a direct sum of von Neumann algebras of Types I, II_1, II_∞, and III (not all types may be present).

If A is a von Neumann algebra on the Hilbert space H, it is said to be a *factor* if $A \cap A' = \mathbf{C}1$, where $1 = \mathrm{id}_H$. Of course, $B(H)$ is a factor, but one has to work harder to get other examples (see Section 6.2). A factor is one and only one of the Types I, II_1, II_∞, III. A factor of Type I is isomorphic to $B(H)$ for some Hilbert space H.

Every von Neumann algebra can be "decomposed" into factors, so these are the building blocks of the theory.

References: [Dix 1], [Tak].

CHAPTER 5

Representations of C*-Algebras

This chapter is concerned with the positive linear functionals, the representations, and the (left and two-sided) ideals of a C*-algebra, and with their inter-relationship. Pure states are introduced and shown to be the extreme points of a certain convex set, and their existence is deduced from the Krein–Milman theorem (see Appendix). From this the existence of irreducible representations is proved by establishing a correspondence between them and the pure states.

We introduce two analogues of the spectrum of an abelian Banach algebra, the space of primitive ideals and (loosely speaking) the space of irreducible representations. These spaces are related to the structure theory of the underlying algebra.

We are also interested in this chapter in the relationship of the representations and ideals of an algebra to the corresponding objects of a subalgebra and a quotient algebra (for the case of a subalgebra this works out nicest if it is hereditary).

The chapter concludes with a brief introduction to the important classes of liminal and postliminal C*-algebras.

5.1. Irreducible Representations and Pure States

If (H, φ) is a representation of a C*-algebra A, we say $x \in H$ is a *cyclic* vector for (H, φ) if x is cyclic for the C*-algebra $\varphi(A)$. If (H, φ) admits a cyclic vector, then we say that it is a *cyclic* representation.

We return now to the GNS construction associated to a state to show that the representations involved are cyclic.

5.1.1. Theorem. *Let A be a C^*-algebra and $\tau \in S(A)$. Then there is a unique vector $x_\tau \in H_\tau$ such that*

$$\tau(a) = \langle a + N_\tau, x_\tau \rangle \qquad (a \in A).$$

Moreover, x_τ is a unit cyclic vector for (H_τ, φ_τ) and

$$\varphi_\tau(a)x_\tau = a + N_\tau \qquad (a \in A).$$

Proof. The function

$$\rho_0 \colon A/N_\tau \to \mathbf{C}, \ a + N_\tau \mapsto \tau(a),$$

is well-defined, linear, and norm-decreasing, so we can extend it to a norm-decreasing linear functional ρ on H_τ. By the Riesz representation theorem, there is a unique vector x_τ in H_τ such that $\rho(y) = \langle y, x_\tau \rangle$ $(y \in H_\tau)$. Thus, x_τ is the unique element of H_τ such that $\tau(a) = \langle a + N_\tau, x_\tau \rangle$ $(a \in A)$.

Let $a, b \in A$. Then $\langle b + N_\tau, \varphi_\tau(a)x_\tau \rangle = \langle a^*b + N_\tau, x_\tau \rangle = \tau(a^*b) = \langle b + N_\tau, a + N_\tau \rangle$, and since this holds for all b, we have $\varphi_\tau(a)x_\tau = a + N_\tau$. Hence, $\varphi_\tau(A)x_\tau$ is dense in H_τ, since it is the space A/N_τ. Therefore, x_τ is cyclic for (H_τ, φ_τ). Consequently, $\varphi_\tau(A)$ acts non-degenerately on H_τ. If $(u_\lambda)_{\lambda \in \Lambda}$ is an approximate unit for A, then $(\varphi_\tau(u_\lambda))$ is one for $\varphi_\tau(A)$, and therefore it converges strongly to id_{H_τ}. Hence, $\|x_\tau\|^2 = \langle x_\tau, x_\tau \rangle = \lim_\lambda \langle \varphi_\tau(u_\lambda)(x_\tau), x_\tau \rangle = \lim_\lambda \tau(u_\lambda) = \|\tau\| = 1$, so x_τ is a unit vector. $\square$

We call the vector x_τ in Theorem 5.1.1 the *canonical* cyclic vector for (H_τ, φ_τ).

If ρ, τ are positive linear functionals on a C^*-algebra A, we write $\rho \leq \tau$ if $\tau - \rho$ is positive. We say τ *majorises* ρ, or ρ is *majorised* by τ, if $\rho \leq \tau$.

5.1.2. Theorem. *Let τ be a state and ρ a positive linear functional on a C^*-algebra A, and suppose that $\rho \leq \tau$. Then there is a unique operator v in $\varphi_\tau(A)'$ such that*

$$\rho(a) = \langle \varphi_\tau(a)vx_\tau, x_\tau \rangle \qquad (a \in A).$$

Moreover, $0 \leq v \leq 1$.

Proof. Define a sesquilinear form σ on A/N_τ by setting $\sigma(a + N_\tau, b + N_\tau) = \rho(b^*a)$ (this is well-defined as $\rho \leq \tau$). Observe that $\|\sigma\| \leq 1$ as $|\rho(b^*a)| \leq \rho(b^*b)^{1/2}\rho(a^*a)^{1/2} \leq \tau(b^*b)^{1/2}\tau(a^*a)^{1/2} = \|b + N_\tau\|\|a + N_\tau\|$. We can therefore extend σ to a bounded sesquilinear form (also denoted σ) on H_τ which also has norm not greater than 1. Hence, there is an operator v on H_τ such that $\langle v(x), y \rangle = \sigma(x, y)$ for all $x, y \in H_\tau$, and $\|v\| \leq 1$. Therefore, $\rho(b^*a) = \sigma(a + N_\tau, b + N_\tau) = \langle v(a + N_\tau), b + N_\tau \rangle = \langle v\varphi_\tau(a)x_\tau, \varphi_\tau(b)x_\tau \rangle$. Consequently, $\langle v(a + N_\tau), a + N_\tau \rangle \geq 0$ for all $a \in A$, so v is positive.

If $a, b, c \in A$, then $\langle \varphi_\tau(a)v(b + N_\tau), c + N_\tau \rangle = \langle v(b + N_\tau), a^*c + N_\tau \rangle = \rho(c^*ab) = \langle v(ab+N_\tau), c+N_\tau \rangle = \langle v\varphi_\tau(a)(b+N_\tau), c+N_\tau \rangle$. Hence, $\varphi_\tau(a)v = v\varphi_\tau(a)$ for all a, so $v \in \varphi_\tau(A)'$. Also, $\rho(a^*b) = \langle v(b + N_\tau), a + N_\tau \rangle = \langle v\varphi_\tau(b)x_\tau, \varphi_\tau(a)x_\tau \rangle = \langle v\varphi_\tau(a^*b)x_\tau, x_\tau \rangle$, so if $(u_\lambda)_{\lambda \in \Lambda}$ is an approximate unit for A, then $\rho(u_\lambda b) = \langle v\varphi_\tau(u_\lambda b)x_\tau, x_\tau \rangle$, and therefore in the limit $\rho(b) = \langle v\varphi_\tau(b)x_\tau, x_\tau \rangle$.

To see uniqueness, suppose that $w \in \varphi_\tau(A)'$ and $\rho(a) = \langle \varphi_\tau(a)wx_\tau, x_\tau \rangle$ $(a \in A)$. Then

$$\langle w\varphi_\tau(a^*b)x_\tau, x_\tau \rangle = \rho(a^*b) = \langle v\varphi_\tau(a^*b)x_\tau, x_\tau \rangle,$$

and therefore

$$\langle w(b + N_\tau), a + N_\tau \rangle = \langle v(b + N_\tau), a + N_\tau \rangle$$

for all a, b. Hence, $w = v$. $\square$

It is an easy exercise to show that we can go in the opposite direction also; that is, if $v \in \varphi_\tau(A)'$ and $0 \leq v \leq 1$, then the equation $\rho(a) = \langle \varphi_\tau(a)vx_\tau, x_\tau \rangle$ defines a positive linear functional ρ on A such that $\rho \leq \tau$.

A representation (H, φ) of a C*-algebra A is *non-degenerate* if the C*-algebra $\varphi(A)$ acts non-degenerately on H.

It is clear that a direct sum of non-degenerate representations is non-degenerate, and also that cyclic representations are non-degenerate. Therefore, the universal representation of A is non-degenerate.

If (H, φ) is a non-degenerate representation for A and $(u_\lambda)_{\lambda \in \Lambda}$ an approximate unit of A, then $(\varphi(u_\lambda))_\lambda$ is an approximate unit of $\varphi(A)$, so the net $(\varphi(u_\lambda))$ converges strongly to id_H.

Let (H, φ) be an arbitrary representation of A. If K is a closed vector subspace of H invariant for $\varphi(A)$, then the map

$$\varphi_K \colon A \to B(K), \quad a \mapsto (\varphi(a))_K,$$

is a *-homomorphism, so the pair (K, φ_K) is a representation of A also. If $K = [\varphi(A)H]$, then K is invariant for $\varphi(A)$ and the representation (K, φ_K) is non-degenerate. Moreover, $\|\varphi(a)\| = \|\varphi_K(a)\|$ $(a \in A)$. We shall often use this device to reduce to the case of a non-degenerate representation.

5.1.3. Theorem. Let (H, φ) be a non-degenerate representation of a C*-algebra A. Then it is a direct sum of cyclic representations of A.

Proof. For each $x \in H$, set $H_x = [\varphi(A)x]$. An easy application of Zorn's lemma shows that there is a maximal set Λ of non-zero elements of H such that the spaces H_x are pairwise orthogonal for $x \in \Lambda$. If $y \in (\cup_{x \in \Lambda} H_x)^\perp$, then for all $x \in \Lambda$ we have $\langle y, \varphi(a^*b)(x) \rangle = 0$, so $\langle \varphi(a)(y), \varphi(b)(x) \rangle = 0$, and therefore the spaces H_y, H_x are orthogonal. Observe that since (H, φ) is non-degenerate, $y \in H_y$. It follows from the maximality of Λ that $y = 0$. Therefore, H is the orthogonal direct sum of the family of Hilbert spaces $(H_x)_{x \in \Lambda}$. Obviously, these spaces are invariant for $\varphi(A)$, and the restriction representation

$$\varphi_x: A \to B(H_x), \quad a \mapsto \varphi(a)_{H_x},$$

has x as a cyclic vector. Since (H, φ) is the direct sum of the representations (H_x, φ_x) the theorem is proved. $\qquad\square$

Two representations (H_1, φ_1) and (H_2, φ_2) of a C*-algebra A are *unitarily equivalent* if there is a unitary $u: H_1 \to H_2$ such that $\varphi_2(a) = u\varphi_1(a)u^*$ ($a \in A$). It is readily verified that unitary equivalence is indeed an equivalence relation.

5.1.4. Theorem. *Suppose that (H_1, φ_1) and (H_2, φ_2) are representations of a C*-algebra A with cyclic vectors x_1 and x_2, respectively. Then there is a unitary $u: H_1 \to H_2$ such that $x_2 = u(x_1)$ and $\varphi_2(a) = u\varphi_1(a)u^*$ for all $a \in A$ if and only if $\langle \varphi_1(a)(x_1), x_1 \rangle = \langle \varphi_2(a)(x_2), x_2 \rangle$ for all $a \in A$.*

Proof. The forward implication is obvious. Suppose, therefore, that we have $\langle \varphi_1(a)(x_1), x_1 \rangle = \langle \varphi_2(a)(x_2), x_2 \rangle$ for all $a \in A$. Define a linear map $u_0: \varphi_1(A)x_1 \to H_2$ by setting $u_0(\varphi_1(a)(x_1)) = \varphi_2(a)(x_2)$. That this is well-defined and isometric follows from the equations

$$\|\varphi_2(a)(x_2)\|^2 = \langle \varphi_2(a^*a)(x_2), x_2 \rangle = \langle \varphi_1(a^*a)(x_1), x_1 \rangle = \|\varphi_1(a)(x_1)\|^2.$$

We extend u_0 to an isometric linear map $u: H_1 \to H_2$, and since $u(H_1) = [\varphi_2(A)x_2] = H_2$, u is a unitary.

If $a, b \in A$, then $u\varphi_1(a)\varphi_1(b)x_1 = \varphi_2(ab)(x_2) = \varphi_2(a)u\varphi_1(b)(x_1)$. Therefore, $u\varphi_1(a) = \varphi_2(a)u$ ($a \in A$). Now $\varphi_2(a)u(x_1) = u\varphi_1(a)(x_1) = \varphi_2(a)(x_2)$, so $\varphi_2(a)(u(x_1) - x_2) = 0$. By non-degeneracy of φ_2, therefore, $u(x_1) = x_2$. $\qquad\square$

A representation (H, φ) of a C*-algebra A is *irreducible* if the algebra $\varphi(A)$ acts irreducibly on H. If two representations are unitarily equivalent, then irreducibility of one implies irreducibility of the other. If H is a one-dimensional Hilbert space, then the zero representation of any C*-algebra on H is irreducible.

5.1.5. Theorem. *Let (H, φ) be a non-zero representation of a C*-algebra A.*

(1) *(H, φ) is irreducible if and only if $\varphi(A)' = \mathbf{C}1$, where $1 = \mathrm{id}_H$.*
(2) *If (H, φ) is irreducible, then every non-zero vector of H is cyclic for (H, φ).*

Proof. Condition (1) is immediate from Theorem 4.1.12.

Suppose that (H, φ) is irreducible, and that x is a non-zero vector of H. The space $[\varphi(A)x]$ is invariant for $\varphi(A)$, and therefore is equal to 0 or H. Because φ is non-zero, there is some element y of H and some element a of A such that $\varphi(a)(y) \neq 0$. Hence, $[\varphi(A)y] = H$, so φ is non-degenerate. It follows that $\varphi(A)x$ is not the zero space, so $[\varphi(A)x] = H$; that is, x is a cyclic vector for (H, φ). $\square$

We say a state τ on a C*-algebra A is *pure* if it has the property that whenever ρ is a positive linear functional on A such that $\rho \leq \tau$, necessarily there is a number $t \in [0, 1]$ such that $\rho = t\tau$.

The set of pure states on A is denoted by PS(A).

5.1.6. Theorem. *Let τ be a state on a C*-algebra A.*

(1) *τ is pure if and only if (H_τ, φ_τ) is irreducible.*
(2) *If A is abelian, then τ is pure if and only if it is a character on A.*

Proof. Suppose that τ is a pure state. Let v be an element of $\varphi_\tau(A)'$ such that $0 \leq v \leq 1$. Then the function

$$\rho: A \to \mathbf{C}, \quad a \mapsto \langle \varphi_\tau(a)v(x_\tau), x_\tau \rangle,$$

is a positive linear functional on A such that $\rho \leq \tau$. Hence, there exists $t \in [0, 1]$ such that $\rho = t\tau$, and therefore $\langle \varphi_\tau(a)v(x_\tau), x_\tau \rangle = \langle t\varphi_\tau(a)(x_\tau), x_\tau \rangle$ for all $a \in A$. Consequently,

$$
\begin{aligned}
\langle v(a + N_\tau), b + N_\tau \rangle &= \langle v\varphi_\tau(a)(x_\tau), \varphi_\tau(b)(x_\tau) \rangle \\
&= \langle v\varphi_\tau(b^*a)(x_\tau), x_\tau \rangle \\
&= \langle t\varphi_\tau(b^*a)(x_\tau), x_\tau \rangle \\
&= \langle t(a + N_\tau), b + N_\tau \rangle
\end{aligned}
$$

for all $a, b \in A$. Therefore, $v = t1$, since A/N_τ is dense in H_τ. It follows that $\varphi_\tau(A)' = \mathbf{C}1$, so (H_τ, φ_τ) is irreducible by Theorem 5.1.5.

Now suppose conversely that (H_τ, φ_τ) is irreducible, and let ρ be a positive linear functional on A such that $\rho \leq \tau$. By Theorem 5.1.2 there is a unique operator v in $\varphi_\tau(A)'$ such that $0 \leq v \leq 1$ and $\rho(a) = \langle \varphi_\tau(a)v(x_\tau), x_\tau \rangle$ for all $a \in A$. But $\varphi_\tau(A)' = \mathbf{C}1$, again by Theorem 5.1.5, so $v = t1$ for some $t \in [0, 1]$. Hence, $\rho = t\tau$, so τ is pure. This proves the equivalence in Condition (1).

Assume now that A is abelian.

If τ is pure, then $\varphi_\tau(A)' = \mathbf{C}1$. But $\varphi_\tau(A) \subseteq \varphi_\tau(A)'$, so $\varphi_\tau(A)$ consists of scalars, and therefore $B(H_\tau) \subseteq \varphi_\tau(A)'$. Hence, $B(H_\tau) = \mathbf{C}1$. Therefore, if $u, v \in B(H_\tau)$ they are scalars, and $\langle uv(x_\tau), x_\tau \rangle = u\langle v(x_\tau), x_\tau \rangle = u\langle x_\tau, x_\tau \rangle \langle v(x_\tau), x_\tau \rangle = \langle u(x_\tau), x_\tau \rangle \langle v(x_\tau), x_\tau \rangle$. Hence, $\tau = \langle \varphi_\tau(\cdot)x_\tau, x_\tau \rangle$ is multiplicative and therefore a character on A.

Now suppose conversely that τ is a character on A, and let ρ be a positive linear functional on A such that $\rho \leq \tau$. If $\tau(a) = 0$, then $\tau(a^*a) = 0$, so $\rho(a^*a) = 0$. Since $|\rho(a)| \leq \rho(a^*a)^{1/2}$, therefore $\rho(a) = 0$. Hence, $\ker(\tau) \subseteq \ker(\rho)$, and it follows from elementary linear algebra that there is a scalar t such that $\rho = t\tau$. Choose $a \in A$ such that $\tau(a) = 1$. Then $\tau(a^*a) = 1$, so $0 \leq \rho(a^*a) = t\tau(a^*a) = t \leq \tau(a^*a) = 1$, and therefore $t \in [0, 1]$. This shows that τ is pure, and the equivalence in Condition (2) is proved.   □

It follows from Theorem 5.1.6 that for an arbitrary abelian C*-algebra A, $\mathrm{PS}(A) = \Omega(A)$. The only thing not obvious is that a character τ on A must have norm 1. To see this, let $(u_\lambda)_{\lambda \in \Lambda}$ be an approximate unit for A. Then (u_λ^2) is also an approximate unit. Hence, $\|\tau\| = \lim_\lambda \tau(u_\lambda^2) = (\lim_\lambda \tau(u_\lambda))^2 = \|\tau\|^2$, so $\|\tau\| = \|\tau\|^2$, and therefore $\|\tau\| = 1$.

5.1.7. Theorem. Let (H, φ) be a representation of a C*-algebra A, and let x be a unit cyclic vector for (H, φ). Then the function

$$\tau: A \to \mathbf{C}, \quad a \mapsto \langle \varphi(a)(x), x \rangle,$$

is a state of A and (H, φ) is unitarily equivalent to (H_τ, φ_τ). Moreover, if (H, φ) is irreducible, then τ is pure.

Proof. Clearly τ is a positive linear functional on A. If $(u_\lambda)_{\lambda \in \Lambda}$ is an approximate unit for A, then because (H, φ) is non-degenerate the net $(\varphi(u_\lambda))_\lambda$ is strongly convergent to id_H. Hence, $\|\tau\| = \lim_\lambda \tau(u_\lambda) = \lim_\lambda \langle \varphi(u_\lambda)(x), x \rangle = \langle x, x \rangle = 1$, so $\tau \in S(A)$. For all $a \in A$,

$$\langle \varphi_\tau(a)(x_\tau), x_\tau \rangle = \tau(a) = \langle \varphi(a)(x), x \rangle,$$

so (H_τ, φ_τ) and (H, φ) are unitarily equivalent by Theorem 5.1.4.

If (H, φ) is irreducible, so is (H_τ, φ_τ), so by Theorem 5.1.6 τ is pure.□

5.1.1. Example. Let H be a non-zero Hilbert space, and $A = K(H)$. We are going to determine the pure states of A. If $x \in H$, then the functional

$$\omega_x: A \to \mathbf{C}, \quad u \mapsto \langle u(x), x \rangle,$$

is positive, and if x is a unit vector, ω_x is a state.

The pure states of A are precisely the states ω_x where x is a unit vector of H.

To prove this, suppose first that x is a unit vector of H, and let $i: A \to B(H)$ be the inclusion map. The representation (H, i) is irreducible, since $A' = \mathbf{C}$ (cf. Remark 4.4.2). Hence, x is a cyclic vector for A, and it follows from Theorem 5.1.7 that the representations $(H_{\omega_x}, \varphi_{\omega_x})$ and (H, i) are unitarily equivalent and ω_x is pure.

Now suppose conversely that τ is a pure state of A. By Theorem 4.2.1 there is a trace-class operator u on H such that $\tau(v) = \text{tr}(uv)$ for all $v \in A$. For any unit vector x of H, the operator $x \otimes x$ is a projection and therefore positive, so $0 \leq \tau(x \otimes x) = \text{tr}(u(x \otimes x)) = \text{tr}(u(x) \otimes x) = \langle u(x), x \rangle$. This shows that the operator u is positive. Since u is a compact normal operator, it is diagonalisable by Theorem 2.4.4; that is, there is an orthonormal basis E for H and there is a family of scalars $(\lambda_e)_{e \in E}$ such that $u(e) = \lambda_e e$ ($e \in E$). Choose $e_0 \in E$. If $v \in A^+$,

$$\tau(v) = \text{tr}(vu) = \sum_{e \in E} \langle vu(e), e \rangle = \sum_{e \in E} \lambda_e \langle v(e), e \rangle \geq \lambda_{e_0} \omega_{e_0}(v).$$

Thus, the pure state τ majorises the positive linear functional $\lambda_{e_0} \omega_{e_0}$, so there exists $t \in [0,1]$ such that $\lambda_{e_0} \omega_{e_0} = t\tau$. Since both ω_{e_0} and τ are of norm one, $\lambda_{e_0} = t$, so $\omega_{e_0} = \tau$; that is, τ is of the required form.

An interesting consequence of our characterisation of the pure states of A is that every non-zero irreducible representation (K, ψ) of A is unitarily equivalent to the identity representation (H, i) of A. To see this, let y be a unit vector in K. The function

$$\rho \colon A \to \mathbf{C}, \quad u \mapsto \langle \psi(u)(y), y \rangle,$$

is a pure state on A, and (K, ψ) is unitarily equivalent to (H_ρ, φ_ρ) by Theorem 5.1.7. Hence, there exists a unit vector x in H such that $\rho = \omega_x$. Thus, (K, ψ) is unitarily equivalent to $(H_{\omega_x}, \varphi_{\omega_x})$, and we have already seen above that $(H_{\omega_x}, \varphi_{\omega_x})$ is unitarily equivalent to (H, i).

5.1.8. Theorem. *If A is a C*-algebra, then the set S of norm-decreasing positive linear functionals on A forms a convex weak* compact set. The extreme points of S are the zero functional and the pure states of A.*

Proof. It is easy to check that S is weak* closed in the closed unit ball of A^*, and therefore weak* compact by the Banach–Alaoglu theorem. Convexity of S is clear.

Let E be the set of extreme points of S.

First we show $0 \in E$: Suppose $0 = t\tau + (1-t)\rho$, where $0 < t < 1$ and $\tau, \rho \in S$. If $a \in A$, then $0 \geq -t\tau(a^*a) = (1-t)\rho(a^*a) \geq 0$. Hence, $\tau = \rho = 0$ on A^+, and therefore on A, so 0 is an extreme point of S.

Next we show that $PS(A) \subseteq E$: Suppose that ρ is a pure state of A, and that $\rho = t\tau + (1-t)\tau'$, where $0 < t < 1$ and $\tau, \tau' \in S$. Then $t\tau$ is a positive linear functional on A majorised by ρ, so there exists $t' \in [0,1]$ such that $t\tau = t'\rho$, because ρ is pure. Since $1 = \|\rho\| = t\|\tau\| + (1-t)\|\tau'\|$, we have $\|\tau\| = \|\tau'\| = 1$. It follows that $t = \|t\tau\| = \|t'\rho\| = t'$, so $\tau = \rho$. Hence, $(1-t)\tau' = (1-t)\rho$, so $\tau' = \rho$. Therefore, $\rho \in E$.

Finally, we suppose that ρ is a non-zero element of E and show that it is a pure state: Since $\rho = \|\rho\|(\rho/\|\rho\|) + (1-\|\rho\|)0$ and $0, \rho/\|\rho\| \in S$, we have $\|\rho\| = 1$, because $\rho \in E$. If τ is a non-zero positive linear functional on A majorised by, and not equal to, ρ, then for $t = \|\tau\| \in (0,1)$ we have $\rho = t(\tau/\|\tau\|) + (1-t)(\rho - \tau)/\|\rho - \tau\|$, since $1 - t = \|\rho - \tau\|$. Hence, $\rho = \tau/\|\tau\|$, since ρ is an extreme point of S. Therefore, $\tau = \|\tau\|\rho$. This proves that ρ is a pure state of A. $\qquad\square$

5.1.9. Corollary. *The set S is the weak* closed convex hull of 0 and the pure states of A.*

Proof. Apply Theorem A.14. $\qquad\square$

5.1.10. Corollary. *If A is a unital C*-algebra, then $S(A)$ is the weak* closed convex hull of the pure states of A.*

Proof. The set $S(A)$ is a non-empty convex weak* compact set, so by Theorem A.14 it is the weak* closed convex hull of its extreme points. It is clear that $S(A)$ is a face of S, where S is as in Theorem 5.1.8, so by that theorem the extreme points of $S(A)$ are the pure states of A. $\qquad\square$

5.1.11. Theorem. *Let a be a positive element of a non-zero C*-algebra A. Then there is a pure state ρ of A such that $\|a\| = \rho(a)$.*

Proof. We may suppose that $a \neq 0$. The function

$$\hat{a}: A^* \to \mathbf{C}, \quad \tau \mapsto \tau(a),$$

is weak* continuous and linear, and by Theorem 3.3.6 $\|a\| = \sup\{\tau(a) \mid \tau \in S\}$, where S is the weak* compact convex set of all norm-decreasing positive linear functionals on A. The set $F = \{\tau \in S \mid \tau(a) = \|a\|\}$ is a weak* compact face of S by Lemma A.13, and therefore has an extreme point ρ by Theorem A.14. Since F is a face in S, the functional ρ is an extreme point of S also. Now $\rho \neq 0$, since $\|a\| = \rho(a)$, and $a \neq 0$. Therefore, ρ is a pure state of A by Theorem 5.1.8. $\qquad\square$

It follows from Theorem 5.1.11 that a non-zero C*-algebra has pure states.

5.1.12. Theorem. *Let A be a C*-algebra, and $a \in A$. Then there is an irreducible representation (H, φ) of A such that $\|a\| = \|\varphi(a)\|$.*

Proof. By the preceding theorem, there is a pure state ρ of A such that $\rho(a^*a) = \|a^*a\|$. By Theorem 5.1.6, the representation (H_ρ, φ_ρ) is irreducible. Since $\|a\|^2 = \rho(a^*a) = \langle \varphi_\rho(a^*a)(x_\rho), x_\rho \rangle = \|\varphi_\rho(a)(x_\rho)\|^2 \leq \|\varphi_\rho(a)\|^2 \leq \|a\|^2$, therefore $\|a\| = \|\varphi_\rho(a)\|$. $\qquad\square$

The characterisation given in Theorem 5.1.8 allows us to prove another extension theorem for positive functionals.

5.1.13. Theorem. *Let B be a C^*-subalgebra of a C^*-algebra A, and let ρ be a pure state on B. Then there is a pure state ρ' on A extending ρ. Moreover, if B is hereditary in A, then ρ' is unique.*

Proof. The set F of all states on A extending ρ is a weak* compact face of the set S of norm-decreasing positive linear functionals on A (by Theorem 3.3.8 F is non-empty). By Theorem A.14 F admits an extreme point, ρ' say. Hence, ρ' is an extreme point of S, and non-zero, so by Theorem 5.1.8 ρ' is a pure state of A. Uniqueness of ρ' when B is hereditary is given by Theorem 3.3.9. $\quad\square$

5.1.14. Theorem. *Let A be a unital C^*-algebra. Suppose that S is a subset of $S(A)$ such that if a hermitian element $a \in A$ satisfies the condition $\tau(a) \geq 0$ for all $\tau \in S$, then necessarily $a \in A^+$. Then the weak* closed convex hull of S is $S(A)$ and the weak* closure of S contains $\mathrm{PS}(A)$.*

Proof. Let C denote the weak* closed convex hull of S. It follows from Theorem 5.1.8 that $\mathrm{PS}(A)$ is the set of extreme points of $S(A)$, so by Theorem A.14 if we show that $S(A) = C$, then $\mathrm{PS}(A)$ is contained in the weak* closure of S.

Suppose that $C \neq S(A)$ and we shall obtain a contradiction. Since the containment $C \subseteq S(A)$ clearly holds, there exists $\tau \in S(A)$ such that $\tau \notin C$. By Theorem A.7 there is a weak* continuous linear functional $\theta: A^* \to \mathbf{C}$ and there is a real number t such that $\mathrm{Re}(\theta(\tau)) > t > \mathrm{Re}(\theta(\rho))$ for all $\rho \in C$. By Theorem A.2 there is an element $a \in A$ such that $\theta = \hat{a}$. If $b = \mathrm{Re}(a)$, then $\mathrm{Re}(\theta(\rho)) = \mathrm{Re}(\rho(a)) = \rho(b)$ for all $\rho \in S(A)$. Since $\rho(t - b) > 0$ for all $\rho \in S$, the hypothesis implies that $t - b \geq 0$. Therefore, $\tau(t - b) \geq 0$, so $t \geq \tau(b)$. But $\tau(b) = \mathrm{Re}(\theta(\tau)) > t$, a contradiction. $\quad\square$

If x is a unit vector in a Hilbert space H, we denote by ω_x the state

$$B(H) \to \mathbf{C}, \quad u \mapsto \langle u(x), x \rangle.$$

5.1.15. Theorem. *Let A be a C^*-algebra and suppose that $(H_\lambda, \varphi_\lambda)_{\lambda \in \Lambda}$ is a family of representations of A. Suppose also that ρ is a pure state of A such that*

$$\cap_{\lambda \in \Lambda} \ker(\varphi_\lambda) \subseteq \ker(\rho).$$

Then ρ belongs to the weak closure in A^* of the set*

$$S = \{\omega_x \varphi_\lambda \mid \lambda \in \Lambda \text{ and } x \in H_\lambda, \ \|x\| = 1\}.$$

Proof. Replacing $(H_\lambda, \varphi_\lambda)$ by the canonically associated non-degenerate representation if necessary, we may suppose that each $(H_\lambda, \varphi_\lambda)$ is non-degenerate. By passing to the quotient of A by the closed ideal $\cap_\lambda \ker(\varphi_\lambda)$ if necessary, we may suppose also that $\cap_\lambda \ker(\varphi_\lambda) = 0$. In this case the direct sum (H, φ) of the representations $(H_\lambda, \varphi_\lambda)$ is faithful. Observe that $\omega_x \varphi_\lambda$ is a state if $\|x\| = 1$.

Denote by $\tilde{\tau}$ the unique state of $\tilde{A}$ extending a state τ of A and by $\tilde{\varphi}_\lambda$ the unique unital $*$-homomorphism from $\tilde{A}$ to $B(H_\lambda)$ extending φ_λ. Then for $\tau = \omega_x \varphi_\lambda \in S$ we have $\tilde{\tau} = \omega_x \tilde{\varphi}_\lambda$.

Suppose that A is non-unital and $a \in A$, $\mu \in \mathbf{C}$, and $a + \mu \in \cap_{\lambda \in \Lambda} \ker(\tilde{\varphi}_\lambda)$. Then for all $b \in A$, we have $ab + \mu b = 0$ because $ab + \mu b \in \cap_{\lambda \in \Lambda} \ker(\varphi_\lambda) = 0$. Thus, if μ were nonzero, then $-a/\mu$ would be a unit for A, which contradicts our assumption. Hence, $\mu = 0$ and therefore $a = 0$. Thus, if A is non-unital, $\cap_\lambda \ker(\tilde{\varphi}_\lambda) = 0$.

From these considerations it follows that to prove the theorem we may suppose that A is unital, replacing A by $\tilde{A}$, φ_λ by $\tilde{\varphi}_\lambda$, and ρ by $\tilde{\rho}$ if necessary ($\tilde{\rho}$ is pure by Theorem 5.1.13).

Suppose then that A is unital and that a is a self-adjoint element of A such that $\tau(a) \geq 0$ for all $\tau \in S$. Then for each $\lambda \in \Lambda$ we have $\langle \varphi_\lambda(a)(x), x \rangle \geq 0$ for all $x \in H_\lambda$, and therefore $\varphi_\lambda(a) \geq 0$. Hence, $\varphi(a) \geq 0$, so $a \geq 0$, as φ is an injective $*$-homomorphism. It now follows from Theorem 5.1.14 that the weak* closure of S contains $\mathrm{PS}(A)$. $\square$

5.2. The Transitivity Theorem

The theorem of the title of this section enables us to relate some topological concepts to purely algebraic ones. For instance, we use it to show that for C*-algebras topological irreducibility of a representation is equivalent to algebraic irreducibility.

We begin with an elementary result.

5.2.1. Lemma. Let H be a Hilbert space and $e_1, \ldots, e_n$, $y_1, \ldots, y_n$ elements of H, where $e_1, \ldots, e_n$ are orthonormal. Then there is an operator $u \in B(H)$ such that

$$u(e_j) = y_j \qquad (j = 1, \ldots, n)$$

and

$$\|u\| \leq \sqrt{2n} \max\{\|y_1\|, \ldots, \|y_n\|\}.$$

Moreover, if there is a self-adjoint operator v on H such that $v(e_j) = y_j$ for $j = 1, \ldots, n$, then we may choose u to be self-adjoint also.

Proof. Set $u = \sum_{j=1}^n y_j \otimes e_j$. Clearly, $u(e_j) = y_j$ $(j = 1, \ldots, n)$. If $x \in H$

and $M = \max_j \|y_j\|$, then

$$\|u(x)\| = \|\sum_{j=1}^{n} \langle x, e_j \rangle y_j\|$$

$$\leq \sum_{j=1}^{n} |\langle x, e_j \rangle| \|y_j\|$$

$$\leq (\sum_{j=1}^{n} |\langle x, e_j \rangle|^2 \sum_{j=1}^{n} \|y_j\|^2)^{1/2}$$

$$\leq \|x\|\sqrt{n}M,$$

so $\|u\| \leq \sqrt{n}M$.

Now suppose that there is a self-adjoint operator v on H such that for all j we have $v(e_j) = y_j$. Then

$$u = \sum_{j=1}^{n} v(e_j) \otimes e_j = v(\sum_{j=1}^{n} e_j \otimes e_j) = vp,$$

where p is the projection $\sum_{j=1}^{n} e_j \otimes e_j$. Because v is hermitian, so is $u' = vp + pv - pvp$, and clearly, $u'(e_j) = y_j$ for all j. Moreover,

$$\|u'\|^2 = \|vp(vp)^* + pv(1-p)(pv(1-p))^*\|$$

$$\leq \|vp\|^2 + \|pv(1-p)\|^2$$

$$\leq \|vp\|^2 + \|pv\|^2$$

$$= 2\|u\|^2 \leq 2nM^2,$$

so $\|u'\| \leq \sqrt{2n}M$. □

The following important result is called the *transitivity* theorem.

5.2.2. Theorem (Kadison). *Let A be a non-zero C*-algebra acting irreducibly on a Hilbert space H, and suppose that $x_1, \ldots, x_n$ and $y_1, \ldots, y_n$ are elements of H and that $x_1, \ldots, x_n$ are linearly independent. Then there exists an operator $u \in A$ such that $u(x_j) = y_j$ for $j = 1, \ldots, n$. If there is a self-adjoint operator v on H such that $v(x_j) = y_j$ for $j = 1, \ldots, n$, then we may choose u to be self-adjoint also. If A contains id_H and there is a unitary v on H such that $v(x_j) = y_j$ for $j = 1, \ldots, n$, then we may choose u to be a unitary also—we may even suppose that $u = e^{iw}$ for some element $w \in A_{sa}$.*

Proof. Suppose first that there is a self-adjoint operator v on H such that $v(x_j) = y_j$ for all j, and we shall show that there exists $u \in A_{sa}$ such that $u(x_j) = y_j$ for all j. We may suppose that $x_1, \ldots, x_n$ are orthonormal (because if we have proved the result in this case, and now suppose that $x_1, \ldots, x_n$ are merely linearly independent, then we may choose an orthonormal basis $e_1, \ldots, e_n$ for $K = \mathbf{C}x_1 + \cdots + \mathbf{C}x_n$, and use the fact that there exists $u \in A_{sa}$ such that $u(e_j) = v(e_j)$ for all j, which implies $u = v$ on K, to get $u(x_j) = v(x_j) = y_j$ for all j). We may also suppose that $\max_j \|y_j\| \leq (2n)^{-1/2}$.

Suppose that $\varepsilon > 0$ and set

$$U_\varepsilon = \{u \in B(H) \mid \max_{1 \leq j \leq n} \|u(x_j)\| < \varepsilon\},$$

so U_ε is a neighbourhood of 0 in the strong topology of $B(H)$. Since A is strongly dense in $B(H)$ by Theorem 4.1.12, it follows from the Kaplansky density theorem, Theorem 4.3.3, that the closed unit ball of A_{sa} is strongly dense in the closed unit ball of $B(H)_{sa}$. Hence, if $w \in B(H)_{sa}$, there is an element $w' \in A_{sa}$ such that $w' - w \in U_\varepsilon$ and $\|w'\| \leq \|w\|$.

By Lemma 5.2.1 there is an element $v_0 \in B(H)_{sa}$ such that $v_0(x_j) = y_j$ for all j, and $\|v_0\| \leq 1$. Hence, there is an element $u_0 \in A_{sa}$ such that $u_0 - v_0 \in U_{1/(2N)}$ (where $N = \sqrt{2n}$) and $\|u_0\| \leq 1$.

We now construct by induction two sequences of self-adjoint operators (u_k) and (v_k) on H, such that $u_k \in A$ and $\|u_k\|, \|v_k\| \leq 2^{-k}$, $u_k - v_k \in U_{2^{-k-1}N^{-1}}$, and for $j = 1, \ldots, n$ we have

$$v_k(x_j) = (v_{k-1} - u_{k-1})(x_j) \qquad (k > 0).$$

Suppose that $u_0, \ldots, u_r$ and $v_0, \ldots, v_r$ have been constructed as above. Then, by Lemma 5.2.1 again, there is a self-adjoint operator v_{r+1} on H such that

$$v_{r+1}(x_j) = (v_r - u_r)(x_j) \qquad (j = 1, \ldots, n)$$

and

$$\|v_{r+1}\| \leq N \max_{1 \leq j \leq n} \|v_r(x_j) - u_r(x_j)\|$$

$$\leq N(\frac{1}{2^{r+1}N}) = \frac{1}{2^{r+1}},$$

so $\|v_{r+1}\| \leq 2^{-r-1}$. Hence, there exists an element $u_{r+1} \in A_{sa}$ such that $u_{r+1} - v_{r+1} \in U_{2^{-r-2}N^{-1}}$ and $\|u_{r+1}\| \leq 2^{-r-1}$. This completes the induction.

Since $\sum_{r=0}^{\infty} \|u_r\| \leq \sum_{r=0}^{\infty} \frac{1}{2^r} < \infty$, the series $\sum_r u_r$ is convergent in A. Set $u = \sum_{r=0}^{\infty} u_r$, so $u \in A_{sa}$. For $j = 1, \ldots, n$ we have

$$y_j - u(x_j) = \lim_{r \to \infty}(y_j - \sum_{k=0}^{r} u_k(x_j)) = \lim_{r \to \infty} v_{r+1}(x_j),$$

since the sum telescopes. Since $\lim_{r\to\infty} v_{r+1}(x_j) = 0$ for each j, we therefore have $y_j = u(x_j)$.

Now we return to the general case; that is, we drop the assumption that there is a hermitian operator v on H such that $v(x_j) = y_j$ for $j = 1, \ldots, n$. However, we may retain the assumption that $x_1, \ldots, x_n$ are orthonormal. By Lemma 5.2.1 there is a (possibly non-hermitian) operator v on H such that $v(x_j) = y_j$ for all j. If v' and v'' are the real and imaginary parts of v, then there are hermitian elements u' and u'' in A such that $u'(x_j) = v'(x_j)$ and $u''(x_j) = v''(x_j)$ for all j, by the first part of this proof. Thus, $u = u' + iu'' \in A$ and $u(x_j) = v(x_j) = y_j$ for all j.

Finally we consider the case where A contains id_H and there exists a unitary v on H such that $v(x_j) = y_j$ for all j. As before, we may suppose that $x_1, \ldots, x_n$ are orthonormal. In this case $y_1, \ldots, y_n$ are also orthonormal. Let K be the linear span of the vectors $x_1, \ldots, x_n, y_1, \ldots, y_n$. Extend $x_1, \ldots, x_n$ (respectively, $y_1, \ldots, y_n$) to an orthonormal basis $x_1, \ldots, x_m$ (respectively, $y_1, \ldots, y_m$) of K. Clearly, there is a unitary $v_0 \in B(K)$ such that $v_0(x_j) = y_j$ for $j = 1, \ldots, m$. Since v_0 is diagonalisable (it is a normal operator on a finite-dimensional Hilbert space), there is an orthonormal basis $e_1, \ldots, e_m$ for K such that $v_0(e_j) = \lambda_j e_j$ for some $\lambda_1, \ldots, \lambda_m \in \mathbf{C}$. Now $|\lambda_j| = 1$, so $\lambda_j = e^{it_j}$ for some $t_j \in \mathbf{R}$. The operator $w' = \sum_{j=1}^m t_j e_j \otimes e_j$ on H is hermitian, and $w'(e_j) = t_j e_j$. Hence, by the first part of this proof there exists $w \in A_{sa}$ such that $w(e_j) = t_j e_j$ for $j = 1, \ldots, m$. Set $u = e^{iw}$. Then u is a unitary in A such that $u(e_j) = e^{it_j} e_j = \lambda_j e_j = v_0(e_j)$, so u equals v_0 on K, and therefore $u(x_j) = v_0(x_j) = y_j$ for $j = 1, \ldots, n$. $\square$

We say a $*$-algebra A acting on a Hilbert space H is *algebraically irreducible* if 0 and H are the only vector subspaces (closed or not) of H that are invariant for A. Obviously, in this case A is topologically irreducible, that is, irreducible in our previous meaning of this word. We say a representation (H, φ) of a C*-algebra A is *algebraically irreducible* if $\varphi(A)$ is algebraically irreducible. Surprisingly, algebraic and topological irreducibility are the same, an important result in the representation theory of C*-algebras:

5.2.3. Theorem. *Let (H, φ) be a representation of a C*-algebra A. Then (H, φ) is algebraically irreducible if and only if it is topologically irreducible.*

Proof. Suppose that (H, φ) is non-zero and topologically irreducible, and let K be a non-zero vector subspace of H invariant for $\varphi(A)$. Let x be a non-zero element of K and y an element of H. Then by Theorem 5.2.2 there exists $u \in A$ such that $\varphi(u)(x) = y$, so $y \in K$. Therefore, $K = H$, so (H, φ) is algebraically irreducible. $\square$

5.2.4. Theorem. *If ρ is a pure state on a C*-algebra A, then $A/N_\rho = H_\rho$.*

Proof. Of course the point here is that A/N_ρ is complete. Since (H_ρ, φ_ρ) is an irreducible representation of A by Theorem 5.1.6, and A/N_ρ is a vector subspace of H_ρ invariant for $\varphi_\rho(A)$, we have $A/N_\rho = 0$ or H_ρ, by the preceding theorem. Since $N_\rho \neq A$, therefore $A/N_\rho = H_\rho$. □

5.3. Left Ideals of C*-Algebras

In this section we show that there is a bijective correspondence between the pure states and the modular maximal left ideals of a C*-algebra. This is used in the next section to analyse the primitive ideals of the algebra.

We begin with a result on hereditary C*-algebras which has its own interest and a nice application in Remark 5.3.1. Moreover, using the correspondence between hereditary C*-subalgebras and closed left ideals, it translates immediately into a key result of this section concerning left ideals (Theorem 5.3.2).

5.3.1. Theorem. Let B_1 and B_2 be hereditary C*-subalgebras of a C*-algebra A. Suppose that $B_1 \subseteq B_2$ and that every positive linear functional τ of A that vanishes on B_1 also vanishes on B_2. Then $B_1 = B_2$.

Proof. We may suppose that A is unital. Let a be a positive element of B_2 and suppose that $\varepsilon > 0$. Then the set

$$F = \{\tau \in S(A) \mid \tau(a) \geq \varepsilon\}$$

is weak* closed in the closed unit ball of A^*, and therefore weak* compact by the Banach–Alaoglu theorem. If $\tau \in F$, then τ does not vanish everywhere on B_2, so it does not vanish everywhere on B_1, either. Choose $a_\tau \in B_1$ such that $\tau(a_\tau) \neq 0$. Then there is a weak* open set U_τ containing τ such that $\rho(a_\tau) \neq 0$ for all $\rho \in U_\tau$. Clearly, the family $(U_\tau)_{\tau \in F}$ forms a weak* open cover of F, so by weak* compactness of F there are finitely many functionals $\tau_1, \ldots, \tau_n \in F$ such that $U_{\tau_1} \cup \ldots \cup U_{\tau_n}$ contains F. Set $b = \sum_{j=1}^n a_{\tau_j}^* a_{\tau_j}$. Then $b \in B_1$, and for any element $\tau \in F$ we have $\tau(b) > 0$ (since there exists some element a_{τ_j} such that $|\tau(a_{\tau_j})| > 0$, and $\tau(b) \geq \tau(a_{\tau_j}^* a_{\tau_j}) \geq |\tau(a_{\tau_j})|^2$). Hence, the weak* continuous linear functional $\hat{b} \colon A^* \to \mathbf{C}$, $\tau \mapsto \tau(b)$, is positive everywhere on F, so (again using weak* compactness of F) there is a positive number M such that $\tau(b) \geq M$ for all $\tau \in F$. Put $c = (\|a\|/M)b$. Then c is a positive element of B_1 and $\tau(c) \geq \|a\| \geq \tau(a)$ $(\tau \in F)$.

Now suppose that τ is an arbitrary state of A. If $\tau(a) < \varepsilon$, then $\tau(c + \varepsilon - a) \geq \tau(\varepsilon - a) > 0$. If $\tau(a) \geq \varepsilon$, then $\tau \in F$ and $\tau(c + \varepsilon - a) \geq \tau(a) + \tau(\varepsilon - a) = \varepsilon > 0$. This shows that for every positive linear functional τ on A, $\tau(c + \varepsilon - a) \geq 0$. Hence, $c + \varepsilon - a \geq 0$ by Theorem 3.4.3. We have therefore shown that for each $\varepsilon > 0$ there is an element c of B_1^+ such that $a \leq c + \varepsilon$. Because B_1 is hereditary in A, by Theorem 3.2.6 $a \in B_1$. Consequently, $B_2^+ \subseteq B_1$, so $B_2 \subseteq B_1$, and therefore $B_2 = B_1$. □

5.3.1. Remark. Let A be a C*-algebra and let a be a self-adjoint element of A such that $\tau(a) > 0$ for all non-zero positive linear functionals τ on A. Then a is positive and $(aAa)^- = A$. Positivity of a is given by Theorem 3.4.3. If the hereditary C*-subalgebra $(aAa)^-$ is not equal to A, then by Theorem 5.3.1 there is a non-zero positive linear functional τ of A which vanishes on $(aAa)^-$, and therefore $\tau(a) = 0$, contradicting our assumption. This shows that $(aAa)^- = A$ as asserted (*cf.* Exercise 3.5).

5.3.2. Theorem. *Let L_1 and L_2 be closed left ideals of a C*-algebra A. Suppose that $L_1 \subseteq L_2$ and that every positive linear functional of A that vanishes on L_1 vanishes on L_2. Then $L_1 = L_2$.*

Proof. By Theorem 3.2.1 $B_1 = L_1 \cap L_1^*$ and $B_2 = L_2 \cap L_2^*$ are hereditary C*-subalgebras of A. If τ is a positive linear functional of A vanishing on B_1, then it is clear from the inequality $|\tau(a)|^2 \leq \|\tau\| \tau(a^*a)$ that τ vanishes on L_1. It follows from the hypothesis that τ vanishes on L_2, and therefore on B_2. Hence, $B_1 = B_2$ by Theorem 5.3.1, and therefore $L_1 = L_2$ by Theorem 3.2.1 again. $\square$

5.3.3. Theorem. *Let L be a proper closed left ideal in a C*-algebra A. Then the set*

$$R = \{N_\rho \mid \rho \in \mathrm{PS}(A) \text{ and } L \subseteq N_\rho\}$$

is non-empty and $L = \cap R$.

Proof. Let S be the set of all norm-decreasing positive linear functionals on A, and let F be the set of all elements τ of S such that $L \subseteq N_\tau$. The functional 0 belongs to F, since $N_0 = A$, so F is non-empty. Also, F is weak* closed, since it is the intersection of the weak* closed sets $\{\tau \in S \mid \tau(a^*a) = 0\}$, where a ranges over L. It follows from the Banach–Alaoglu theorem that F is weak* compact. It is easily checked that F is a face of S. By Theorem A.14 the set E of extreme points of F is non-empty and F is the weak* closed convex hull of E. By Theorem 5.1.8, $E \subseteq \{0\} \cup \mathrm{PS}(A)$. If $E = \{0\}$, then $F = 0$, so for all $\tau \in S$ that vanish on L we have $\tau = 0$ (since in this case $L \subseteq N_\tau$ and therefore $\tau \in F$). By Theorem 5.3.2 L equals A, contradicting the assumption that L is proper. This argument shows that $E \neq \{0\}$ and therefore E intersects $\mathrm{PS}(A)$, so R is non-empty.

Now set $L_1 = \cap R$, so L_1 is a closed left ideal of A containing L. If τ is a non-zero element of E, then N_τ contains L and therefore contains L_1 by the definition of R. Thus, if $a \in L_1$, then $\tau(a^*a) = 0$ for all $\tau \in E$, so $\tau(a^*a) = 0$ for all $\tau \in F$, as F is the weak* closed convex hull of E. Hence, every functional in F vanishes on L_1. Since every functional in S vanishing on L is an element of F, we conclude from Theorem 5.3.2 that $L = L_1$. $\square$

If τ is a positive linear functional on a C*-algebra A, then it is easily checked that the set $N_\tau + N_\tau^* = \{a + b^* \mid a, b \in N_\tau\}$ is contained in $\ker(\tau)$.

5.3.4. Theorem. *If τ is a state on a C*-algebra A, then τ is pure if and only if $\ker(\tau) = N_\tau + N_\tau^*$.*

Proof. If τ is pure, then the representation (H_τ, φ_τ) is irreducible. Suppose that a is a hermitian element of $\ker(\tau)$ which is not in N_τ. Then $a + N_\tau$ and x_τ are orthogonal elements of H_τ (since $\langle a + N_\tau, x_\tau \rangle = \tau(a) = 0$), so if p is the projection of H_τ onto $\mathbf{C}(a + N_\tau)$, we have $p(a + N_\tau) = a + N_\tau$ and $p(x_\tau) = 0 + N_\tau$. By the transitivity theorem, Theorem 5.2.2, there is a hermitian element $b \in A$ such that $\varphi(b)(a + N_\tau) = a + N_\tau$ and $\varphi(b)(x_\tau) = 0 + N_\tau$. Hence, the elements $c = ba - a$ and b belong to N_τ. Since a is self-adjoint, $a = ba - c = a^*b - c^* \in N_\tau + N_\tau^*$. This shows that the hermitian elements of $\ker(\tau)$ belong to $N_\tau + N_\tau^*$, so $\ker(\tau) \subseteq N_\tau + N_\tau^*$ (since $\ker(\tau)$ is self-adjoint), and therefore $\ker(\tau) = N_\tau + N_\tau^*$.

Now suppose conversely that $\ker(\tau) = N_\tau + N_\tau^*$. Suppose also that ρ is a positive linear functional on A majorised by τ. Then $N_\tau \subseteq N_\rho$, and therefore $\ker(\tau) = N_\tau + N_\tau^* \subseteq N_\rho + N_\rho^* \subseteq \ker(\rho)$. By elementary linear algebra, there is a scalar t such that $\rho = t\tau$. If $\rho \neq 0$, then there exists $a \in A^+$ such that $\rho(a) > 0$, and therefore t is a positive number. Moreover, if $(u_\lambda)_{\lambda \in \Lambda}$ is an approximate unit for A, then $t = \|\rho\| = \lim_\lambda \rho(u_\lambda) \leq \lim_\lambda \tau(u_\lambda) = 1$, so $t \in [0,1]$. This shows that τ is a pure state of A. $\quad\square$

5.3.5. Theorem. *If A is a non-zero C*-algebra, then the correspondence $\tau \mapsto N_\tau$ is a bijection from $\mathrm{PS}(A)$ onto the set R of all modular maximal left ideals of A.*

Proof. Suppose that $\tau, \rho \in \mathrm{PS}(A)$ are such that $N_\tau \subseteq N_\rho$. Then by Theorem 5.3.4 $\ker(\tau) = N_\tau + N_\tau^* \subseteq N_\rho + N_\rho^* = \ker(\rho)$, so there is a scalar t such that $\rho = t\tau$. Obviously, t is positive and by equating norms we get $t = 1$, so $\rho = \tau$. This shows that the map $\rho \mapsto N_\rho$ is injective.

If $\tau \in \mathrm{PS}(A)$, then $H_\tau = A/N_\tau$ by Theorem 5.2.4, so there is an element $u \in A$ such that $x_\tau = u + N_\tau$. Then for all $a \in A$ we have $a + N_\tau = \varphi_\tau(a)(x_\tau) = au + N_\tau$, so $a - au \in N_\tau$, and therefore N_τ is modular. Now suppose that L is a proper left ideal of A containing N_τ. Since L is modular (as N_τ is), $\bar{L}$ is a proper left ideal of A, by Remark 1.3.1. Hence, by Theorem 5.3.3 there is a pure state ρ of A such that $\bar{L} \subseteq N_\rho$. Therefore, $N_\tau \subseteq N_\rho$, so $\tau = \rho$ by the first part of this proof. Hence, $L = N_\tau$, and this shows that $N_\tau \in R$.

Finally, suppose that L is an arbitrary element of R. Since by Remark 1.3.1 L is a proper closed left ideal of A, there is a pure state τ of A such that $L \subseteq N_\tau$, again using Theorem 5.3.3. By maximality of L, therefore, $L = N_\tau$. $\quad\square$

5.3.2. Remark. If A is a non-zero abelian C*-algebra, then for $\tau \in \Omega(A) = \mathrm{PS}(A)$ we have $N_\tau = \ker(\tau)$, so Theorem 5.3.5 asserts that the correspondence $\tau \mapsto \ker(\tau)$ is a bijection from $\Omega(A)$ onto the set of all modular maximal ideals of A.

5.4. Primitive Ideals

For abelian C*-algebras the ideal structure is investigated in terms of the modular maximal ideals, that is, the kernels of the characters, and in terms of the topology of the spectrum. In the non-abelian case, the role of the modular maximal ideals is taken over by the primitive ideals. There are a number of candidates for the position of analogue of the character space. The most obvious of these is set of primitive ideals, which we endow with a suitable topology. Another analogue is obtained in terms of unitary equivalence classes of non-zero irreducible representations.

We begin with a simple result from pure algebra that allows us to define primitive ideals.

5.4.1. Theorem. *Let L be a modular left ideal in an algebra A. Then there is a largest ideal I of A contained in L, namely*

$$I = \{a \in A \mid aA \subseteq L\}.$$

Proof. It is clear that $I = \{a \in A \mid aA \subseteq L\}$ is an ideal of A. Since L is modular, there is an element $u \in A$ such that $a - au \in L$ for all $a \in A$. If $a \in I$, then au and $a - au \in L$, so $a \in L$. Therefore, $I \subseteq L$. If J is an ideal of A contained in L, then for all $a \in J$ we have $aA \subseteq J \subseteq L$, so $J \subseteq I$. $\quad\square$

If L is a modular maximal left ideal in an algebra A, we call the ideal I in Theorem 5.4.1 the *primitive* ideal of A associated to L. We denote by $\mathrm{Prim}(A)$ the set of primitive ideals of A.

5.4.1. Remark. If τ is a pure state on a C*-algebra A, then $\ker(\varphi_\tau)$ is the ideal associated to the modular left ideal N_τ, as in Theorem 5.4.1, since

$$\ker(\varphi_\tau) = \{a \in A \mid \varphi_\tau(a)(A/N_\tau) = 0\}$$
$$= \{a \in A \mid aA \subseteq N_\tau\}.$$

5.4.2. Theorem. *An ideal I of a C*-algebra A is primitive if and only if there exists a non-zero irreducible representation (H, φ) of A such that $I = \ker(\varphi)$.*

Proof. If I is primitive, then there is a modular maximal left ideal of A to which I is associated, and by Theorem 5.3.5 this left ideal is of the form N_ρ for some $\rho \in \mathrm{PS}(A)$. Hence, $I = \ker(\varphi_\rho)$ by Remark 5.4.1. Since (H_ρ, φ_ρ) is a non-zero irreducible representation of A, the forward implication of the theorem is proved.

Now suppose conversely that $I = \ker(\varphi)$, where (H, φ) is some non-zero irreducible representation of A. By Theorem 5.1.5 (H, φ) has a unit cyclic vector, x say. The function

$$\rho: A \to \mathbf{C}, \quad a \mapsto \langle \varphi(a)(x), x \rangle,$$

is a pure state on A and the representations (H, φ) and (H_ρ, φ_ρ) are unitarily equivalent by Theorem 5.1.7. Hence, $I = \ker(\varphi_\rho)$ is the primitive ideal associated to the modular maximal left ideal N_ρ.   □

If S is a subset of a C*-algebra A, we let $\mathrm{hull}(S)$ denote the set of primitive ideals of A containing S. If R is a non-empty set of primitive ideals of A, we denote by $\ker(R)$ the intersection of the ideals in R. We set $\ker(\emptyset) = A$.

5.4.3. Theorem. *If I is a proper modular ideal of a C*-algebra A, then* $\mathrm{hull}(I)$ *is non-empty. If I is a proper closed ideal in A, then*

$$I = \ker(\mathrm{hull}(I));$$

that is, I is the intersection of the primitive ideals that contain it.

Proof. If I is a proper modular ideal of A, an application of Zorn's lemma shows that there is a modular maximal left ideal L of A containing I. If J is the associated primitive ideal, then $I \subseteq J$, since $I \subseteq L$. Therefore, $J \in \mathrm{hull}(I)$, so $\mathrm{hull}(I) \neq \emptyset$.

Now suppose that I is a proper closed ideal of A. By Theorems 5.3.3 and 5.3.5, the set R of modular maximal left ideals of A that contain I is non-empty and $I = \cap R$. If L is a modular maximal left ideal of A with associated primitive ideal J, then $I \subseteq J$ if and only if $I \subseteq L$. Hence, $\mathrm{hull}(I) \neq \emptyset$, and $I \subseteq \ker(\mathrm{hull}(I)) \subseteq \cap R = I$, so $I = \ker(\mathrm{hull}(I))$.   □

It follows from Theorem 5.4.3 that a modular maximal ideal of a C*-algebra is primitive. For abelian C*-algebras the two concepts coincide.

5.4.4. Theorem. *Let A be an abelian C*-algebra and I an ideal of A. Then I is primitive if and only if it is modular maximal.*

Proof. Suppose that I is primitive. By Remark 5.4.1 and Theorem 5.3.5, there exists $\rho \in \mathrm{PS}(A)$ such that $I = \ker(\varphi_\rho)$. By Theorem 5.1.6 ρ is a character on A, so $N_\rho = \ker(\rho)$, and therefore $I = \ker(\rho)$. Hence, I is a modular maximal ideal of A (*cf.* Remark 5.3.2).   □

A C*-algebra A is *primitive* if its zero ideal is primitive, that is, if A has a faithful non-zero irreducible representation. The primitive C*-algebras, as with the simple C*-algebras, are thought of as the basic building blocks of the theory, and it is important to construct examples of these algebras. We present a few here. More will be given at various points as we proceed.

If H is a non-zero Hilbert space, then the identity representation of $B(H)$ on H is irreducible by Theorem 4.1.12, since $B(H)$ is strongly dense in itself. Therefore, $B(H)$ is primitive.

The Toeplitz algebra $\mathbf{A}$ is primitive because it acts irreducibly on the Hardy space H^2 by Theorem 3.5.5.

Since every non-zero C*-algebra admits a pure state and therefore a non-zero irreducible representation, it follows that every non-zero simple C*-algebra is primitive, because the kernel of every non-zero irreducible representation is the zero ideal in this case.

Not all primitive C*-algebras are simple. An easy counterexample is provided by $B(H)$, where H is an infinite-dimensional Hilbert space.

An abelian C*-algebra is almost never primitive. To be precise, a non-zero abelian C*-algebra A is primitive if and only if $A = \mathbf{C}$. The backward implication is trivial. To see the forward implication, suppose that A admits a faithful irreducible representation (H, φ). Since A is abelian, $\varphi(A) \subseteq \varphi(A)'$, and since (H, φ) is irreducible, $\varphi(A)' = \mathbf{C}1$ by Theorem 5.1.5. Therefore, $\varphi(A) = \mathbf{C}1$, so $A = \mathbf{C}$.

5.4.2. Remark. A closed ideal I in a C*-algebra A is *prime* if whenever J_1 and J_2 are closed ideals of A such that $J_1 J_2 \subseteq I$, we necessarily have $J_1 \subseteq I$ or $J_2 \subseteq I$. If I is a prime ideal in A, and S_1, S_2 are subsets of A such that $S_1 A S_2$ is contained in I, then S_1 or S_2 is contained in I. To see this, set $J_1 = A S_1 A$ and $J_2 = A S_2 A$. Then J_1, J_2 are closed ideals in A such that $J_1 J_2 \subseteq I$, so J_1 or J_2 is contained in I (because I is prime). But $S_j \subseteq A S_j A$ (this follows from the existence of an approximate unit for A), so S_1 or S_2 is contained in I.

We say that A is a *prime* C*-algebra if the zero ideal of A is prime. Equivalently, every pair of non-zero closed ideals of A has non-zero intersection (the equivalence holds because $I \cap J = IJ$ for all closed ideals I, J of A). It is immediate from the following theorem that a primitive C*-algebra is prime.

5.4.5. Theorem. *If I is a primitive ideal of a C*-algebra A, then I is prime.*

Proof. First, suppose that $\rho \in \mathrm{PS}(A)$, and let L_1, L_2 be left ideals of A such that $L_1 L_2 \subseteq N_\rho$. We claim that L_1 or L_2 is contained in N_ρ. For suppose that $L_2 \nsubseteq N_\rho$. Then $\varphi_\rho(L_2) x_\rho$ is a non-zero vector subspace of H_ρ invariant for (H_ρ, φ_ρ), and therefore by (algebraic) irreducibility of this representation, $\varphi_\rho(L_2) x_\rho = H_\rho$. Hence, $x_\rho = \varphi_\rho(a)(x_\rho)$ for some element $a \in L_2$. If b is an arbitrary element of L_1, then $\varphi_\rho(b)(x_\rho) = \varphi_\rho(ba)(x_\rho) =$

$ba + N_\rho = 0 + N_\rho$, since $L_1 L_2 \subseteq N_\rho$. Hence, $b \in N_\rho$. This shows that $L_1 \subseteq N_\rho$.

Now suppose that J_1 and J_2 are ideals of A such that $J_1 J_2 \subseteq I$. There is a pure state ρ of A such that $I = \ker(\varphi_\rho)$, so $J_1 J_2 \subseteq N_\rho$, and therefore, by what we have just shown, $J_1 \subseteq N_\rho$ or $J_2 \subseteq N_\rho$. Hence, J_1 or J_2 is contained in I. □

Just as we put a topology on the character space of an abelian Banach algebra, we now endow the set of primitive ideals of a C*-algebra with a topology that reflects the ideal structure of the algebra.

5.4.6. Theorem. *If A is a C*-algebra, then there is a unique topology on* $\mathrm{Prim}(A)$ *such that for each subset R the set* $\mathrm{hull}(\ker(R))$ *is the closure R.*

Proof. If $R \subseteq \mathrm{Prim}(A)$, set $R' = \mathrm{hull}(\ker(R))$. Clearly, $R \subseteq R'$ and $\ker(R) = \ker(R')$, so $R' = (R')'$. Also, $\emptyset' = \emptyset$.

If R_1 and R_2 are subsets of $\mathrm{Prim}(A)$, then $(R_1 \cup R_2)' = R_1' \cup R_2'$. To show this we may suppose that R_1 and R_2 are non-empty. Set $I_1 = \cap R_1$ and $I_2 = \cap R_2$. Then $\cap(R_1 \cup R_2) = I_1 \cap I_2 = I_1 I_2$, so for any ideal $I \in \mathrm{Prim}(A)$, we have $I \in (R_1 \cup R_2)' \Leftrightarrow I_1 I_2 \subseteq I \Leftrightarrow I_1 \subseteq I$ or $I_2 \subseteq I$ (since I is prime by Theorem 5.4.5) $\Leftrightarrow I \in R_1' \cup R_2'$. Thus, $(R_1 \cup R_2)' = R_1' \cup R_2'$, as asserted.

If $(R_\lambda)_{\lambda \in \Lambda}$ is an arbitrary family of subsets of $\mathrm{Prim}(A)$, then

$$(\cap_{\lambda \in \Lambda} R_\lambda')' = \cap_{\lambda \in \Lambda} R_\lambda'.$$

This is so because for each index $\mu \in \Lambda$, $\cap_{\lambda \in \Lambda} R_\lambda' \subseteq R_\mu'$, so $(\cap_{\lambda \in \Lambda} R_\lambda')' \subseteq R_\mu'$, and therefore $(\cap_{\lambda \in \Lambda} R_\lambda')' \subseteq \cap_{\lambda \in \Lambda} R_\lambda'$, and the reverse inclusion is obvious.

Now define τ to be the collection of all sets $\mathrm{Prim}(A) \setminus R'$, where R ranges over all subsets of $\mathrm{Prim}(A)$. From what we have just shown, it is easily checked that τ is a topology on $\mathrm{Prim}(A)$, for which R' is the closure of R for each $R \subseteq \mathrm{Prim}(A)$, and there can be only one such topology. □

The topology on $\mathrm{Prim}(A)$ in Theorem 5.4.6 is called the *Jacobson* or *hull-kernel* topology on $\mathrm{Prim}(A)$.

Recall that a T_0-*space* is a topological space Ω such that for every pair of distinct points of Ω there is an open set containing one of the points and not the other.

5.4.7. Theorem. *Let A be a C*-algebra.*

(1) $\mathrm{Prim}(A)$ *is a T_0-space.*

(2) *The correspondence $I \mapsto \mathrm{hull}(I)$ is a bijection from the set of closed ideals of A onto the set of closed subsets of* $\mathrm{Prim}(A)$.

(3) *If I_1, I_2 are closed ideals of A, then $I_1 \subseteq I_2$ if and only if $\mathrm{hull}(I_2) \subseteq \mathrm{hull}(I_1)$.*

Proof. If I is a closed ideal of A, then $I = \ker(\text{hull}(I))$ by Theorem 5.4.3, so $\text{hull}(I) = \text{hull}(\ker(\text{hull}(I)))$. Hence, $\text{hull}(I)$ is closed in $\text{Prim}(A)$. Conditions (2) and (3) are immediate from these observations. If I_1 and I_2 are distinct points of $\text{Prim}(A)$, then one is not contained in the other; say $I_1 \nsubseteq I_2$. Hence, $I_2 \notin \text{hull}(I_1)$, and since $\text{hull}(I_1)$ is closed, this shows that $\text{Prim}(A)$ is a T_0-space; that is, Condition (1) holds. $\square$

5.4.8. Theorem. *If A is a unital C*-algebra, then $\text{Prim}(A)$ is compact.*

Proof. Let $(C_\lambda)_{\lambda \in \Lambda}$ be a family of closed sets in $\text{Prim}(A)$ with the finite intersection property, and for each index λ, let I_λ be the closed ideal of A such that $\text{hull}(I_\lambda) = C_\lambda$. For each non-empty finite subset $F = \{\lambda_1, \ldots, \lambda_n\}$ of Λ, the ideal $I_F = I_{\lambda_1} + \cdots + I_{\lambda_n}$ is proper because $C_{\lambda_1} \cap \ldots \cap C_{\lambda_n}$ is non-empty. Hence, $1 \notin I_F$ and the ideal $J = \cup_F I_F$ (where F ranges over all non-empty finite subsets of Λ) is proper. It follows from Theorem 5.4.3 that there is a primitive ideal I containing J. Hence, $I_\lambda \subseteq I$, so $I \in C_\lambda$, for all $\lambda \in \Lambda$. Thus, $\cap_{\lambda \in \Lambda} C_\lambda$ is non-empty. Therefore, the space $\text{Prim}(A)$ is compact. $\square$

The converse of the preceding theorem is false. An easy counter-example is provided by $K(H)$, where H is an infinite-dimensional Hilbert space. In this case the primitive ideal space is a point space (since $K(H)$ is simple), but $K(H)$ is non-unital.

We introduce now another topological space which is also an analogue of the character space of an abelian algebra:

If A is a non-zero C*-algebra, we denote by $\hat{A}$ the set of unitary equivalence classes of non-zero irreducible representations of A. If (H, φ) is a non-zero irreducible representation of A, we denote its equivalence class in $\hat{A}$ by $[H, \varphi]$, and we set $\ker[H, \varphi] = \ker(\varphi)$. The surjection

$$\theta \colon \hat{A} \to \text{Prim}(A), \quad [H, \varphi] \mapsto \ker[H, \varphi],$$

is the *canonical* map from $\hat{A}$ to $\text{Prim}(A)$. We endow $\hat{A}$ with the weakest topology making θ continuous, and call the topological space $\hat{A}$ the *spectrum* of A. By elementary topology θ is open and closed. If R is a subset of A, let $\text{hull}'(R) = \theta^{-1}(\text{hull}(R))$. Then the correspondence $I \mapsto \text{hull}'(I)$ is a bijection from the closed ideals of A onto the closed subsets of $\hat{A}$.

Note that if A is a unital C*-algebra, then it follows from Theorem 5.4.8 that $\hat{A}$ is compact.

The proof of the following is a short, easy exercise:

5.4.9. Theorem. *Let A be a C*-algebra. The following conditions are equivalent:*

(1) $\hat{A}$ *is a T_0-space.*

(2) *Any two non-zero irreducible representations of A with the same kernel are unitarily equivalent.*
(3) *The canonical map $\hat{A} \to \mathrm{Prim}(A)$ is a homeomorphism.*

We have seen that the closed ideals of a C*-algebra A correspond to the closed subsets of the primitive ideal space and of the spectrum. They also correspond to certain subsets of $\mathrm{PS}(A)$, a result that is very useful, as we shall see in the theory of tensor products (Chapter 6).

Let A be a C*-algebra and $\rho \in \mathrm{PS}(A)$. If u is a unitary in $\tilde{A}$, then the linear functional

$$\rho^u : A \to \mathbf{C}, \quad a \mapsto \rho(uau^*),$$

is a pure state of A. We say that a subset S of $\mathrm{PS}(A)$ is *unitarily invariant* if whenever $\rho \in S$ and u is a unitary of $\tilde{A}$ we have $\rho^u \in S$. In this case we set

$$S^\perp = \{a \in A \mid \rho(a) = 0 \ (\rho \in S)\}.$$

If I is a closed ideal in A, we set

$$I^\perp = \{\rho \in \mathrm{PS}(A) \mid \rho(a) = 0 \ (a \in I)\}.$$

It is clear that $I^\perp$ is a unitarily invariant weak* closed subset of $\mathrm{PS}(A)$, and it follows from the next theorem that $S^\perp$ is a closed ideal in A.

5.4.10. Theorem. *Let A be a C*-algebra.*
(1) *If S is a non-empty unitarily invariant subset of $\mathrm{PS}(A)$, then*

$$S^\perp = \cap_{\rho \in S} \ker(\varphi_\rho).$$

If in addition S is relatively weak closed in $\mathrm{PS}(A)$, then $S = (S^\perp)^\perp$.*
(2) *If I is a closed ideal of A, then $I = (I^\perp)^\perp$.*
(3) *The map $S \mapsto S^\perp$ from the set of relatively weak* closed unitarily invariant subsets S of $\mathrm{PS}(A)$ to the set of closed ideals of A is a bijection.*

Proof. Suppose that S is a non-empty unitarily invariant subset of $\mathrm{PS}(A)$. If $a \in \cap_{\rho \in S} \ker(\varphi_\rho)$, then for each $\rho \in S$ we have $\rho(a) = \langle \varphi_\rho(a)(x_\rho), x_\rho \rangle = 0$, so $a \in S^\perp$. Thus, $\cap_{\rho \in S} \ker(\varphi_\rho) \subseteq S^\perp$. If these sets are not equal, then there is an element $a \in S^\perp$ such that for some $\rho \in S$ we have $\varphi_\rho(a) \neq 0$. Hence, there exists a unit vector $x \in H_\rho$ such that $\langle \varphi_\rho(a)(x), x \rangle \neq 0$. Since ρ is pure, (H_ρ, φ_ρ) is an irreducible representation, and therefore if φ denotes the unique unital *-homomorphism from $\tilde{A}$ to $B(H_\rho)$ extending φ_ρ, the representation (H_ρ, φ) of $\tilde{A}$ is also irreducible. It follows from Theorem 5.2.2 that there exists a unitary u in $\tilde{A}$ such that $\varphi(u)x = x_\rho$. Now

$$
\begin{aligned}
\rho^u(a) &= \langle \varphi_\rho(uau^*)(x_\rho), x_\rho \rangle \\
&= \langle \varphi_\rho(a)\varphi(u^*)(x_\rho), \varphi(u^*)(x_\rho) \rangle \\
&= \langle \varphi_\rho(a)x, x \rangle \\
&\neq 0,
\end{aligned}
$$

so $\rho^u \notin S$, contradicting the unitary invariance of S. We therefore conclude that $\cap_{\rho \in S} \ker(\varphi_\rho) = S^\perp$.

Now suppose that in addition S is relatively weak* closed in $PS(A)$. It is clear that $S \subseteq (S^\perp)^\perp$. To prove the reverse inclusion, suppose that $\tau \in (S^\perp)^\perp$. By applying Theorem 5.1.15 to the family $(H_\rho, \varphi_\rho)_{\rho \in S}$ of representations of A, we see that τ is a weak* limit of states of the form $\omega_x \varphi_\rho$ with $\rho \in S$ and x a unit vector of H_ρ—the symbol ω_x denotes the state

$$B(H_\rho) \to \mathbf{C}, \quad u \mapsto \langle u(x), x \rangle.$$

It is clear from the first part of this proof that for any such state $\omega_x \varphi_\rho$, there is a unitary u in $\tilde{A}$ such that $\omega_x \varphi_\rho = \rho^u$, and therefore $\omega_x \varphi_\rho \in S$ by unitary invariance of S. Hence, $\tau \in S$, since S is relatively weak* closed in $PS(A)$. Therefore, $S = (S^\perp)^\perp$.

Let I be a proper closed ideal of A. By Theorem 5.3.3 $I = \cap_{\rho \in S} N_\rho$, where $S = \{\rho \in PS(A) \mid I \subseteq N_\rho\}$. The set S is non-empty and clearly weak* closed in $PS(A)$ and unitarily invariant. Moreover, since $\ker(\varphi_\rho)$ is the largest ideal contained in N_ρ (cf. Remark 5.4.1), we have $I \subseteq N_\rho$ if and only if $I \subseteq \ker(\varphi_\rho)$, and therefore $I = \cap_{\rho \in S} \ker(\varphi_\rho) = S^\perp$. It follows that $I^\perp = (S^\perp)^\perp = S$, so $(I^\perp)^\perp = S^\perp = I$. This proves the theorem. $\square$

5.5. Extensions and Restrictions of Representations

As is indicated by the title, we are concerned in this section with extending and restricting representations, usually with the aim of getting representations of the same type as the ones with which we started. We also investigate the relationships between the primitive ideal space and the spectrum of a C*-algebra and the corresponding spaces for its hereditary C*-subalgebras and quotient algebras.

Let (H, φ) be a representation of a C*-algebra A, and suppose that B is a C*-subalgebra of A and that K is a closed vector subspace of H invariant for $\varphi(B)$. Then the map

$$\psi \colon B \to B(K), \quad b \mapsto \varphi(b)_K,$$

is a *-homomorphism. We denote the representation (K, ψ) by $(H, \varphi)_{B,K}$.

5.5.1. Theorem. *Let B be a C*-subalgebra of a C*-algebra A and suppose that (K, ψ) is a non-degenerate representation of B. Then there is a non-degenerate representation (H, φ) of A and a closed vector subspace K' of H invariant for $\varphi(B)$ such that (K, ψ) is unitarily equivalent to $(H, \varphi)_{B,K'}$. If (K, ψ) is cyclic (respectively irreducible), we may take (H, φ) cyclic (respectively irreducible).*

Proof. We may assume that ψ is non-zero. First suppose that (K, ψ) is cyclic, and let y be a unit cyclic vector. The function

$$\tau_0 \colon B \to \mathbf{C}, \quad b \mapsto \langle \psi(b)(y), y \rangle,$$

is a state of B, and therefore extends to a state τ of A, by Theorem 3.3.8. Set $(H, \varphi) = (H_\tau, \varphi_\tau)$. If $(u_\lambda)_{\lambda \in \Lambda}$ is an approximate unit for B, then the net of positive operators $(\varphi(u_\lambda))$ is increasing and bounded above, and therefore by Theorem 4.1.1 it converges strongly to a positive operator, p say, on H. Clearly, $p\varphi(b) = \varphi(b)p = \varphi(b)$ for all $b \in B$. Let $x = p(x_\tau)$ and $K' = [\varphi(B)x]$. Then K' is a closed vector subspace of H invariant for $\varphi(B)$. Set $(K', \psi') = (H, \varphi)_{B,K'}$. Then x is a cyclic vector for the representation (K', ψ') ($x \in K'$, since $x = p(x_\tau) = \lim_\lambda \varphi(u_\lambda)(x)$). For each $b \in B$, we have $\langle \psi'(b)(x), x \rangle = \langle \varphi(b)(x_\tau), x_\tau \rangle = \tau(b) = \tau_0(b) = \langle \psi(b)(y), y \rangle$. Hence, by Theorem 5.1.4 the representations (K, ψ) and (K', ψ') are unitarily equivalent.

Now suppose that (K, ψ) is irreducible. Then τ_0 is a pure state by Theorem 5.1.7. Applying Theorem 5.1.13, we may suppose that τ is pure also. It then follows that $(H, \varphi) = (H_\tau, \varphi_\tau)$ is irreducible.

Finally, suppose only that (K, ψ) is non-degenerate. In this case, by Theorem 5.1.3, we can write (K, ψ) as a direct sum of a family $((K_\lambda, \psi_\lambda))_\lambda$ of cyclic representations of B. For each index λ there is a cyclic representation $(H_\lambda, \varphi_\lambda)$ of A, and a closed vector subspace K'_λ of H_λ invariant for $\varphi_\lambda(B)$ such that $(K_\lambda, \psi_\lambda)$ is unitarily equivalent to $(H_\lambda, \varphi_\lambda)_{B,K'_\lambda}$. Hence, if (H, φ) is the direct sum of the representations $(H_\lambda, \varphi_\lambda)$, it is a non-degenerate representation of A, and the orthogonal direct sum K' of the spaces K'_λ is a closed vector subspace of H invariant for $\varphi(B)$. It is easily checked that (K, ψ) is unitarily equivalent to $(H, \varphi)_{B,K'}$. $\square$

If (H, φ) is a representation for a C*-algebra A, and B a C*-subalgebra of A, we denote the representation $(H, \varphi)_{B,[\varphi(B)H]}$ by $(H, \varphi)_B$, and call it the *restriction* of the representation (H, φ) to B.

5.5.2. Theorem. *Let B be a hereditary C*-subalgebra of a C*-algebra A, and suppose that (H, φ) is an irreducible representation of A. Then $(H, \varphi)_B$ is an irreducible representation of B. Moreover, $\varphi(B)H$ is closed.*

Proof. We may suppose that $\varphi \neq 0$. Let p be the orthogonal projection of H onto $[\varphi(B)H]$, and let $(u_\lambda)_{\lambda \in \Lambda}$ be an approximate unit for B. Then it is easily checked that $(\varphi(u_\lambda))$ converges strongly to p on H. Suppose that x and y are elements of $p(H)$ and that x is non-zero. By (algebraic) irreducibility of (H, φ), there is an element $a \in A$ such that $\varphi(a)(x) = y$. Set $b_\lambda = u_\lambda a u_\lambda$, so $b_\lambda \in B$ (because B is hereditary in A). Then, since

$$\|\varphi(b_\lambda)(x) - y\| \leq \|\varphi(u_\lambda a u_\lambda)(x) - \varphi(u_\lambda)(y)\| + \|\varphi(u_\lambda)(y) - y\|$$
$$\leq \|\varphi(a)\varphi(u_\lambda)x - y\| + \|\varphi(u_\lambda)y - y\|,$$

and $\lim_\lambda \varphi(u_\lambda)(x) = x$, and $\lim_\lambda \varphi(u_\lambda)(y) = y$, therefore $\lim_\lambda \varphi(b_\lambda)(x) = y$. Hence, $y \in [\varphi(B)x]$. This argument shows that $(H, \varphi)_B$ is (topologically) irreducible. By Theorem 5.2.3 $(H, \varphi)_B$ is therefore algebraically irreducible, and since $\varphi(B)H$ is an invariant vector space for $(H, \varphi)_B$, it is equal to $[\varphi(B)H]$.   □

5.5.3. Corollary. *Let B be a non-zero hereditary C*-subalgebra of a C*-algebra A, and ρ a pure state of A. Then there exists $t \in [0, 1]$ and there is a pure state ρ' of B such that $\rho_B = t\rho'$, where ρ_B is the restriction of ρ to B.*

Proof. We may suppose that $\rho_B \neq 0$. Let p be the projection of H onto $\varphi_\rho(B)H$, so $p \in \varphi_\rho(B)'$, since $\varphi_\rho(B)H$ is invariant for $\varphi_\rho(B)$. If $y = p(x_\rho)$, then for all $b \in B$ we have $\rho(b) = \langle \varphi_\rho(b)(x_\rho), x_\rho \rangle = \langle p\varphi_\rho(b)(x_\rho), x_\rho \rangle = \langle \varphi_\rho(b)(y), y \rangle$. Hence, the irreducible representation $(H, \varphi_\rho)_B$ is non-zero, and y is also non-zero. By Theorem 5.1.7 the function

$$\rho' : B \to \mathbf{C}, \quad b \mapsto \langle \varphi_\rho(b)y, y \rangle / \|y\|^2,$$

is a pure state of B, since $y/\|y\|$ is a unit cyclic vector for $(H, \varphi_\rho)_B$. Clearly, $\rho_B = t\rho'$, where $t = \|y\|^2$. Since $\|y\| = \|p(x_\rho)\| \leq \|x_\rho\| = 1$, therefore $t \in [0, 1]$.   □

5.5.4. Lemma. *Suppose that B is a non-zero hereditary C*-subalgebra of a C*-algebra A, and that I is a primitive ideal of A not containing B. Then $I \cap B$ is a primitive ideal of B. Moreover, if J is a closed ideal of A such that $J \cap B \subseteq I$, then $J \subseteq I$.*

Proof. The characterisation of primitive ideals given in Theorem 5.4.2 implies that $I = \ker(\varphi)$ for some non-zero irreducible representation (H, φ) of A. Now $I \cap B$ is the kernel of the representation $(H, \varphi)_B$ of B, and this representation is non-zero because $B \not\subseteq I$. Since $(H, \varphi)_B$ is irreducible (Theorem 5.5.2), we can apply Theorem 5.4.2 again to infer that $I \cap B$ is a primitive ideal in B.

Suppose now that J is a closed ideal of A such that $J \cap B \subseteq I$, and we shall show that $J \subseteq I$. Because B is hereditary in A, we have $BAB \subseteq B$; hence, $BJB \subseteq J \cap B \subseteq I$, and therefore $(BJ)A(JB) \subseteq I$. Since I is prime (Theorem 5.4.5), it follows from Remark 5.4.2 that one of the sets BJ or JB is contained in I. Hence, BAJ or JAB is contained in I, so again applying Remark 5.4.2, B or J is contained in I. The containment $B \subseteq I$ is ruled out by hypothesis, so $J \subseteq I$.   □

If B is a non-zero hereditary C*-subalgebra of a C*-algebra A, we call the map

$$\mathrm{Prim}(A) \setminus \mathrm{hull}(B) \to \mathrm{Prim}(B), \quad I \mapsto I \cap B,$$

the *canonical* map from $\mathrm{Prim}(A) \setminus \mathrm{hull}(B)$ to $\mathrm{Prim}(B)$. Similarly, the map

$$\hat{A} \setminus \mathrm{hull}'(B) \to \hat{B}, \quad [H, \varphi] \mapsto [(H, \varphi)_B],$$

is the *canonical* map from $\hat{A} \setminus \mathrm{hull}'(B)$ to $\hat{B}$. (By Theorem 5.4.3, the intersection of the primitive ideals of A is the zero ideal. Hence, $\mathrm{hull}(B) \neq \mathrm{Prim}(A)$ and $\mathrm{hull}'(B) \neq \hat{A}$.)

5.5.5. Theorem. *Suppose that B is a non-zero hereditary C^*-subalgebra of a C^*-algebra A. Then the following diagram commutes where the maps are the canonical ones:*

$$
\begin{array}{ccc}
\hat{A} \setminus \mathrm{hull}'(B) & \to & \hat{B} \\
\downarrow & & \downarrow \\
\mathrm{Prim}(A) \setminus \mathrm{hull}(B) & \to & \mathrm{Prim}(B).
\end{array}
$$

Moreover, the horizontal maps are homeomorphisms.

Proof. Commutativity of the diagram is clear.

Denote by θ and θ' the upper and lower horizontal maps, respectively. First we show that θ is injective. Suppose that $\theta[H_1, \varphi_1] = \theta[H_2, \varphi_2]$. If $(K_j, \psi_j) = (H_j, \varphi_j)_B$ $(j = 1, 2)$, then (K_1, ψ_1) and (K_2, ψ_2) are non-zero unitarily equivalent irreducible representations of B. Let $u \colon K_1 \to K_2$ be a unitary such that $\psi_2(b) = u\psi_1(b)u^*$ $(b \in B)$, and choose unit vectors x_1, x_2 of K_1, K_2, respectively, such that $x_2 = u(x_1)$. Then $\varphi_1(A)x_1 = H_1$ and $\varphi_2(A)x_2 = H_2$ (because φ_1 and φ_2 are algebraically irreducible). Theorem 5.1.7 implies that the functions

$$\rho_j \colon A \to \mathbf{C}, \quad a \mapsto \langle \varphi_j(a)(x_j), x_j \rangle, \quad (j = 1, 2)$$

are pure states of A such that (H_j, φ_j) is unitarily equivalent to $(H_{\rho_j}, \varphi_{\rho_j})$ $(j = 1, 2)$. Moreover, ρ_j extends the function

$$\rho_j' \colon B \to \mathbf{C}, \quad b \mapsto \langle \psi_j(b)(x_j), x_j \rangle,$$

which is a pure state of B, since (K_j, ψ_j) is irreducible. However, for each $b \in B$ we have

$$\rho_2'(b) = \langle \psi_2(b)u(x_1), u(x_1) \rangle = \langle \psi_1(b)(x_1), x_1 \rangle = \rho_1'(b),$$

so $\rho_2' = \rho_1'$. It follows from Theorem 3.3.9 that $\rho_2 = \rho_1$, and therefore (H_2, φ_2) and (H_1, φ_1) are unitarily equivalent. Therefore, θ is injective.

To show surjectivity of θ, suppose that $[K, \psi] \in \hat{B}$. By Theorem 5.5.1 there is an irreducible representation (H, φ) of A and a closed vector subspace K' of H invariant for $\varphi(B)$ such that (K, ψ) and $(H, \varphi)_{B, K'}$ are unitarily equivalent. Since (H, φ) is clearly non-zero, $[H, \varphi] \in \hat{A}$. Choose x in

K' such that $\varphi(B)x = K'$ (this is possible, since $(H,\varphi)_{B,K'}$ is irreducible).
Then there is an element $b_0 \in B$ such that $x = \varphi(b_0)(x)$. Now $\varphi(A)x = H$,
as (H,φ) is irreducible, so for any $y \in H$ and $b \in B$ there exists an element
$a \in A$ such that $\varphi(b)(y) = \varphi(b)\varphi(a)(x) = \varphi(bab_0)(x) \in \varphi(B)x$. This shows
that $\varphi(B)H = K'$. Therefore, $(H,\varphi)_{B,K'} = (H,\varphi)_B$, so $[H,\varphi] \notin \mathrm{hull}'(B)$
and $\theta[H,\varphi] = [K,\psi]$. We have therefore shown that θ is a bijection. It
follows directly from commutativity of the diagram in the statement of
the theorem that θ' is surjective, and injectivity of θ' is immediate from
Lemma 5.5.4.

We now show θ' is a homeomorphism (from which it is an elementary
exercise to show that θ is also a homeomorphism).

Suppose that C is a non-empty closed set of $\mathrm{Prim}(B)$. Then $C = \mathrm{hull}_B(I)$ for some closed ideal I of B (the subscript indicates we are looking
at the hull relative to B). Now

$$(\theta')^{-1}(C) = \{J \in \mathrm{Prim}(A) \mid I \subseteq J \cap B \text{ and } B \not\subseteq J\}$$
$$= \mathrm{hull}_A(I) \cap (\mathrm{Prim}(A) \setminus \mathrm{hull}_A(B)),$$

so $(\theta')^{-1}(C)$ is closed in $\mathrm{Prim}(A) \setminus \mathrm{hull}_A(B)$. Therefore, θ' is continuous.

Now we show that θ' is a closed map. Suppose that C is a non-empty
closed set in $\mathrm{Prim}(A) \setminus \mathrm{hull}_A(B)$, so $C = \mathrm{hull}_A(I) \setminus \mathrm{hull}_A(B)$ for some
proper closed ideal I of A. If $J \in \mathrm{hull}_B(I \cap B)$, then $J = J' \cap B$ for
some $J' \in \mathrm{Prim}(A) \setminus \mathrm{hull}_A(B)$, because θ' is surjective. Hence, $I \cap B \subseteq
J'$. It follows from Lemma 5.5.4 that $I \subseteq J'$. Therefore, $J = \theta'(J') \in
\theta'(C)$. Hence, $\mathrm{hull}_B(I \cap B) \subseteq \theta'(C)$, and the reverse inclusion is obvious,
so $\mathrm{hull}_B(I \cap B) = \theta'(C)$. Consequently, θ' is a closed map.   □

5.5.6. Corollary. *Suppose that B is a non-zero hereditary C*-subalgebra
of a primitive C*-algebra A. Then B also is primitive.*

Proof. It follows immediately from Theorem 5.5.5 that since $0 \in \mathrm{Prim}(A) \setminus
\mathrm{hull}(B)$, therefore $0 \in \mathrm{Prim}(B)$.   □

If the assumption in Corollary 5.5.6 that B is hereditary is dropped,
then the conclusion may fail. For instance, if H is a Hilbert space of
dimension greater than 1, then the primitive C*-algebra $B(H)$ contains
non-trivial, and therefore non-primitive, abelian C*-subalgebras.

Let $\psi: A \to B$ be a surjective $*$-homomorphism of C*-algebras, and
suppose that $I = \ker(\psi)$. If (H,φ) is a representation of B, then $(H,\varphi\psi)$
is a representation of A, and clearly $(H,\varphi\psi)$ is irreducible if (H,φ) is irreducible. If $B \neq 0$, we therefore have a well-defined map

$$\hat{B} \to \mathrm{hull}'(I), \quad [H,\varphi] \mapsto [H,\varphi\psi],$$

which we call the *canonical* map from $\hat{B}$ to hull$'(I)$. Note that $\ker(\varphi\psi) = \psi^{-1}(\ker(\varphi))$. It follows from Theorem 5.4.2 that if $J \in \mathrm{Prim}(B)$, then $\psi^{-1}(J) \in \mathrm{hull}(I)$. The map

$$\mathrm{Prim}(B) \to \mathrm{hull}(I), \quad J \mapsto \psi^{-1}(J),$$

is the *canonical* map from $\mathrm{Prim}(B)$ to hull(I). It is an easy exercise to show that these two canonical maps are bijective.

5.5.7. Theorem. *Suppose that the map $\psi \colon A \to B$ is a non-zero surjective $*$-homomorphism from the C*-algebra A onto the C*-algebra B, and suppose that $I = \ker(\psi)$. Then the following diagram commutes, where all the maps are canonical:*

$$
\begin{array}{ccc}
\hat{B} & \to & \mathrm{hull}'(I) \\
\downarrow & & \downarrow \\
\mathrm{Prim}(B) & \to & \mathrm{hull}(I).
\end{array}
$$

Moreover, the horizontal maps are homeomorphisms.

Proof. We shall show only that the second horizontal map, which we denote by θ, is a homeomorphism, because the rest is then routine. To show that θ is continuous, let C be a closed set in hull(I). For some closed ideal I_0 of A containing I, we have $C = \mathrm{hull}_A(I_0)$, so

$$
\begin{aligned}
\theta^{-1}(C) &= \{J \in \mathrm{Prim}(B) \mid I_0 \subseteq \psi^{-1}(J)\} \\
&= \{J \in \mathrm{Prim}(B) \mid \psi(I_0) \subseteq J\},
\end{aligned}
$$

and therefore $\theta^{-1}(C)$ is closed in $\mathrm{Prim}(B)$. Hence, θ is continuous.

To show that θ is a closed map, suppose that D is a closed set of $\mathrm{Prim}(B)$. Then there is a closed ideal J_0 in B such that $D = \mathrm{hull}_B(J_0)$. If $J \in \mathrm{hull}_A(\psi^{-1}(J_0))$, then $I \subseteq J$, so $J = \psi^{-1}(J')$ for some primitive ideal J' of B (because θ is bijective). Clearly, $J_0 \subseteq J'$, so $J = \theta(J') \in \theta(D)$, and therefore $\mathrm{hull}_A(\psi^{-1}(J_0)) \subseteq \theta(D)$. Since the reverse inclusion also holds, we have $\mathrm{hull}_A(\psi^{-1}(J_0)) = \theta(D)$, so $\theta(D)$ is closed in hull(I), and therefore θ is a closed map. $\qquad\square$

5.6. Liminal and Postliminal C*-Algebras

The algebras of the title of this section form the best-behaved class of C*-algebras. Their theory is deep and very well-developed, but we shall have space here only to touch upon the elements of the subject.

A C*-algebra A is said to be *liminal* if for every non-zero irreducible representation (H, φ) of A we have $\varphi(A) = K(H)$ (equivalently, invoking Theorem 2.4.9, $\varphi(A) \subseteq K(H)$). Liminal algebras are also called *CCR algebras*. CCR is an acronym for *completely continuous representations* (an older terminology for a compact operator is a *completely continuous* operator).

5.6.1. *Example.* If A is an abelian C*-algebra, then it is liminal. For if (H, φ) is a non-zero irreducible representation of A, then $\varphi(A)' = \mathbf{C}1$, and $\varphi(A) \subseteq \varphi(A)'$, since A is abelian. Hence, $\varphi(A) = \mathbf{C}1$, so H is one-dimensional, since $\varphi(A)$ has no non-trivial invariant vector subspaces. Therefore, $\varphi(A) = K(H)$.

5.6.2. *Example.* Every finite-dimensional C*-algebra A is liminal. For if (H, φ) is a non-zero irreducible representation of A, then $H = \varphi(A)x$ for some non-zero vector $x \in H$, so H is finite-dimensional and therefore $\varphi(A) \subseteq K(H) = B(H)$.

5.6.3. *Example.* If H is a Hilbert space, $K(H)$ is a liminal C*-algebra. This follows immediately from Example 5.1.1, where it is shown that every non-zero irreducible representation of $K(H)$ is unitarily equivalent to the identity representation of $K(H)$ on H.

Not every C*-algebra is liminal. The algebra $B(H)$ for H infinite-dimensional provides an easy counterexample (consider the identity representation of $B(H)$ on H).

5.6.1. Theorem. *If A is a liminal C*-algebra, then its C*-subalgebras and its quotient C*-algebras are liminal also.*

Proof. Suppose that B is a C*-subalgebra of A. If (K, ψ) is a non-zero irreducible representation of B, then by Theorem 5.5.1 there is an irreducible representation (H, φ) of A and a closed vector subspace K' of H such that (K, ψ) is unitarily equivalent to $(K', \psi') = (H, \varphi)_{B,K'}$. Clearly, $\varphi \neq 0$ because $\psi \neq 0$. By hypothesis, $\varphi(A) = K(H)$, and therefore $\psi'(b)$ is a compact operator for all $b \in B$ (restrictions of compact operators are compact). Therefore, $\psi(B)$ consists of compact operators on K. Consequently, B is liminal.

Now let C be a quotient C*-algebra of A, and let $\pi: A \to C$ be the quotient *-homomorphism. Let (H, φ) be a non-zero irreducible representation of C. Then $(H, \varphi\pi)$ is a non-zero irreducible representation of A, and therefore $\varphi\pi(A) = K(H)$ from the hypothesis; that is, $\varphi(C) = K(H)$. Therefore, C is liminal. $\square$

If I is a closed ideal in a liminal C*-algebra A, then it follows from Theorem 5.6.1 that I and A/I are both liminal. The converse is false. To see this, let us first observe that a unital liminal C*-algebra A has only finite-dimensional irreducible representations. For if (H, φ) is a non-zero irreducible representation of A, then it is non-degenerate, and therefore $\varphi(1) = \mathrm{id}_H$. Hence, id_H is compact, and therefore $\dim(H) < \infty$. Now if H is any infinite-dimensional Hilbert space, then the C*-algebra $A = K(H) + \mathbf{C}1$ is unital and has an infinite-dimensional non-zero irreducible representation, namely, the identity representation on H. Hence, A is not liminal. But $K(H)$ is a liminal ideal of A such that $A/K(H) = \mathbf{C}$ is liminal.

A C*-algebra A is said to be *postliminal* if for every non-zero irreducible representation (H, φ) of A we have $K(H) \subseteq \varphi(A)$ (equivalently, by Theorem 2.4.9, $K(H) \cap \varphi(A) \neq 0$). The postliminal C*-algebras are also called *GCR algebras* and *Type I* C*-algebras. Unfortunately, the Type I terminology conflicts with the terminology for von Neumann algebras because Type I von Neumann algebras are not Type I C*-algebras in general.

Every liminal C*-algebra is obviously postliminal.

5.6.2. Theorem. *Let I be a closed ideal in a C*-algebra A. Then A is postliminal if and only if I and A/I are postliminal.*

Proof. First suppose that A is postliminal. If (K, ψ) is a non-zero irreducible representation of I, then by Theorem 5.5.1 there is an irreducible representation (H, φ) of A and a closed vector subspace K' of H invariant for $\varphi(I)$ such that (K, ψ) is unitarily equivalent to $(K', \psi') = (H, \varphi)_{I,K'}$. Clearly, $\varphi \neq 0$, since $\psi \neq 0$. Since A is postliminal, $K(H) \subseteq \varphi(A)$. Observe that if u is a compact operator on K', then it is the restriction of a compact operator v on H (for example, take $v = u$ on K' and $v = 0$ on $K'^{\perp}$). Suppose now that x is a non-zero element of K' and let u be the projection of K' onto $\mathbf{C}x$. Since (K', ψ') is algebraically irreducible and $\psi'(I)x$ is a non-zero vector subspace of K' invariant for $\psi'(I)$, we have $K' = \psi'(I)x$. Hence, $x = \psi'(b)x$ for some $b \in I$. Moreover, since v is compact, we have $v = \varphi(a)$ for some $a \in A$, because A is postliminal. Consequently, $u = \psi'(b)u = \psi'(b)\varphi(a)_{K'} = \varphi(ba)_{K'} = \psi'(ba)$ (as $ba \in I$). Therefore, $\psi'(I)$ contains the rank-one projections of K', and therefore the compact operators on K'. Since (K, ψ) and (K', ψ') are unitarily equivalent, $\psi(I)$ contains the compact operators on K. This shows that I is postliminal. The proof that A/I is also postliminal when A is postliminal is completely straightforward and left as an exercise.

Now we suppose that I and A/I are postliminal and we show that A is also. Let (H, φ) be a non-zero irreducible representation of A. If $\ker(\varphi)$ contains I, then there is a *-homomorphism $\psi \colon A/I \to B(H)$ such that $\varphi = \psi\pi$, where π is the quotient map from A to A/I. Clearly, (H, ψ) is a non-zero irreducible representation of A/I. Since A/I is postliminal, $K(H) \subseteq \psi(A/I)$; that is, $K(H) \subseteq \varphi(A)$. Suppose now that $\ker(\varphi)$ does not contain I. Then the representation $(H', \varphi') = (H, \varphi)_I$ of I is non-zero. It is also irreducible, by Theorem 5.5.2. Since I is postliminal, this implies that $K(H') \subseteq \varphi'(I)$. It is easily checked that for $b \in I$ the operator $\varphi(b)$ is compact if and only if $\varphi'(b)$ is compact. Hence, there is an element b of I such that $\varphi(b)$ is non-zero and compact. Thus, whether $\ker(\varphi)$ does or does not contain I, there is an element a of A such that $\varphi(a)$ is non-zero and compact. This shows that A is postliminal. $\square$

A consequence of this result is that a C*-algebra A is postliminal if and only if its unitisation $\tilde{A}$ is postliminal.

5.6.4. Example. Let **A** denote the Toeplitz algebra (Section 3.5). From Theorem 3.5.10, its commutator ideal is $K(H^2)$, and this is liminal as we saw in Example 5.6.3. Moreover, the quotient $\mathbf{A}/K(H^2)$ is *-isomorphic to $C(\mathbf{T})$ (Theorem 3.5.11), so it is abelian and therefore liminal (*cf.* Example 5.6.1). Hence, **A** is postliminal by Theorem 5.6.2. However, **A** is not liminal, as is seen by observing that the identity representation of **A** on H^2 is irreducible (by Theorem 3.5.5) and not finite-dimensional.

A C*-algebra is said to be *elementary* if it is *-isomorphic to $K(H)$ for some Hilbert space H. It is easily checked that a simple postliminal C*-algebra is elementary. An elementary C*-algebra is unital if and only if it is finite-dimensional, so an infinite-dimensional unital simple C*-algebra is not postliminal. In particular, it follows from Theorem 4.1.16 that if H is a separable infinite-dimensional Hilbert space, then the corresponding Calkin algebra is a simple C*-algebra which is not postliminal. An application of Theorem 5.6.2 shows that $B(H)$ therefore cannot be postliminal, either.

5.6.3. Theorem. *Let (H, φ) and (H', φ') be non-zero irreducible representations of a postliminal C*-algebra A. Then they are unitarily equivalent if and only if their kernels are the same.*

Proof. We show only the backward implication because the forward implication is trivial. Suppose then that $\ker(\varphi) = \ker(\varphi')$. Observe that the map

$$\psi \colon \varphi(A) \to \varphi'(A), \quad \varphi(a) \mapsto \varphi'(a),$$

is well-defined and a *-isomorphism. Since A is postliminal, $K(H) \subseteq \varphi(A)$. We show that $\psi(K(H)) \subseteq K(H')$: Let p be a rank-one projection in $B(H)$. Then $pB(H)p = \mathbf{C}p$, so if $q = \psi(p)$, we have $qK(H')q \subseteq \psi(\mathbf{C}p) = \mathbf{C}q$, since $K(H') \subseteq \varphi'(A)$. From this it is easily verified that q is a rank-one projection on H'. Since the rank-one projections in $B(H)$ have closed linear span $K(H)$, we have $\psi(K(H)) \subseteq K(H')$, and the reverse inclusion holds by symmetry, so the restriction $\psi \colon K(H) \to K(H')$ is a *-isomorphism. It follows from Theorem 2.4.8 that there exists a unitary $u \colon H \to H'$ such that $\psi(v) = uvu^*$ for all $v \in K(H)$.

Let $a \in A$ and $w \in K(H')$. Then there exists $v \in K(H)$ such that $w = uvu^*$, and there exists $b \in A$ such that $v = \varphi(b)$. Since $\varphi(a)v \in K(H)$, we have $\psi(\varphi(a))\psi(v) = \psi(\varphi(a)v) = \psi(\varphi(ab)) = u\varphi(ab)u^*$ $= (u\varphi(a)u^*)(u\varphi(b)u^*)$. Hence, $\psi(\varphi(a))w = (u\varphi(a)u^*)w$. Now $K(H')$ is an essential ideal in $B(H')$ (Example 3.1.2), so this argument shows that $\psi(\varphi(a)) = u\varphi(a)u^*$; that is, $\varphi'(a) = u\varphi(a)u^*$. Thus, u implements a unitary equivalence between the representations (H, φ) and (H', φ'). $\square$

5.6.4. Theorem. *If A is a non-zero postliminal C*-algebra, then the canonical map $\hat{A} \to \mathrm{Prim}(A)$ is a homeomorphism.*

Proof. This is immediate from Theorems 5.4.9 and 5.6.3. □

5. Exercises

1. Let τ be a pure state on a C*-algebra A, and y a unit vector in H_τ such that $\tau(a) = \langle \varphi_\tau(a)(y), y \rangle$ for all $a \in A$. Show that there is a scalar λ of modulus one such that $y = \lambda x_\tau$.

2. Let H be a Hilbert space and x a unit vector of H. Show that the functional

$$\omega_x \colon B(H) \to \mathbf{C}, \quad u \mapsto \langle u(x), x \rangle,$$

is a pure state of $B(H)$. Show that not all pure states of $B(H)$ are of this form if H is separable and infinite-dimensional.

3. Give an example to show that a quotient C*-algebra of a primitive C*-algebra need not be primitive.

4. If I is a primitive ideal of a C*-algebra A, show that $M_n(I)$ is a primitive ideal of $M_n(A)$. (Thus, if A is primitive, so is $M_n(A)$.)

5. Let A be a C*-algebra. Show the following conditions are equivalent:
(a) A is prime.
(b) If $aAb = 0$, then a or $b = 0$ $(a, b \in A)$.

6. Let S be a set of C*-subalgebras of a C*-algebra A that is *upwards-directed*, that is, if $B, C \in S$, then there exists $D \in S$ such that $B, C \subseteq D$. Show that $(\cup S)^-$ is a C*-subalgebra of A.

Suppose that all the algebras in S are prime and that $A = (\cup S)^-$. Show that A is prime.

7. If A is a C*-algebra, its *centre* C is the set of elements of A commuting with every element of A. Show that C is a C*-subalgebra of A. Show that if A is simple, then $C = 0$ if A is non-unital and $C = \mathbf{C}1$ if A is unital.

8. Let S be an upwards-directed set of closed ideals in a C*-algebra A (*cf.* Exercise 5.6 for the term *upwards-directed*). Suppose that $A = (\cup S)^-$, and that all of the algebras in S are postliminal. Show that A is postliminal.

9. Let A be a C*-algebra. If I, J are postliminal ideals in A (that is, closed ideals that are postliminal C*-algebras), show that $I + J$ is postliminal also. Deduce from this and Exercise 5.8 that there is a largest postliminal ideal I in A (which may, of course, be the zero ideal). Show that A/I has no non-zero postliminal ideals.

5. Addenda

If G is an ordered group, then the C*-algebra $T(G)$ generated by all generalised Toeplitz operators with symbol in $C(\hat{G})$ is primitive [Mur] (*cf.* Addenda 3).

A representation (H, φ) of a C*-algebra A is said to be *Type* I if the von Neumann algebra $\varphi(A)''$ is Type I (*cf.* Addenda 4). We say A itself is *Type* I (as a C*-algebra) if all its representations are Type I. A C*-algebra is Type I if and only if it is postliminal.

If A is a C*-algebra, a *composition series* for A is a family $(I_\beta)_{\beta \leq \alpha}$ of closed ideals I_β of A indexed by the ordinals β less than or equal to a fixed ordinal α, and such that
(a) $I_0 = 0$, $I_\alpha = A$;
(b) I_γ is contained in I_β if $\gamma < \beta \leq \alpha$;
(c) if β is a limit ordinal, $\beta \leq \alpha$, we have $I_\beta = (\cup_{\gamma < \beta} I_\gamma)^-$.
    The following conditions are equivalent:
(a) A is a postliminal C*-algebra.
(b) A has a composition series $(I_\beta)_{\beta \leq \alpha}$ such that $I_{\beta+1}/I_\beta$ is postliminal for all β.
(c) A has a composition series $(I_\beta)_{\beta \leq \alpha}$ such that $I_{\beta+1}/I_\beta$ is liminal for all β.

A C*-subalgebra of a postliminal C*-algebra is postliminal also.

References: [Dix 2], [Ped].

CHAPTER 6

Direct Limits and Tensor Products

This chapter is concerned with a number of techniques for constructing new C*-algebras from old. In Section 6.1 we introduce direct limits, and in Section 6.2 we use them to exhibit examples of AF-algebras, particularly examples of simple AF-algebras. The AF-algebras form a large class, which is relatively easy to analyse in that it is closely associated with the class of finite-dimensional C*-algebras, but which is nevertheless highly non-trivial. Some of these algebras play an important role in mathematical physics.

The second fundamental construction of this chapter is the tensor product. Again, this is a device for getting new C*-algebras, but is also a powerful tool in the general theory.

Finally, we introduce the nuclear C*-algebras, whose distinguishing feature is that they behave very well with respect to tensor products. A key result is a theorem of Takesaki, which asserts that abelian C*-algebras are nuclear.

6.1. Direct Limits of C*-Algebras

Although our principal aim in this section is to construct direct limits of C*-algebras, we begin with direct limits of groups, because these will be needed in Chapter 7 in connection with K-theory.

If $(G_n)_{n=1}^\infty$ is a sequence of groups, and if for each n we have a homomorphism $\tau_n \colon G_n \to G_{n+1}$, then we call $(G_n, \tau_n)_{n=1}^\infty$ a *direct sequence of groups*. Given such a sequence and positive integers $n \leq m$, we set $\tau_{nn} = \mathrm{id}_{G_n}$ and we define $\tau_{nm} \colon G_n \to G_m$ inductively on m by setting $\tau_{n,m+1} = \tau_m \tau_{nm}$. If $n \leq m \leq k$, we have $\tau_{nk} = \tau_{mk} \tau_{nm}$.

If G' is a group and we have homomorphisms $\rho^n \colon \to G'$ such that the diagram

$$
\begin{array}{ccc}
G_n & \xrightarrow{\ \tau_n\ } & G_{n+1} \\[2pt]
& \rho^n \searrow \quad \downarrow \rho^{n+1} & \\[4pt]
& G' &
\end{array}
$$

commutes for each n, that is, $\rho^n = \rho^{n+1}\tau_n$, then $\rho^n = \rho^m\tau_{nm}$ for all $m \geq n$.

The product $\prod_{k=1}^{\infty} G_k$ is a group with the pointwise-defined operation, and if we let G' be the set of all elements $(x_k)_k$ in $\prod_{k=1}^{\infty} G_k$ such that there exists N for which $x_{k+1} = \tau_k(x_k)$ for all $k \geq N$, then G' is a subgroup of $\prod_{k=1}^{\infty} G_k$. Let e_k be the unit of G_k. The set F of all $(x_k)_k \in \prod_{k=1}^{\infty} G_k$ such that there exists N for which $x_k = e_k$ for all $k \geq N$ is a normal subgroup of G', and we denote the quotient group G'/F by G. We call G the *direct limit* of the sequence $(G_n, \tau_n)_{n=1}^{\infty}$, and where no ambiguity can result we sometimes write $\varinjlim G_n$ for G.

If $x \in G_n$, define $\hat{\tau}^n(x)$ to be the sequence (x_k) where $x_k = e_k$ for $k < n$, and $x_{n+k} = \tau_{n,n+k}(x)$ for all $k \geq 0$. Then $\hat{\tau}^n(x) \in G'$, and the map

$$\tau^n : G_n \to G, \quad x \mapsto \hat{\tau}^n(x)F,$$

is a homomorphism, called the *natural* homomorphism from G_n to G. It is straightforward to check that the diagram

$$
\begin{array}{ccc}
G_n & \xrightarrow{\ \tau_n\ } & G_{n+1} \\
 & {\scriptstyle \tau^n}\searrow & \big\downarrow {\scriptstyle \tau^{n+1}} \\
 & & G
\end{array}
$$

commutes for each n, and that G is the union of the increasing sequence $(\tau^n(G_n))_{n=1}^{\infty}$.

6.1.1. Theorem. *Let G be the direct limit of the direct sequence of groups $(G_n, \tau_n)_{n=1}^{\infty}$, and let $\tau^n : G_n \to G$ be the natural map for each n.*

(1) *If $x \in G_n$ and $y \in G_m$ and $\tau^n(x) = \tau^m(y)$, then there exists $k \geq n, m$ such that $\tau_{nk}(x) = \tau_{mk}(y)$.*

(2) *If G' is a group, and for each n there is a homomorphism $\rho^n : G_n \to G'$ such that the diagram*

$$
\begin{array}{ccc}
G_n & \xrightarrow{\ \tau_n\ } & G_{n+1} \\
 & {\scriptstyle \rho^n}\searrow & \big\downarrow {\scriptstyle \rho^{n+1}} \\
 & & G'
\end{array}
$$

commutes, then there is a unique homomorphism $\rho : G \to G'$ such that for each n the diagram

$$
\begin{array}{ccc}
G_n & \xrightarrow{\ \tau^n\ } & G \\
 & {\scriptstyle \rho^n}\searrow & \big\downarrow {\scriptstyle \rho} \\
 & & G'
\end{array}
$$

commutes.

Proof. Condition (1) follows directly from the definitions.

Assume we have G' and ρ^n as in Condition (2). If $x \in G_n$ and $y \in G_m$ and $\tau^n(x) = \tau^m(y)$, then by Condition (1) there exists $k \geq n, m$ such that $\tau_{nk}(x) = \tau_{mk}(y)$. Hence, $\rho^n(x) = \rho^k \tau_{nk}(x) = \rho^k \tau_{mk}(y) = \rho^m(y)$. Thus, we can well-define a map $\rho \colon G \to G'$ by setting $\rho(\tau^n(x)) = \rho^n(x)$. It is easily checked that ρ is a homomorphism, and by definition $\rho \tau^n = \rho^n$ for all n. Uniqueness of ρ is clear. $\square$

6.1.1. Remark. From Condition (1) of the preceding theorem, if $\tau^n(x) = e$, then there exists $k \geq n$ such that $\tau_{nk}(x) = e$, a result we shall be using frequently. (The symbol e denotes the unit of the relevant group.)

Let A be a *-algebra. A C*-*seminorm* on A is a seminorm p on A such that for all $a, b \in A$ we have $p(ab) \leq p(a)p(b)$, $p(a^*) = p(a)$ and $p(a^*a) = p(a)^2$. If in addition p is in fact a norm, we call p a C*-*norm*.

If $\varphi \colon A \to B$ is a *-homomorphism from a *-algebra into a C*-algebra, then the function

$$p \colon A \to \mathbf{R}^+, \quad a \mapsto \|\varphi(a)\|,$$

is a C*-seminorm on A, and if φ is injective, p is a C*-norm.

If p is any C*-seminorm on a *-algebra A, the set $N = p^{-1}\{0\}$ is a self-adjoint ideal of A, and we get a C*-norm on the quotient *-algebra A/N by setting $\|a + N\| = p(a)$. If B denotes the Banach space completion of A/N with this norm, it is easily checked that the multiplication and involution operations extend uniquely to operations of the same type on B so as to make B a C*-algebra. We call B the *enveloping* C*-algebra of the pair (A, p), and the map

$$i \colon A \to B, \quad a \mapsto a + N,$$

the *canonical* map from A to B. Of course, $i(A)$ is a dense *-subalgebra of B.

If p is a C*-norm, we refer to B more simply as the C*-*completion* of A. In this case A is a dense *-subalgebra of B.

Let $(A_n)_{n=1}^\infty$ be a sequence of C*-algebras and suppose that for each n we have a *-homomorphism $\varphi_n \colon A_n \to A_{n+1}$. Then we call $(A_n, \varphi_n)_{n=1}^\infty$ a *direct sequence of C*-algebras*. The product $\prod_{k=1}^\infty A_k$ is a *-algebra with the pointwise-defined operations, and if A' denotes the set of all elements $a = (a_k)_k$ of $\prod_{k=1}^\infty A_k$ such that there is an integer N for which $a_{k+1} = \varphi_k(a_k)$ for all $k \geq N$, then A' is a *-subalgebra of $\prod_{k=1}^\infty A_k$. Note that $\|a_{k+1}\| \leq \|a_k\|$ if $k \geq N$ (since φ_k is norm-decreasing), so the sequence $(\|a_k\|)_k$ is eventually decreasing (and of course bounded below). It therefore converges, and we set $p(a) = \lim_{k \to \infty} \|a_k\|$. It is straightforward to verify that

$$p \colon A' \to \mathbf{R}^+, \quad a \mapsto p(a),$$

is a C*-seminorm on A'. We denote the enveloping C*-algebra of (A',p) by A, and call it the *direct limit* of the sequence $(A_n,\varphi_n)_{n=1}^{\infty}$. If no ambiguity results, we sometimes write $\varinjlim A_n$ for A.

Similar to the group case, if $a \in A_n$, we define $\hat{\varphi}^n(a)$ in A' to be the sequence $(a_k)_k$ such that $a_1,\ldots,a_{n-1}$ are zero and $a_{n+k} = \varphi_{n,n+k}(a)$ for all $k \geq 0$. If $i: A' \to A$ is the canonical map, then the map

$$\varphi^n: A_n \to A, \quad a \mapsto i(\hat{\varphi}^n(a)),$$

is a $*$-homomorphism, called the *natural* map from A_n to A. A routine argument shows that for all n the diagram

$$
\begin{array}{ccc}
A_n & \xrightarrow{\varphi_n} & A_{n+1} \\
& {\scriptstyle \varphi^n}\searrow \quad \downarrow {\scriptstyle \varphi^{n+1}} & \\
& A &
\end{array}
$$

commutes, and that the union of the increasing sequence of C*-subalgebras $(\varphi^n(A_n))_n$ is a dense $*$-subalgebra of A. Also,

$$\|\varphi^n(a)\| = \lim_{k\to\infty} \|\varphi_{n,n+k}(a)\| \tag{1}$$

if $a \in A_n$.

6.1.2. Theorem. *Let A be the direct limit of the direct sequence of C*-algebras $(A_n,\varphi_n)_{n=1}^{\infty}$, and suppose that $\varphi^n: A_n \to A$ is the natural map for each n.*

(1) *If $a \in A_n$, $b \in A_m$, $\varepsilon > 0$ and $\varphi^n(a) = \varphi^m(b)$, then there exists $k \geq n,m$ such that $\|\varphi_{nk}(a) - \varphi_{mk}(b)\| < \varepsilon$.*

(2) *If B is a C*-algebra and there is a $*$-homomorphism $\psi^n: A_n \to B$ for each n such that the diagram*

$$
\begin{array}{ccc}
A_n & \xrightarrow{\varphi_n} & A_{n+1} \\
& {\scriptstyle \psi^n}\searrow \quad \downarrow {\scriptstyle \psi^{n+1}} & \\
& B &
\end{array}
$$

commutes, then there is a unique $*$-homomorphism $\psi: A \to B$ such that for each n the diagram

$$
\begin{array}{ccc}
A_n & \xrightarrow{\varphi^n} & A \\
& {\scriptstyle \psi^n}\searrow \quad \downarrow {\scriptstyle \psi} & \\
& B &
\end{array}
$$

commutes.

Proof. Condition (1) follows from Eq. (1) above.

Suppose that B and ψ^n are as in Condition (2). Let $a \in A_n$ and $b \in A_m$, and suppose that $\varphi^n(a) = \varphi^m(b)$. If $\varepsilon > 0$, then by Condition (1) there exists $k \geq n, m$ such that $\|\varphi_{nk}(a) - \varphi_{mk}(b)\| < \varepsilon$. Consequently, $\|\psi^n(a) - \psi^m(b)\| = \|\psi^k(\varphi_{nk}(a) - \varphi_{mk}(b))\| \leq \|\varphi_{nk}(a) - \varphi_{mk}(b)\| < \varepsilon$. Letting $\varepsilon \to 0$, we therefore have $\psi^n(a) = \psi^m(b)$. This shows that we can well-define a map ψ from $C = \cup_{n=1}^{\infty} \varphi^n(A_n)$ to B by setting $\psi(\varphi^n(a)) = \psi^n(a)$. If k is any integer, then $\|\psi^n(a)\| = \|\psi^{n+k}\varphi_{n,n+k}(a)\| \leq \|\varphi_{n,n+k}(a)\|$, and therefore $\|\psi(\varphi^n(a))\| = \|\psi^n(a)\| \leq \lim_{k\to\infty} \|\varphi_{n,n+k}(a)\| = \|\varphi^n(a)\|$. Thus, ψ is norm-decreasing, and it is easily seen to be a *-homomorphism. Since C is a dense *-subalgebra of A, we can extend ψ to a *-homomorphism $\psi: A \to B$, and $\psi\varphi^n = \psi^n$ for all n. Uniqueness of ψ follows from density of C in A.   □

6.1.2. Remark. Retaining the notation of the preceding theorem, if $a \in A_n$ and $\varphi^n(a) = 0$ and $\varepsilon > 0$, then by Condition (1) there exists $k \geq n$ such that $\|\varphi_{nk}(a)\| < \varepsilon$.

6.1.3. Remark. Let A be a C*-algebra and let $(A_n)_{n=1}^{\infty}$ be an increasing sequence of C*-subalgebras of A whose union is dense in A. Let $\varphi_n: A_n \to A_{n+1}$ be the inclusion map. A straightforward application of Theorem 6.1.2, Condition (2), shows that A is (*-isomorphic to) the direct limit of the direct sequence $(A_n, \varphi_n)_{n=1}^{\infty}$.

6.1.3. Theorem. *Let S be a non-empty set of simple C*-subalgebras of a C*-algebra A. Suppose that S is upwards-directed (that is, if $B, C \in S$, then there exists $D \in S$ such that D contains B and C), and $\cup S$ is dense in A. Then A is simple also.*

Proof. To show that A is simple it suffices to show that if $\pi: A \to B$ is a surjective *-homomorphism onto a non-zero C*-algebra B, then it is injective. If $C \in S$, then the restriction of π to C is either zero, or it is injective, and therefore isometric. Since π is not the zero map on $\cup S$, it follows easily from the upwards-directed property of S that π is not the zero map on any non-zero $C \in S$. Hence, π is isometric on $\cup S$, and therefore, by continuity, π is isometric on A.   □

6.1.4. Theorem. *Suppose that $(A_n, \varphi_n)_{n=1}^{\infty}$ is a direct sequence of simple C*-algebras. Then the direct limit $\varinjlim A_n$ is simple, also.*

Proof. Let $\varphi^n: A_n \to A$ be the natural map, where $A = \varinjlim A_n$. Then the set $S = \{\varphi^n(A_n) \mid n \geq 1\}$ is an upwards-directed family of simple C*-subalgebras of A whose union is dense in A, so by Theorem 6.1.3 A is simple.   □

6.2. Uniformly Hyperfinite Algebras

The C*-algebras of the title form an interesting class, since they are highly non-trivial yet accessible to detailed analysis. Before introducing them, however, we shall need to consider some preliminary material.

We begin by characterising the finite-dimensional simple C*-algebras, since uniformly hyperfinite algebras are defined in terms of these algebras.

6.2.1. Remark. A non-zero finite-dimensional C*-algebra is simple if and only if it is of the form $M_n(\mathbf{C})$ for some n. To see this, suppose that A is a non-zero simple finite-dimensional C*-algebra. By Example 5.6.2 A is liminal. Hence, if (H, φ) is any non-zero irreducible representation of A, then $\varphi(A) = K(H)$ and H is therefore finite-dimensional. Moreover, since $\ker(\varphi)$ is a proper closed ideal of A, it is the zero ideal. Therefore, if $n = \dim(H)$, then A is *-isomorphic to $K(H) = B(H)$, which in turn is *-isomorphic to $M_n(\mathbf{C})$.

We shall need the following lemmas in this section and also at various points in the sequel.

6.2.1. Lemma. *Let p, q be projections in a unital C*-algebra A and suppose that $\|q - p\| < 1$. Then there exists a unitary u in A such that $q = upu^*$ and $\|1 - u\| \leq \sqrt{2}\|q - p\|$, namely, $u = v|v|^{-1}$, where $v = 1 - p - q + 2qp$.*

Proof. If $v = 1 - p - q + 2qp$, then $v^*v = 1 - (q - p)^2$ by computation. Because $v^* = 1 - p - q + 2pq$, we have also $vv^* = 1 - (p - q)^2$, so $vv^* = v^*v$; that is, v is normal. Now $\|q - p\| < 1$ implies that $\|(q - p)^2\| = \|q - p\|^2 < 1$, so $1 - (q - p)^2 = v^*v$ is invertible. Hence, v is invertible by normality. Consequently, $u = v|v|^{-1}$ is a unitary. Now $vp = qp = qv$, so $pv^* = v^*q$, and $pv^*v = v^*qv = v^*vp$. It follows that p commutes with $|v|$ and therefore with $|v|^{-1}$. Hence, $up = qu$; that is, $q = upu^*$.

Since $\mathrm{Re}(v) = 1 - (q - p)^2 = |v|^2$, we have $\mathrm{Re}(u) = \mathrm{Re}(v)|v|^{-1} = |v|$. Therefore, $\|1 - u\|^2 = \|(1 - u^*)(1 - u)\| = 2\|1 - \mathrm{Re}(u)\| = 2\|1 - |v|\| \leq 2\|1 - |v|^2\|$ (because $1 - t \leq 1 - t^2$ for all $t \in [0, 1]$). Since $1 - |v|^2 = (q - p)^2$, therefore $\|1 - u\|^2 \leq 2\|q - p\|^2$, so $\|1 - u\| \leq \sqrt{2}\|q - p\|$. $\square$

If Ω is a locally compact Hausdorff space and $f \in C_0(\Omega)$, extend f to a continuous function $\tilde{f}$ on the one-point compactification $\tilde{\Omega}$ of Ω by setting $\tilde{f}(\infty) = 0$, where ∞ is the "point at infinity." If $\delta > |f(\omega)|$ for all $\omega \in \Omega$, then $\delta > \|f\|_\infty$, since $\|f\|_\infty = \|\tilde{f}\|_\infty$, and $\delta > \|\tilde{f}\|_\infty$ because the continuous function $\omega \mapsto |\tilde{f}(\omega)|$ attains its upper bound on the compact space $\tilde{\Omega}$.

6.2.2. Lemma. *Let a be a self-adjoint element of a C*-algebra A such that $\|a - a^2\| < 1/4$. Then there is a projection $p \in A$ such that $\|a - p\| < 1/2$.*

Proof. We may suppose that A is abelian and may therefore suppose that $A = C_0(\Omega)$ for some locally compact Hausdorff space Ω. The hypothesis implies that $1/2$ is not in the range of $|a|$, and therefore the set $S = |a|^{-1}(1/2, \infty)$ is open and compact in Ω (it is compact since it is equal to $\{\omega \in \Omega \mid |a(\omega)| \geq 1/2\}$). Hence, $p = \chi_S$ is a projection in A. Since $|a(\omega) - \chi_S(\omega)| < 1/2$ for all $\omega \in \Omega$, therefore by the observation preceding this lemma, $\|a - p\|_\infty < 1/2$. $\qquad\square$

A positive linear functional on a C*-algebra A is *tracial* if $\tau(a^*a) = \tau(aa^*)$ for all $a \in A$. Equivalently, $\tau(ab) = \tau(ba)$ for all $a, b \in A$. To see the equivalence, let τ be tracial and let b, c be self-adjoint elements of A. Then if $a = b + ic$, we have $a^*a = b^2 + c^2 + i(bc - cb)$ and $aa^* = b^2 + c^2 + i(cb - bc)$. Since $\tau(a^*a) = \tau(aa^*)$, we get $\tau(bc - cb) = \tau(cb - bc) = -\tau(bc - cb)$, so $\tau(bc - cb) = 0$. Thus, $\tau(bc) = \tau(cb)$ for all $b, c \in A_{sa}$ and, therefore, for all $b, c \in A$.

A tracial positive linear functional τ is *faithful* if $\tau(a^*a) = 0$ implies that $a = 0$.

6.2.1. Example. The function

$$\mathrm{tr} \colon M_n(\mathbf{C}) \to \mathbf{C}, \quad (\lambda_{ij}) \mapsto \tfrac{1}{n} \sum_{i=1}^{n} \lambda_{ii},$$

is a faithful tracial state on $M_n(\mathbf{C})$. In fact, this is the only tracial state on $M_n(\mathbf{C})$. To show this, we need only show that all tracial states take the same value on the rank-one projections, since these span $M_n(\mathbf{C})$. But this will follow easily if we show that all rank-one projections are unitarily equivalent. Supposing then that p and q are rank-one projections, there exist unit vectors e and f such that $p = e \otimes e$ and $q = f \otimes f$. Since there exists a unitary $u \in M_n(\mathbf{C})$ such that $u(e) = f$, we have $q = u(e) \otimes u(e) = u(e \otimes e)u^* = upu^*$.

6.2.2. Remark. If H is an infinite-dimensional Hilbert space, then $K(H)$ does not admit a tracial state. Suppose the contrary, and let τ be a tracial state on $K(H)$. Observe that all the rank-one projections on H are unitarily equivalent (same proof as in Example 6.2.1), and therefore τ takes on the same (positive) value, t say, on all rank-one projections. Now let E be an orthonormal basis for H. If $e_1, \ldots, e_n \in E$ and p is the projection $\sum_{i=1}^{n} e_i \otimes e_i$, then $\tau(p) \leq 1$ and $\tau(p) = nt$, so $n \leq 1/t$. Thus, $1/t$ is an upper bound for the integers, an absurdity. This shows that $K(H)$ has no tracial state, as claimed.

6.2.3. Remark. If τ is a tracial positive linear functional of a C*-algebra A, then N_τ is an ideal of A. For, we know that N_τ is a left ideal, because

τ is a positive functional, and the tracial condition implies that $N_\tau = N_\tau^*$, from which N_τ is an ideal as claimed.

If A is simple, then any non-zero tracial positive linear functional τ on A is faithful, since in this case the proper closed ideal N_τ of A is the zero ideal.

6.2.4. Remark. Let $(A_n)_{n=1}^\infty$ be an increasing sequence of C*-subalgebras of a C*-algebra A such that $A = (\cup_{n=1}^\infty A_n)^-$. Suppose that A is unital and that all the algebras A_n contain the unit of A. Then if each algebra A_n admits a unique tracial state, τ_n say, A also admits a unique tracial state. We show this: The restriction of τ_{n+1} to A_n is a tracial state ($\tau_{n+1}(1) = 1$, so the restriction has norm one). Therefore, by the uniqueness assumption, τ_{n+1} is equal to τ_n on A_n. Define τ on the *-subalgebra $\cup_{n=1}^\infty A_n$ of A by setting $\tau(a) = \tau_n(a)$ if $a \in A_n$. It is easily checked that this gives a norm-decreasing linear function from $\cup_n A_n$ to $\mathbf{C}$, so we can extend to get a bounded linear functional $\tau: A \to \mathbf{C}$. It is clear that τ is a tracial state on A; that it is the unique such state follows from the uniqueness assumption on the algebras A_n.

A *uniformly hyperfinite algebra* or *UHF algebra* is a unital C*-algebra A which has an increasing sequence $(A_n)_{n=1}^\infty$ of finite-dimensional simple C*-subalgebras each containing the unit of A such that $\cup_{n=1}^\infty A_n$ is dense in A.

Since A_n is simple and finite-dimensional, it is *-isomorphic to some $M_k(\mathbf{C})$, so it admits a unique tracial state. It follows from Remark 6.2.4 that A has a unique tracial state, τ say. By Theorem 6.1.3 A is simple.

Let n, d be positive integers. We call the unital *-homomorphism

$$\varphi: M_n(\mathbf{C}) \to M_{dn}(\mathbf{C}), \quad a \mapsto \begin{pmatrix} a & & 0 \\ & \ddots & \\ 0 & & a \end{pmatrix},$$

the *canonical* map from $M_n(\mathbf{C})$ to $M_{dn}(\mathbf{C})$ ($\varphi(a)$ has d blocks of a down the main diagonal, and everywhere else it is zero).

Denote by $\mathbf{S}$ the set of all functions $s: \mathbf{N} \setminus \{0\} \to \mathbf{N} \setminus \{0\}$. If $s \in \mathbf{S}$, define $s! \in \mathbf{S}$ by setting

$$s!(n) = s(1)s(2)\ldots s(n) \qquad (n \geq 1).$$

Let $\varphi_n: M_{s!(n)}(\mathbf{C}) \to M_{s!(n+1)}(\mathbf{C})$ be the canonical map (obviously, $s!(n)$ divides $s!(n+1)$). We denote by M_s the direct limit of the direct sequence $(M_{s!(n)}(\mathbf{C}), \varphi_n)_{n=1}^\infty$. Since the C*-algebras $M_{s!(n)}(\mathbf{C})$ are finite-dimensional simple C*-algebras, it is clear that M_s is a UHF algebra.

Define a function ε_s from the set of prime numbers to $\mathbf{N} \cup \{+\infty\}$ by setting, for each prime r,

$$\varepsilon_s(r) = \sup\{m \in \mathbf{N} \mid r^m \text{ divides some } s!(n)\}.$$

6.2.3. Theorem. Let $s, s' \in S$ and suppose that $M_s, M_{s'}$ are *-isomorphic. Then $\varepsilon_s = \varepsilon_{s'}$.

Proof. Let $\pi: M_s \to M_{s'}$ be a *-isomorphism, let τ and τ' be the unique tracial states of M_s and $M_{s'}$, respectively, and let $\varphi^n: M_{s!(n)}(\mathbf{C}) \to M_s$ and $\psi^n: M_{s'!(n)}(\mathbf{C}) \to M_{s'}$ be the natural *-homomorphisms. Clearly, $\tau'\pi$ is a tracial state on M_s, so by uniqueness of the tracial state, $\tau = \tau'\pi$.

To prove the theorem, it suffices to show that $\varepsilon_s \leq \varepsilon_{s'}$, since the reverse inequality will then follow by symmetry. Therefore, it is enough to show that for each positive integer n there is a positive integer m such that $s!(n)$ divides $s'!(m)$. (For then if r is a prime and k a positive integer such that r^k divides $s!(n)$, then r^k divides $s'!(m)$, and this shows that $\varepsilon_s(r) \leq \varepsilon_{s'}(r)$.)

Suppose then that n is a positive integer. Let p be a rank-one projection in $M_{s!(n)}(\mathbf{C})$. Since $\tau\varphi^n$ is the unique tracial state on $M_{s!(n)}(\mathbf{C})$, we have $\tau(\varphi^n(p)) = 1/s!(n)$. Now $\pi(\varphi^n(p))$ is a projection in $M_{s'}$, so there is a positive integer m and a self-adjoint element $a \in M_{s'!(m)}(\mathbf{C})$ such that

$$\|\pi(\varphi^n(p)) - \psi^m(a)\| < 1/8 \qquad \text{and} \qquad \|\pi(\varphi^n(p)) - \psi^m(a^2)\| < 1/8$$

(this uses the density of $\cup_{k=1}^{\infty} \psi^k(M_{s'!(k)}(\mathbf{C}))$ in $M_{s'}$). Hence,

$$
\begin{aligned}
\|a - a^2\| &= \|\psi^m(a) - \psi^m(a^2)\| \\
&\leq \|\psi^m(a) - \pi(\varphi^n(p))\| + \|\pi(\varphi^n(p)) - \psi^m(a^2)\| \\
&< 1/4.
\end{aligned}
$$

It follows from Lemma 6.2.2 that there is a projection q in $M_{s'!(m)}(\mathbf{C})$ such that $\|a - q\| < 1/2$. Hence,

$$
\begin{aligned}
\|\pi(\varphi^n(p)) - \psi^m(q)\| &\leq \|\pi(\varphi^n(p)) - \psi^m(a)\| + \|\psi^m(a) - \psi^m(q)\| \\
&< 1/8 + 1/2 < 1.
\end{aligned}
$$

Applying Lemma 6.2.1, the projections $\pi(\varphi^n(p))$ and $\psi^m(q)$ are unitarily equivalent in $M_{s'}$, and therefore $\tau'(\psi^m(q)) = \tau'(\pi(\varphi^n(p))) = \tau(\varphi^n(p)) = 1/s!(n)$. But $\tau'\psi^m$ is the unique tracial state on $M_{s'!(m)}(\mathbf{C})$, so $\tau'\psi^m(q)$ must be of the form $d/s'!(m)$ for some positive integer d. Therefore, $s'!(m) = ds!(n)$, so $s!(n)$ divides $s'!(m)$. $\square$

6.2.4. Corollary. There exists an uncountable number of UHF algebras that are not *-isomorphic.

Proof. Let (r_n) be the sequence of prime numbers. If $s \in \mathbf{S}$, let $\bar{s} \in \mathbf{S}$ be defined by $\bar{s}(n) = r_n^{s_n}$. Then $\varepsilon_{\bar{s}}(r_n) = s_n$. It follows that if $s, s' \in \mathbf{S}$ and $\varepsilon_{\bar{s}} = \varepsilon_{\bar{s}'}$ then $s = s'$. Since the set $\mathbf{S}$ is uncountable, therefore by Theorem 6.2.3, $(M_{\bar{s}})_{s \in \mathbf{S}}$ is an uncountable family of UHF algebras that are not *-isomorphic. $\qquad\square$

Now we present an application of these algebras to the theory of von Neumann algebras.

A *factor* on the Hilbert space H is a von Neumann algebra A on H such that $A \cap A' = \mathbf{C} \operatorname{id}_H$ (*cf.* Addenda 4). If H is a Hilbert space, then $B(H)$ is a factor. It is harder to give other examples (although examples exist in abundance). We shall presently exhibit an example of an infinite-dimensional factor not *-isomorphic to any $B(H)$.

A von Neumann algebra on a Hilbert space H is *hyperfinite* if it has a weakly dense C*-subalgebra that is a UHF algebra and whose unit is id_H.

6.2.2. Example. If H is a separable Hilbert space, then $B(H)$ is hyperfinite. This is clear if H is finite-dimensional. To prove it for the infinite-dimensional case, let A be an infinite-dimensional UHF algebra, and let (H, φ) be a non-zero irreducible representation of A. Since A is simple, φ is a *-isomorphism of A onto $\varphi(A)$, so $\varphi(A)$ is a UHF algebra. Now if x is any non-zero vector of H, then it is cyclic for (H, φ) (by irreducibility), so $H = [\varphi(A)x]$, and since A is separable, this shows that H is separable. Clearly, H is infinite-dimensional since A is. Because (H, φ) is irreducible, $\varphi(A)' = \mathbf{C}$, so $\varphi(A)'' = B(H)$. Therefore, $\varphi(A)$ is a UHF subalgebra of $B(H)$ containing the identity map id_H and is weakly dense, so $B(H)$ is a hyperfinite algebra.

6.2.5. Theorem. *Let A be a UHF algebra, and let τ be its unique tracial state. Then the von Neumann algebra $\varphi_\tau(A)''$ is a hyperfinite factor admitting a faithful tracial state.*

Proof. Let $B = \varphi_\tau(A)$. Since A is unital and the representation (H_τ, φ_τ) is non-degenerate (it is cyclic), we have $\varphi(1) = \operatorname{id}_{H_\tau}$. The representation (H_τ, φ_τ) is faithful because A is simple, so B is *-isomorphic to A and is therefore a UHF algebra. Hence, the von Neumann algebra B'' is hyperfinite.

The tracial condition gives the equation

$$\langle uu'(x_\tau), x_\tau \rangle = \langle u'u(x_\tau), x_\tau \rangle$$

for all $u, u' \in B$, and weak density of B in B'' implies that this equation holds for all $u, u' \in B''$ also. Hence, the function

$$\omega: B'' \to \mathbf{C}, \quad u \mapsto \langle u(x_\tau), x_\tau \rangle,$$

is a tracial state on B''.

To show that ω is faithful, we show that the vector x_τ is separating for B''. Suppose that $u \in B''$ and $u(x_\tau) = 0$. If $v \in B$, then $\|uv(x_\tau)\|^2 = \omega(v^*u^*uv) = \omega(vv^*u^*u)$ by the tracial condition, so $\|uv(x_\tau)\|^2 = \langle vv^*u^*u(x_\tau), x_\tau \rangle = 0$. Hence, $u[Bx_\tau] = 0$, and since x_τ is cyclic for B, this shows that $u = 0$. Thus, x_τ is a separating vector for B'', as claimed.

Now let p be a projection in $B' \cap B''$. The function

$$\omega': B'' \to \mathbf{C}, \quad u \mapsto \omega(pu),$$

is a weakly continuous tracial positive linear functional on B''. Hence, when we restrict it to the UHF algebra B, it is a constant t times the unique tracial state ω_B of B; that is, $\omega'(v) = t\omega(v)$ for all $v \in B$. Therefore, by weak density of B in B'' and weak continuity of ω' and ω, we have $\omega' = t\omega$. Hence, $t = \omega(p)$, so $0 = \omega'(1 - p) = \omega(p)\omega(1 - p)$. Since p and $1 - p$ are positive and ω is faithful, this implies that either p or $1 - p$ is zero. We have therefore shown that the only projections in the von Neumann algebra $B' \cap B''$ are the trivial ones. Since a von Neumann algebra is the closed linear span of its projections, this shows that $B' \cap B'' = \mathbf{C}$, and therefore B'' is a factor. $\square$

If we suppose in the preceding theorem that A is an infinite-dimensional UHF algebra, then the von Neumann algebra $\varphi_\tau(A)''$ is a factor that is not $*$-isomorphic to $B(H)$ for any Hilbert space H. (By Remark 6.2.2 $B(H)$ has no faithful tracial state if H is infinite-dimensional.)

A more general class than the UHF algebras, but which is similarly defined in terms of finite-dimensional algebras, is the class of AF-algebras.

An *AF-algebra* is a C*-algebra that contains an increasing sequence $(A_n)_{n=1}^\infty$ of finite-dimensional C*-subalgebras such that $\cup_{n=1}^\infty A_n$ is dense in A.

6.2.3. Example. If H is a separable Hilbert space, then $K(H)$ is an AF-algebra. To show this, we may suppose that H is infinite-dimensional. Let $(e_n)_{n=1}^\infty$ be an orthonormal basis for H, and let p_n be the projection $\sum_{i=1}^n e_i \otimes e_i$. The sequence (p_n) is an approximate unit for $K(H)$ (cf. Example 3.1.1), so $K(H) = (\cup_{n=1}^\infty p_n K(H) p_n)^-$. If $u \in K(H)$, then $p_n u p_n = \sum_{i,j=1}^n (e_i \otimes e_i) u (e_j \otimes e_j) = \sum_{i,j=1}^n \langle u(e_j), e_i \rangle e_i \otimes e_j$, so the C*-algebra $p_n K(H) p_n$ is finite-dimensional. This shows that $K(H)$ is an AF-algebra.

6.2.4. Example. If A is a direct limit of a direct sequence $(A_n, \varphi_n)_{n=1}^\infty$ of C*-algebras, where the A_n are finite-dimensional, then A is an AF-algebra. For in this case the sequence of algebras $(\varphi^n(A_n))_n$ is increasing, its union is dense in A, and $\varphi^n(A_n)$ is finite-dimensional for each n.

The "converse" is also true. More precisely, if A is an AF-algebra, then it is $*$-isomorphic to a direct limit of finite-dimensional C*-algebras (by Remark 6.1.3).

A finite-dimensional C*-algebra is the linear span of its projections, as we observed in Section 2.4, so an AF-algebra is the closed linear span of its projections. A consequence is that an abelian AF-algebra has totally disconnected spectrum.

6.2.6. Theorem. *If I is a closed ideal in an AF-algebra A, then I and A/I are AF-algebras.*

Proof. Suppose $(A_n)_{n=1}^{\infty}$ is an increasing sequence of finite-dimensional C*-subalgebras of A whose union is dense in A, and let $\pi \colon A \to A/I$ be the quotient $*$-homomorphism. Then $(\pi(A_n))_n$ is an increasing sequence of finite-dimensional C*-subalgebras of A/I and the union of these algebras is dense in A/I, so A/I is an AF-algebra.

Set $I_n = I \cap A_n$, and let $J = (\cup_n I_n)^-$. Then J is a closed ideal of A contained in I, and since $(I_n)_n$ is an increasing sequence of finite-dimensional C*-subalgebras of I, we shall have shown that I is an AF-algebra if we show that $I = J$. To prove this consider the well-defined $*$-homomorphism

$$\varphi \colon A/J \to A/I, \quad a + J \mapsto a + I.$$

We shall prove that $I = J$ by showing that φ is isometric, and to see this it suffices to show that φ is isometric on the C*-subalgebras $(A_n + J)/J$, since these form an increasing sequence whose union is a dense $*$-subalgebra of A/J. Denote by $\psi \colon (A_n + J)/J \to A_n/(A_n \cap J)$ and $\theta \colon (A_n + I)/I \to A_n/(A_n \cap I)$ the canonical $*$-isomorphisms (*cf.* Remark 3.1.3), and by $i \colon (A_n + I)/I \to A/I$ the inclusion. Since $A_n \cap I = A_n \cap J$, and since the restriction of φ to $(A_n + J)/J$ is the composition $i\theta^{-1}\psi$ of isometric maps, φ is isometric on $(A_n + J)/J$. $\qquad\square$

It is evident from Theorem 6.2.6 that a C*-algebra A is an AF-algebra if and only if $\tilde{A}$ is an AF-algebra.

We shall have more to say about AF-algebras in later sections.

6.3. Tensor Products of C*-Algebras

If H and K are vector spaces, we denote by $H \otimes K$ their algebraic tensor product. This is linearly spanned by the elements $x \otimes y$ ($x \in H$, $y \in K$). (There is a conflict between the tensor notation and our use of $x \otimes y$ to denote a rank-one operator on a Hilbert space, but the context will always resolve the ambiguity.)

One reason why tensor products are useful is that they turn bilinear maps into linear maps. More precisely, if $\varphi \colon H \times K \to L$ is a bilinear map, where H, K and L are vector spaces, then there is a unique linear map $\varphi' \colon H \otimes K \to L$ such that $\varphi'(x \otimes y) = \varphi(x, y)$ for all $x \in H$ and $y \in K$.

If τ, ρ are linear functionals on the vector spaces H, K, respectively, then there is a unique linear functional $\tau \otimes \rho$ on $H \otimes K$ such that

$$(\tau \otimes \rho)(x \otimes y) = \tau(x)\rho(y) \qquad (x \in H, \ y \in K),$$

since the function

$$H \times K \to \mathbf{C}, \ \ (x, y) \mapsto \tau(x)\rho(y),$$

is bilinear.

Suppose that $\sum_{j=1}^{n} x_j \otimes y_j = 0$, where $x_j \in H$ and $y_j \in K$. If $y_1, \ldots, y_n$ are linearly independent, then $x_1 = \cdots = x_n = 0$. For, in this case, there exist linear functionals $\rho_j \colon K \to \mathbf{C}$ such that $\rho_j(y_i) = \delta_{ij}$ (Kronecker delta). If $\tau \colon H \to \mathbf{C}$ is linear, we have $0 = (\tau \otimes \rho_j)(\sum_{i=1}^{n} x_i \otimes y_i) = \sum_{i=1}^{n} \tau(x_i)\rho_j(y_i) = \sum_{i=1}^{n} \tau(x_i)\delta_{ij} = \tau(x_j)$. Thus, $\tau(x_j) = 0$ for arbitrary τ, and this shows that $x_1 = \cdots = x_n = 0$.

Similarly, if $\sum_{j=1}^{n} x_j \otimes y_j = 0$ and $x_1, \ldots, x_n$ are linearly independent, then $y_1 = \cdots = y_n = 0$.

If $u \colon H \to H'$ and $v \colon K \to K'$ are linear maps between vector spaces, then by elementary linear algebra there exists a unique linear map

$$u \otimes v \colon H \otimes K \to H' \otimes K'$$

such that $(u \otimes v)(x \otimes y) = u(x) \otimes v(y)$ for all $x \in H$ and all $y \in K$.

The map $(u, v) \mapsto u \otimes v$ is bilinear.

If H and K are normed, then there are in general many possible norms on $H \otimes K$ which are related in a suitable manner to the norms on H and K, and indeed it is this very lack of uniqueness that creates the difficulties of the theory, as we shall see in the case that H and K are C*-algebras.

When the spaces are Hilbert spaces, however, matters are simple.

6.3.1. Theorem. *Let H and K be Hilbert spaces. Then there is a unique inner product $\langle \cdot, \cdot \rangle$ on $H \otimes K$ such that*

$$\langle x \otimes y, x' \otimes y' \rangle = \langle x, x' \rangle \langle y, y' \rangle \qquad (x, x' \in H, \ y, y' \in K).$$

Proof. If τ and ρ are conjugate-linear maps from H and K, respectively, to $\mathbf{C}$, then there is a unique conjugate-linear map $\tau \otimes \rho$ from $H \otimes K$ to $\mathbf{C}$ such that $(\tau \otimes \rho)(x \otimes y) = \tau(x)\rho(y)$ for $x \in H$ and $y \in K$. (Observe that $\bar{\tau}$ and $\bar{\rho}$ are linear, and set $\tau \otimes \rho = (\bar{\tau} \otimes \bar{\rho})^-$.)

If x is an element of a Hilbert space, let τ_x be the conjugate-linear functional defined by setting $\tau_x(y) = \langle x, y \rangle$.

Let X be the vector space of all conjugate-linear functionals on $H \otimes K$. The map

$$H \times K \to X, \quad (x, y) \mapsto \tau_x \otimes \tau_y,$$

is bilinear, so there is a unique linear map $M: H \otimes K \to X$ such that $M(x \otimes y) = \tau_x \otimes \tau_y$ for all x and y. The map

$$\langle \cdot, \cdot \rangle : (H \otimes K)^2 \to \mathbf{C}, \quad (z, z') \mapsto M(z)(z'),$$

is a sesquilinear form on $H \otimes K$ such that

$$\langle x \otimes y, x' \otimes y' \rangle = \langle x, x' \rangle \langle y, y' \rangle \qquad (x, x' \in H, \ y, y' \in K).$$

That it is the unique such sesquilinear form is clear.

If $z \in H \otimes K$, then $z = \sum_{j=1}^{n} x_j \otimes y_j$ for some $x_1, \ldots, x_n \in H$ and $y_1, \ldots, y_n \in K$. Let $e_1, \ldots, e_m$ be an orthonormal basis for the linear span of $y_1, \ldots, y_n$. Then $z = \sum_{j=1}^{m} x'_j \otimes e_j$ for some $x'_1, \ldots, x'_m \in H$, and therefore,

$$\langle z, z \rangle = \sum_{i,j=1}^{m} \langle x'_i \otimes e_i, x'_j \otimes e_j \rangle$$

$$= \sum_{i,j=1}^{m} \langle x'_i, x'_j \rangle \langle e_i, e_j \rangle$$

$$= \sum_{j=1}^{m} \| x'_j \|^2.$$

Thus, $\langle \cdot, \cdot \rangle$ is positive, and if $\langle z, z \rangle = 0$, then $x'_j = 0$ for $j = 1, \ldots, m$, so $z = 0$. Therefore, $\langle \cdot, \cdot \rangle$ is an inner product. $\square$

If H and K are as in Theorem 6.3.1, we shall always regard $H \otimes K$ as a pre-Hilbert space with the above inner product. The Hilbert space completion of $H \otimes K$ is denoted by $H \hat{\otimes} K$, and called the *Hilbert space tensor product* of H and K. Note that

$$\| x \otimes y \| = \| x \| \| y \|.$$

It is an elementary exercise to show that if E_1 and E_2 are orthonormal bases for H and K, respectively, then $E_1 \otimes E_2 = \{x \otimes y \mid x \in E_1, \ y \in E_2\}$ is an orthonormal basis for $H \hat{\otimes} K$.

If H', K' are closed vector subspaces of H, K, respectively, then the inclusion map $H' \otimes K' \to H \hat{\otimes} K$ is isometric when $H' \otimes K'$ has its canonical inner product. It follows that we may regard $H' \hat{\otimes} K'$ as a closed vector subspace of $H \hat{\otimes} K$.

6.3.2. Lemma. *Let H, K be Hilbert spaces and suppose that $u \in B(H)$ and $v \in B(K)$. Then there is a unique operator $u \hat{\otimes} v \in B(H \hat{\otimes} K)$ such that*

$$(u \hat{\otimes} v)(x \otimes y) = u(x) \otimes v(y) \qquad (x \in H, \ y \in K).$$

Moreover, $\|u \hat{\otimes} v\| = \|u\| \|v\|$.

Proof. The map $(u, v) \mapsto u \otimes v$ is bilinear, so to show that $u \otimes v \colon H \otimes K \to H \otimes K$ is bounded, we may assume that u and v are unitaries, since the unitaries span the C*-algebras $B(H)$ and $B(K)$. If $z \in H \otimes K$, then we may write $z = \sum_{j=1}^{n} x_j \otimes y_j$, where $y_1, \ldots, y_n$ are orthogonal. Hence,

$$\|(u \otimes v)(z)\|^2 = \|\sum_{j=1}^{n} u(x_j) \otimes v(y_j)\|^2$$

$$= \sum_{j=1}^{n} \|u(x_j) \otimes v(y_j)\|^2$$

$$\text{(since } v(y_1), \ldots, v(y_n) \text{ are orthogonal)}$$

$$= \sum_{j=1}^{n} \|u(x_j)\|^2 \|v(y_j)\|^2$$

$$= \sum_{j=1}^{n} \|x_j\|^2 \|y_j\|^2$$

$$= \|z\|^2.$$

Consequently, $\|u \otimes v\| = 1$.

Thus, for all operators u, v on H, K, respectively, the linear map $u \otimes v$ is bounded on $H \otimes K$ and hence has an extension to a bounded linear map $u \hat{\otimes} v$ on $H \hat{\otimes} K$.

It is easily verified that the maps

$$B(H) \to B(H \hat{\otimes} K), \quad u \mapsto u \hat{\otimes} \operatorname{id}_K,$$

and

$$B(K) \to B(H \hat{\otimes} K), \quad v \mapsto \operatorname{id}_H \hat{\otimes} v,$$

are injective *-homomorphisms and therefore isometric. Hence, $\|u \hat{\otimes} \operatorname{id}\| = \|u\|$ and $\|\operatorname{id} \hat{\otimes} v\| = \|v\|$, so $\|u \hat{\otimes} v\| = \|(u \hat{\otimes} \operatorname{id})(\operatorname{id} \hat{\otimes} v)\| \le \|u\| \|v\|$. If ε is a sufficiently small positive number, and if $u, v \ne 0$, then there are unit vectors x and y such that $\|u(x)\| > \|u\| - \varepsilon > 0$ and $\|v(y)\| > \|v\| - \varepsilon > 0$. Hence,

$$\|(u \hat{\otimes} v)(x \otimes y)\| = \|u(x)\| \|v(y)\|$$
$$> (\|u\| - \varepsilon)(\|v\| - \varepsilon),$$

so $\|u \hat{\otimes} v\| > (\|u\| - \varepsilon)(\|v\| - \varepsilon)$. Letting $\varepsilon \to 0$, we get $\|u \hat{\otimes} v\| \ge \|u\| \|v\|$. $\square$

6.3.1. *Remark.* Let H and K be Hilbert spaces and suppose that $u, u' \in B(H)$ and $v, v' \in B(K)$. It is routine to show that

$$(u \,\hat{\otimes}\, v)(u' \,\hat{\otimes}\, v') = uu' \,\hat{\otimes}\, vv'$$

and

$$(u \,\hat{\otimes}\, v)^* = u^* \,\hat{\otimes}\, v^*.$$

If $u_1, \ldots, u_n$ are operators on H and $v_1, \ldots, v_n$ are linearly independent operators on K such that $\sum_{j=1}^n u_j \,\hat{\otimes}\, v_j = 0$, then $u_1 = \cdots = u_n = 0$. For if $x \in H$, choose orthonormal vectors $e_1, \ldots, e_m$ in H such that

$$\mathbf{C}u_1(x) + \cdots + \mathbf{C}u_n(x) \subseteq \mathbf{C}e_1 + \cdots + \mathbf{C}e_m.$$

Then there are scalars λ_{ij} such that $u_j(x) = \sum_{i=1}^m \lambda_{ij} e_i$ for $1 \leq j \leq n$. If $y \in K$, we have

$$0 = \sum_{j=1}^n \left(\sum_{i=1}^m \lambda_{ij} e_i \right) \otimes v_j(y) = \sum_{i=1}^m e_i \otimes \sum_{j=1}^n \lambda_{ij} v_j(y),$$

so $\sum_{j=1}^n \lambda_{ij} v_j(y) = 0$. This shows that $\sum_{j=1}^n \lambda_{ij} v_j = 0$, and therefore, by linear independence of $v_1, \ldots, v_n$, we get $\lambda_{ij} = 0$ for all i and j. It follows that $u_1(x) = \cdots = u_n(x) = 0$.

If A and B are algebras, there is a unique multiplication on $A \otimes B$ such that

$$(a \otimes b)(a' \otimes b') = aa' \otimes bb'$$

for all $a, a' \in A$ and $b, b' \in B$. We show this: Let L_a denote left multiplication by a, and let X be the vector space of all linear maps on $A \otimes B$. If $a \in A$ and $b \in B$, then $L_a \otimes L_b \in X$, and the map

$$A \times B \to X, \quad (a, b) \mapsto L_a \otimes L_b,$$

is bilinear. Hence, there is a unique linear map $M \colon A \otimes B \to X$ such that $M(a \otimes b) = L_a \otimes L_b$ for all a and b. The bilinear map

$$(A \otimes B)^2 \to A \otimes B, \quad (c, d) \mapsto cd = M(c)(d),$$

is readily seen to be the required unique multiplication on $A \otimes B$.

We call $A \otimes B$ endowed with this multiplication the *algebra tensor product* of the algebras A and B.

If A and B are *-algebras, then there is a unique involution on $A \otimes B$ such that $(a \otimes b)^* = a^* \otimes b^*$ for all a and b. The existence of such an involution is easily seen if we show that

$$\sum_{j=1}^n a_j \otimes b_j = 0 \Rightarrow \sum_{j=1}^n a_j^* \otimes b_j^* = 0.$$

Choose linearly independent elements $c_1, \ldots, c_m$ in B having the same linear span as $b_1, \ldots, b_n$. Then $b_j = \sum_{i=1}^{m} \lambda_{ij} c_i$ for unique scalars λ_{ij}. Since $\sum_{j=1}^{n} a_j \otimes b_j = 0$, we have $\sum_{i,j} \lambda_{ij} a_j \otimes c_i = 0$, and therefore $\sum_{j=1}^{n} \lambda_{ij} a_j = 0$ for $i = 1, \ldots, m$, because $c_1, \ldots, c_m$ are linearly independent. Hence, $\sum_{j=1}^{n} \bar{\lambda}_{ij} a_j^* = 0$, and therefore,

$$
\sum_{j=1}^{n} a_j^* \otimes b_j^* = \sum_{j=1}^{n} \sum_{i=1}^{m} a_j^* \otimes \bar{\lambda}_{ij} c_i^*
$$
$$
= \sum_{i=1}^{m} \left(\sum_{j=1}^{n} \bar{\lambda}_{ij} a_j^* \right) \otimes c_i^*
$$
$$
= \sum_{i=1}^{m} 0 \otimes c_i^*
$$
$$
= 0.
$$

We call $A \otimes B$ with the above involution the *-algebra tensor product* of A and B.

If A, B are *-subalgebras of *-algebras A', B', respectively, we may clearly regard $A \otimes B$ as a *-subalgebra of $A' \otimes B'$.

6.3.2. Remark. Let $\varphi \colon A \to C$ and $\psi \colon B \to C$ be *-homomorphisms from *-algebras A and B into a *-algebra C such that every element of $\varphi(A)$ commutes with every element of $\psi(B)$. Then there is a unique *-homomorphism $\pi \colon A \otimes B \to C$ such that

$$
\pi(a \otimes b) = \varphi(a)\psi(b) \qquad (a \in A, \ b \in B).
$$

This follows from the observation that the map

$$
A \times B \to C, \quad (a, b) \mapsto \varphi(a)\psi(b),
$$

is bilinear and so induces a linear map $\pi \colon A \otimes B \to C$, which is easily seen to have the required properties.

If A, A', B, B' are *-algebras and $\varphi \colon A \to A'$ and $\psi \colon B \to B'$ are *-homomorphisms, then $\varphi \otimes \psi \colon A \otimes B \to A' \otimes B'$ is a *-homomorphism. The proof is a routine exercise.

We shall use the next result to show that there is at least one C*-norm on $A \otimes B$ if A and B are C*-algebras.

6.3.3. Theorem. *Suppose that (H, φ) and (K, ψ) are representations of the C*-algebras A and B, respectively. Then there exists a unique *-homomorphism $\pi \colon A \otimes B \to B(H \hat{\otimes} K)$ such that*

$$
\pi(a \otimes b) = \varphi(a) \hat{\otimes} \psi(b) \qquad (a \in A, \ b \in B).
$$

Moreover, if φ and ψ are injective, so is π.

Proof. The maps

$$\varphi': A \to B(H \hat{\otimes} K), \quad a \mapsto \varphi(a) \hat{\otimes} \mathrm{id}_K,$$

and

$$\psi': B \to B(H \hat{\otimes} K), \quad b \mapsto \mathrm{id}_H \otimes \psi(b),$$

are *-homomorphisms, and the elements of $\varphi'(A)$ commute with the elements of $\psi'(B)$. Hence, by Remark 6.3.2, there is a unique *-homomorphism $\pi: A \otimes B \to B(H \hat{\otimes} K)$ such that $\pi(a \otimes b) = \varphi'(a)\psi'(b) = \varphi(a) \hat{\otimes} \psi(b)$ ($a \in A,\ b \in B$).

Now suppose that the representations (H, φ) and (K, ψ) are faithful, and let $c \in \ker(\pi)$. We can write c in the form $c = \sum_{j=1}^n a_j \otimes b_j$, where the elements $b_1, \ldots, b_n$ are linearly independent. Then $\psi(b_1), \ldots, \psi(b_n)$ are linearly independent, because ψ is injective, and $\sum_{j=1}^n \varphi(a_j) \hat{\otimes} \psi(b_j) = 0$. Hence, $\varphi(a_1) = \cdots = \varphi(a_n) = 0$ (Remark 6.3.1). Since φ is injective, $a_1 = \cdots = a_n = 0$, so $c = 0$. Thus, $\ker(\pi) = 0$. $\qquad\square$

We denote π in Theorem 6.3.3 by $\varphi \hat{\otimes} \psi$.

Let A and B be C*-algebras with universal representations (H, φ) and (K, ψ), respectively. By Theorem 6.3.3 there is a unique injective *-homomorphism $\pi: A \otimes B \to B(H \hat{\otimes} K)$ such that $\pi(a \otimes b) = \varphi(a) \hat{\otimes} \psi(b)$ for all a and b. Hence, the function

$$\|.\|_*: A \otimes B \to \mathbf{R}^+, \quad c \mapsto \|\pi(c)\|,$$

is a C*-norm on $A \otimes B$, called the *spatial* C*-norm. Note that $\|a \otimes b\|_* = \|a\|\|b\|$. We call the C*-completion of $A \otimes B$ with respect to $\|.\|_*$ the *spatial tensor product* of A and B, and denote it by $A \otimes_* B$.

In general, there may be more than one C*-norm on $A \otimes B$.

If γ is a C*-norm on $A \otimes B$, we denote the C*-completion of $A \otimes B$ with respect to γ by $A \otimes_\gamma B$.

6.3.4. Lemma. *Let A, B be C*-algebras and let γ be a C*-norm on $A \otimes B$. Then for $a' \in A$ and $b' \in B$ the maps*

$$\varphi: A \to A \otimes_\gamma B, \quad a \to a \otimes b',$$

and

$$\psi: B \to A \otimes_\gamma B, \quad b \mapsto a' \otimes b,$$

are continuous.

Proof. Since φ is a linear map between Banach spaces, we may invoke the closed graph theorem. Thus, to show that φ is continuous we need only show that if a sequence (a_n) converges to 0 in A and the sequence $(\varphi(a_n))$ converges to c in $A \otimes_\gamma B$, then $c = 0$. We may suppose that a_n and b' are positive (by replacing a_n by $a_n^* a_n$ and b' by $b'^* b'$ if necessary). Hence, $c \geq 0$. If τ is a positive linear functional on $A \otimes_\gamma B$, then the linear functional

$$\rho \colon A \to \mathbf{C}, \quad a \mapsto \tau(a \otimes b'),$$

is positive, hence continuous. Consequently, $\tau(c) = \lim_{n\to\infty} \tau(a_n \otimes b') = \lim_{n\to\infty} \rho(a_n) = 0$, since $\lim_{n\to\infty} a_n = 0$. Since τ is an arbitrary positive linear functional on $A \otimes_\gamma B$, Theorem 3.3.6 implies that $c = 0$. Therefore, φ is continuous. Similar reasoning shows that ψ is continuous. $\qquad\square$

6.3.5. Theorem. Let A, B be non-zero C*-algebras and suppose that γ is a C*-norm on $A \otimes B$. Let (H, π) be a non-degenerate representation of $A \otimes_\gamma B$. Then there exist unique *-homomorphisms $\varphi \colon A \to B(H)$ and $\psi \colon B \to B(H)$ such that

$$\pi(a \otimes b) = \varphi(a)\psi(b) = \psi(b)\varphi(a) \qquad (a \in A, \ b \in B).$$

Moreover, the representations (H, φ) and (H, ψ) are non-degenerate.

Proof. Let $H_0 = \pi(A \otimes B)H$. If $z \in H_0$, it can be written in the form

$$z = \sum_{i=1}^n \pi(a_i \otimes b_i)(x_i).$$

Suppose we have two such expressions; that is, z can be written

$$z = \sum_{i=1}^n \pi(a_i \otimes b_i)(x_i) = \sum_{j=1}^m \pi(c_j \otimes d_j)(y_j), \qquad (1)$$

where $a_i, c_j \in A$, $b_i, d_j \in B$, and $x_i, y_j \in H$. If $(v_\mu)_{\mu \in M}$ is an approximate unit for B and $a \in A$, then

$$\pi(a \otimes v_\mu)(z) = \sum_{i=1}^n \pi(aa_i \otimes v_\mu b_i)(x_i) = \sum_{j=1}^m \pi(ac_j \otimes v_\mu d_j)(y_j),$$

so in the limit (using Lemma 6.3.4),

$$\lim_\mu \pi(a \otimes v_\mu)(z) = \sum_{i=1}^n \pi(aa_i \otimes b_i)(x_i) = \sum_{j=1}^m \pi(ac_j \otimes d_j)(y_j).$$

We can therefore well-define a map $\varphi(a): H_0 \to H_0$ by setting $\varphi(a)(z) = \sum_{i=1}^{n} \pi(aa_i \otimes b_i)(x_i)$, if z is as in Eq. (1). Since $\varphi(a)(z) = \lim_{\mu} \pi(a \otimes v_{\mu})(z)$, it is clear that $\varphi(a)$ is linear. By Lemma 6.3.4 there exists a positive number M (depending on a) such that $\|\pi(a \otimes b)\| \leq M\|b\|$ ($b \in B$), so $\|\varphi(a)(z)\| \leq M\|z\|$. Hence, $\varphi(a)$ is bounded. Since H_0 is dense in H (because (H, π) is non-degenerate), we can therefore extend $\varphi(a)$ uniquely to a bounded linear map on H, also denoted by $\varphi(a)$.

Suppose now that (u_{λ}) is an approximate unit for A. Reasoning as before, for all $b \in B$ we have $\lim_{\lambda} \pi(u_{\lambda} \otimes b)(z) = \sum_{i=1}^{n} \pi(a_i \otimes bb_i)(x_i)$, if z is as in Eq. (1). We can therefore well-define a map $\psi(b): H_0 \to H_0$ by setting $\psi(b)(z) = \sum_{i=1}^{n} \pi(a_i \otimes bb_i)(x_i) = \lim_{\lambda} \pi(u_{\lambda} \otimes b)(z)$. The linear map $\psi(b)$ is bounded and extends uniquely to a bounded linear map on H, also denoted by $\psi(b)$.

A routine verification shows that the maps

$$\varphi: A \to B(H), \quad a \mapsto \varphi(a),$$

and

$$\psi: B \to B(H), \quad b \mapsto \psi(b),$$

are $*$-homomorphisms, and that $\pi(a \otimes b) = \varphi(a)\psi(b) = \psi(b)\varphi(a)$.

Now suppose that $\varphi': A \to B(H)$ and $\psi': B \to B(H)$ are another pair of $*$-homomorphisms such that $\pi(a \otimes b) = \varphi'(a)\psi'(b) = \psi'(b)\varphi'(a)$ for all a and b. Suppose that $z \in H$ is such that $\varphi'(a)(z) = 0$ ($a \in A$). Then $\pi(a \otimes b)(z) = 0$ for all $a \in A$ and $b \in B$, and therefore $\pi(c)(z) = 0$ for all $c \in A \otimes_{\gamma} B$. By non-degeneracy of (H, π), we have $z = 0$. Thus, φ' is non-degenerate. Similarly, ψ' is non-degenerate.

In particular, φ and ψ are non-degenerate.

If (u_{λ}) and (v_{μ}) are approximate units as above, then the nets $(\varphi'(u_{\lambda}))$ and $(\psi'(v_{\mu}))$ converge strongly to id_H (by non-degeneracy of φ' and ψ'). Hence, $(\pi(a \otimes v_{\mu}))_{\mu}$ converges strongly to both $\varphi'(a)$ and $\varphi(a)$ for all $a \in A$, so $\varphi' = \varphi$. Similarly, $(\pi(u_{\lambda} \otimes b))_{\lambda}$ converges strongly to both $\psi'(b)$ and $\psi(b)$ for all $b \in B$, so $\psi' = \psi$. $\square$

We denote the maps φ and ψ in the preceding theorem by π_A and π_B, respectively.

6.3.6. Corollary. Let A and B be C^*-algebras and let γ be a C^*-seminorm on $A \otimes B$. Then

$$\gamma(a \otimes b) \leq \|a\|\|b\| \qquad (a \in A, \ b \in B).$$

Proof. Let $\delta = \max(\gamma, \|.\|_*)$, so δ is a C^*-norm on $A \otimes B$. Let (H, π) be the universal representation of $A \otimes_{\delta} B$. This is faithful and non-degenerate, so Theorem 6.3.5 applies. If $a \in A$ and $b \in B$, then $\pi(a \otimes b) = \pi_A(a)\pi_B(b)$.

Hence, $\delta(a \otimes b) = \|\pi(a \otimes b)\| \leq \|\pi_A(a)\|\|\pi_B(b)\| \leq \|a\|\|b\|$, so $\gamma(a \otimes b) \leq \delta(a \otimes b) \leq \|a\|\|b\|$. $\qquad\qquad\qquad\qquad\qquad\qquad\qquad\qquad\qquad\qquad\qquad\qquad\square$

Let A and B be C*-algebras. Denote by Γ the set of all C*-norms γ on $A \otimes B$. We define $\|c\|_{max} = \sup_{\gamma \in \Gamma} \gamma(c)$ for each $c \in A \otimes B$. If $c = \sum_{j=1}^n a_j \otimes b_j$ with $a_j \in A$ and $b_j \in B$, then for any $\gamma \in \Gamma$ we have $\gamma(c) \leq \sum_{j=1}^n \gamma(a_j \otimes b_j) \leq \sum_{j=1}^n \|a_j\|\|b_j\|$ by Corollary 6.3.6. Hence, $\|c\|_{max} < \infty$.

It is readily verified that

$$\|.\|_{max} : A \otimes B \to \mathbf{R}^+, \quad c \mapsto \|c\|_{max},$$

is a C*-norm, called the *maximal* C*-norm. We denote by $A \otimes_{max} B$ the C*-completion of $A \otimes B$ under this norm, and call $A \otimes_{max} B$ the *maximal tensor product* of A and B.

If γ is a C*-seminorm, then $\gamma \leq \|.\|_{max}$ (because $\max(\gamma, \|.\|_*) \leq \|.\|_{max}$, since $\max(\gamma, \|.\|_*)$ is a C*-norm).

The maximal tensor product has a very useful universal property:

6.3.7. Theorem. Let $A, B,$ and C be C*-algebras and suppose that $\varphi : A \to C$ and $\psi : B \to C$ are *-homomorphisms such that every element of $\varphi(A)$ commutes with every element of $\psi(B)$. Then there is a unique *-homomorphism $\pi : A \otimes_{max} B \to C$ such that

$$\pi(a \otimes b) = \varphi(a)\psi(b) \qquad (a \in A, \ b \in B).$$

Proof. Uniqueness is clear. By Remark 6.3.2 there is a *-homomorphism $\pi : A \otimes B \to C$ satisfying the equation in the statement of the theorem. The function

$$\gamma : A \otimes B \to \mathbf{R}^+, \quad c \mapsto \|\pi(c)\|,$$

is a C*-seminorm. Hence, $\gamma(c) \leq \|c\|_{max}$ for all $c \in A \otimes B$. Therefore, π is a norm-decreasing *-homomorphism, and so extends to a norm-decreasing *-homomorphism on $A \otimes_{max} B$. $\qquad\qquad\qquad\qquad\qquad\qquad\square$

We say a C*-algebra A is *nuclear* if, for each C*-algebra B, there is only one C*-norm on $A \otimes B$.

6.3.3. Remark. If a *-algebra A admits a complete C*-norm $\|.\|$, then it is the only C*-norm on A. For if γ is another C*-norm on A, and B denotes the C*-completion of A with respect to γ, then the inclusion $(A, \|.\|) \to (B, \gamma)$ is an injective *-homomorphism and therefore isometric, so $\gamma = \|.\|$.

6.3.1. Example. For each $n \geq 1$, the C*-algebra $M_n(\mathbf{C})$ is nuclear. The reason is that for each C*-algebra A, the *-algebra $M_n(\mathbf{C}) \otimes A$ admits a complete C*-norm. This is seen by showing that the unique linear map $\pi \colon M_n(\mathbf{C}) \otimes A \to M_n(A)$, such that $\pi((\lambda_{ij})_{ij} \otimes a) = (\lambda_{ij}a)_{ij}$ for $(\lambda_{ij}) \in M_n(\mathbf{C})$ and $a \in A$, is a *-isomorphism (this is a routine exercise).

We are going to show that all finite-dimensional C*-algebras are nuclear and for this we shall need to determine the structure theory for such algebras. This is given in the following.

6.3.8. Theorem. *If A is a non-zero finite-dimensional C*-algebra, it is *-isomorphic to $M_{n_1}(\mathbf{C}) \oplus \cdots \oplus M_{n_k}(\mathbf{C})$ for some integers $n_1, \ldots, n_k$.*

Proof. If A is simple, the result is immediate from Remark 6.2.1. We prove the general result by induction on the dimension m of A. The case $m = 1$ is obvious. Suppose the result holds for all dimensions less than m. We may suppose that A is not simple, and so contains a non-zero proper closed ideal I, and we may take I to be of minimum dimension. In this case I has no non-trivial ideals, so I is *-isomorphic to $M_{n_1}(\mathbf{C})$ for some integer n_1. Hence, I has a unit p, so $I = Ap$ and p commutes with all the elements of A. Also, $A(1 - p)$ is a C*-subalgebra of A and the map

$$A \to Ap \oplus A(1 - p), \quad a \mapsto (ap, a(1 - p)),$$

is a *-isomorphism. Since the algebra $A(1 - p)$ has dimension less than m, it is *-isomorphic to $M_{n_2}(\mathbf{C}) \oplus \cdots \oplus M_{n_k}(\mathbf{C})$ for some $n_2, \ldots, n_k$ (inductive hypothesis). Hence, A is *-isomorphic to $M_{n_1}(\mathbf{C}) \oplus \cdots \oplus M_{n_k}(\mathbf{C})$.   □

6.3.9. Theorem. *A finite-dimensional C*-algebra is nuclear.*

Proof. Let A be a finite-dimensional C*-algebra, which we may suppose to be the direct sum $A = M_{n_1}(\mathbf{C}) \oplus \cdots \oplus M_{n_k}(\mathbf{C})$. Let B be an arbitrary C*-algebra. A routine verification shows that the unique linear map

$$\pi \colon A \otimes B \to (M_{n_1}(\mathbf{C}) \otimes B) \oplus \cdots \oplus (M_{n_k}(\mathbf{C}) \otimes B),$$

such that

$$\pi((a_1, \ldots, a_k) \otimes b) = (a_1 \otimes b, \ldots, a_k \otimes b)$$

for all $a_j \in M_{n_j}(\mathbf{C})$ and $b \in B$, is a *-isomorphism. Hence, $A \otimes B$ admits a complete C*-norm, so it admits only one C*-norm. This shows that A is nuclear.   □

The next result suggests that nuclear algebras exist in abundance.

6.3.10. Theorem. Let S be a non-empty set of C*-subalgebras of a C*-algebra A which is upwards-directed (that is, if $B, C \in S$, then there exists $D \in S$ such that $B, C \subseteq D$). Suppose that $\cup S$ is dense in A and that all the algebras in S are nuclear. Then A is nuclear.

Proof. Let B be an arbitrary C*-algebra and suppose β, γ are C*-norms on $A \otimes B$. Set $C = \cup_{D \in S} D \otimes B$ (we may regard $D \otimes B$ as a *-subalgebra of $A \otimes B$ for each $D \in S$). Then C is a *-subalgebra of $A \otimes B$ and it is clear that C is dense in the C*-algebras $A \otimes_\beta B$ and $A \otimes_\gamma B$. Now $\beta = \gamma$ on $D \otimes B$ for each $D \in S$, by nuclearity of D, so $\beta = \gamma$ on C, and therefore the identity map on C extends to a *-isomorphism $\pi \colon A \otimes_\beta B \to A \otimes_\gamma B$. If $a \in A$ and $b \in B$, then there is a sequence (a_n) in $\cup S$ such that $a = \lim_{n \to \infty} a_n$ in A, so $a \otimes b = \lim_{n \to \infty} a_n \otimes b$ in $A \otimes_\beta B$ and in $A \otimes_\gamma B$, and therefore $\pi(a \otimes b) = \lim_{n \to \infty} \pi(a_n \otimes b) = \lim_{n \to \infty} a_n \otimes b = a \otimes b$, where convergence is with respect to γ. Therefore, $\pi = \mathrm{id}$ on $A \otimes B$. Hence, for any $c \in A \otimes B$, we have $\gamma(c) = \gamma(\pi(c)) = \beta(c)$, so $\gamma = \beta$ on $A \otimes B$. Therefore, the algebra A is nuclear. $\square$

6.3.2. Example. If H is a Hilbert space, $K(H)$ is a nuclear C*-algebra. To see this, suppose that $e_1, \ldots, e_n$ are orthonormal vectors in H. If $p = \sum_{j=1}^{n} e_j \otimes e_j$, then p is a projection, $p \in K(H)$, and the map

$$\varphi \colon M_n(\mathbf{C}) \to p K(H) p, \quad (\lambda_{ij}) \mapsto \sum_{i,j=1}^{n} \lambda_{ij} e_i \otimes e_j,$$

is a *-isomorphism. We show surjectivity only: If $u \in p K(H) p$, then

$$u = pup$$
$$= \sum_{i,j=1}^{n} (e_i \otimes e_i) u (e_j \otimes e_j)$$
$$= \sum_{i,j=1}^{n} \langle u(e_j), e_i \rangle e_i \otimes e_j$$
$$= \varphi((\langle u(e_j), e_i \rangle)_{ij}).$$

Since $p K(H) p$ is finite-dimensional, it is nuclear (Theorem 6.3.9).

If E is an orthonormal basis for H, let I be the set of all finite non-empty subsets of E made into an upwards-directed poset by setting $i \leq j$ if $i \subseteq j$. For $i \in I$, let $p_i = \sum_{x \in i} x \otimes x$ and $A_i = p_i K(H) p_i$. Each A_i is therefore a nuclear C*-algebra, and the set $S = \{A_i \mid i \in I\}$ is upwards-directed.

If u is a finite-rank operator on H, we can write $u = \sum_{k=1}^{m} x_k \otimes y_k$ for some $x_k, y_k \in H$. Therefore,

$$u p_i = \sum_{k=1}^{m} x_k \otimes p_i(y_k),$$

so $\lim_i u p_i = \sum_{k=1}^m x_k \otimes y_k = u$, since $\lim_i p_i(y) = y$ $(y \in H)$. From this it follows by norm-density of the finite-rank operators in $K(H)$ that $\lim_i u p_i = u$ for all $u \in K(H)$. Therefore, $\lim_i p_i u p_i = u$. Hence, $\cup S$ is dense in $K(H)$. By Theorem 6.3.10 $K(H)$ is a nuclear C*-algebra.

6.3.11. Theorem. *All AF-algebras are nuclear.*

Proof. If A is an AF-algebra, then it has an increasing sequence $(A_n)_{n=1}^\infty$ of finite-dimensional C*-subalgebras such that $\cup_n A_n$ is dense in A. Each A_n is nuclear by Theorem 6.3.9, so A is nuclear by Theorem 6.3.10.   □

6.4. Minimality of the Spatial C*-Norm

As the title of the section suggests, we show here that the spatial C*-norm on a tensor product of C*-algebras is the least C*-norm. Along the way we shall also show the important result—due to Takesaki—that abelian C*-algebras are nuclear.

We begin with a result on approximate units that will be used at a number of points in the sequel.

6.4.1. Lemma. *Let A and B be C*-algebras and suppose that γ is a C*-norm on $A \otimes B$. Then $A \otimes_\gamma B$ admits an approximate unit of the form $(u_\lambda \otimes v_\lambda)_{\lambda \in \Lambda}$, where $(u_\lambda)_{\lambda \in \Lambda}$ and $(v_\lambda)_{\lambda \in \Lambda}$ are approximate units for A and B, respectively.*

Proof. Let $(u_\lambda)_{\lambda \in \Lambda}$ and $(v_\mu)_{\mu \in M}$ be approximate units for A and B, respectively. Write $(\lambda, \mu) \leq (\lambda', \mu')$ in $\Lambda \times M$ if $\lambda \leq \lambda'$ and $\mu \leq \mu'$. The relation $\leq$ is reflexive, transitive, and upwards-directed, and if we set $u'_{(\lambda,\mu)} = u_\lambda$ and $v'_{(\lambda,\mu)} = v_\mu$, then a routine argument shows that the net $(u'_{(\lambda,\mu)} \otimes v'_{(\lambda,\mu)})_{(\lambda,\mu) \in \Lambda \times M}$ is an approximate unit for $A \otimes_\gamma B$ of the required type.   □

6.4.1. Remark. Suppose that A and B are C*-algebras and suppose also that $(H_\lambda, \varphi_\lambda)_{\lambda \in \Lambda}$ and $(K_\mu, \psi_\mu)_{\mu \in M}$ are families of representations of A and B, respectively. Set $(H, \varphi) = \oplus_{\lambda \in \Lambda}(H_\lambda, \varphi_\lambda)$ and $(K, \psi) = \oplus_{\mu \in M}(K_\mu, \psi_\mu)$. It is readily verified that there is a unique unitary

$$ u \colon H \hat{\otimes} K \to \bigoplus_{\substack{\lambda \in \Lambda \\ \mu \in M}} H_\lambda \hat{\otimes} K_\mu, $$

such that for all $x = (x_\lambda)_{\lambda \in \Lambda} \in H$ and $y = (y_\mu)_{\mu \in M} \in K$,

$$ u(x \otimes y) = (x_\lambda \otimes y_\mu)_{\lambda, \mu}. $$

For each element $c \in A \otimes B$,

$$(\varphi \,\hat{\otimes}\, \psi)(c) = u^*\left(\bigoplus_{\substack{\lambda \in \Lambda \\ \mu \in M}} (\varphi_\lambda \,\hat{\otimes}\, \psi_\mu)(c)\right)u.$$

(To see this, show it first for $c = a \otimes b$.) It follows that

$$\|(\varphi \,\hat{\otimes}\, \psi)(c)\| = \sup_{\substack{\lambda \in \Lambda \\ \mu \in M}} \|(\varphi_\lambda \,\hat{\otimes}\, \psi_\mu)(c)\|.$$

6.4.2. Theorem. *If A, B are non-zero C*-algebras and $c \in A \otimes B$, then*

$$\|c\|_* = \sup_{\substack{\tau \in S(A) \\ \rho \in S(B)}} \|(\varphi_\tau \,\hat{\otimes}\, \varphi_\rho)(c)\|.$$

Proof. If (H, φ) and (K, ψ) are the universal representations of A and B, respectively, then $\|c\|_* = \|(\varphi \,\hat{\otimes}\, \psi)(c)\|$ by definition of the norm $\|.\|_*$. Since $(H, \varphi) = \oplus_{\tau \in S(A)}(H_\tau, \varphi_\tau)$ and $(K, \psi) = \oplus_{\rho \in S(B)}(H_\rho, \varphi_\rho)$, the theorem follows from Remark 6.4.1. $\qquad\qquad\qquad\qquad\qquad\qquad\qquad\qquad\qquad\square$

6.4.3. Corollary. *If τ, ρ are states on C*-algebras A, B, respectively, then $\tau \otimes \rho$ is continuous on $A \otimes B$ with respect to the spatial C*-norm.*

Proof. If $c \in A \otimes B$, then

$$(\tau \otimes \rho)(c) = \langle (\varphi_\tau \,\hat{\otimes}\, \varphi_\rho)(c)(x_\tau \otimes x_\rho), x_\tau \otimes x_\rho \rangle \qquad\qquad (1)$$

(to see this, first show it for $c = a \otimes b$). Since $\|(\varphi_\tau \,\hat{\otimes}\, \varphi_\rho)(c)\| \le \|c\|_*$ by Theorem 6.4.2, we have $|(\tau \otimes \rho)(c)| \le \|c\|_*$ by Eq. (1). Thus, $\tau \otimes \rho$ is continuous with respect to $\|.\|_*$. $\qquad\qquad\qquad\qquad\qquad\qquad\qquad\qquad\square$

6.4.2. Remark. If (H, φ) and (K, ψ) are representations of C*-algebras A and B that are unitarily equivalent to representations (H', φ') and (K', ψ'), respectively, then there exists a unitary $u: H \,\hat{\otimes}\, K \to H' \,\hat{\otimes}\, K'$ such that for all $c \in A \otimes B$ we have $(\varphi' \,\hat{\otimes}\, \psi')(c) = u(\varphi \,\hat{\otimes}\, \psi)(c)u^*$. (The proof is routine.)

6.4.4. Theorem. *If (H, φ) and (K, ψ) are arbitrary representations of C*-algebras A and B, respectively, and $c \in A \otimes B$, then*

$$\|(\varphi \,\hat{\otimes}\, \psi)(c)\| \le \|c\|_*.$$

Proof. Let $H' = [\varphi(A)H]$ and $K' = [\psi(B)K]$. Then

$$H' \,\hat{\otimes}\, K' = [(\varphi \,\hat{\otimes}\, \psi)(A \otimes B)(H \,\hat{\otimes}\, K)],$$

and a routine verification shows that

$$(\varphi \mathbin{\hat{\otimes}} \psi)(c)_{H' \hat{\otimes} K'} = (\varphi_{H'} \mathbin{\hat{\otimes}} \psi_{K'})(c)$$

for all $c \in A \otimes B$. Therefore,

$$\|(\varphi \mathbin{\hat{\otimes}} \psi)(c)\| = \|(\varphi \mathbin{\hat{\otimes}} \psi)(c)_{H' \hat{\otimes} K'}\| = \|(\varphi_{H'} \mathbin{\hat{\otimes}} \psi_{K'})(c)\|.$$

Thus to prove the theorem we may suppose that φ and ψ are non-degenerate (replacing them with the non-degenerate representations $\varphi_{H'}$ and $\psi_{K'}$ if necessary). Hence, we may write φ and ψ as direct sums of cyclic representations. By Theorem 5.1.7 each non-zero cyclic representation of a C*-algebra is unitarily equivalent to a representation of the form (H_τ, φ_τ) for some state τ of the algebra. Replacing (H, φ) and (K, ψ) by unitarily equivalent representations if necessary, we may suppose that $(H, \varphi) = \oplus_{\lambda \in \Lambda}(H_{\tau_\lambda}, \varphi_{\tau_\lambda})$ for some index set Λ with $\tau_\lambda \in S(A)$ for all λ, and likewise we may suppose that $(K, \psi) = \oplus_{\mu \in M}(H_{\rho_\mu}, \varphi_{\rho_\mu})$ for an index set M with $\rho_\mu \in S(B)$ for all μ (we can do this by Remark 6.4.2). For all $c \in A \otimes B$,

$$\|(\varphi \mathbin{\hat{\otimes}} \psi)(c)\| = \sup_{\substack{\lambda \in \Lambda \\ \mu \in M}} \|(\varphi_{\tau_\lambda} \mathbin{\hat{\otimes}} \varphi_{\rho_\mu})(c)\|$$

by Remark 6.4.1, and therefore,

$$\|(\varphi \mathbin{\hat{\otimes}} \psi)(c)\| \leq \sup_{\substack{\tau \in S(A) \\ \rho \in S(B)}} \|(\varphi_\tau \mathbin{\hat{\otimes}} \varphi_\rho)(c)\| = \|c\|_*,$$

by Theorem 6.4.2.   □

We shall use the following elementary observation in the proof of Theorem 6.4.5.

6.4.3. Remark. If p is a rank-one projection of $M_n(\mathbf{C})$, then there exist scalars $\lambda_1, \ldots, \lambda_n$ such that $p = (\lambda_i \bar{\lambda}_j)_{ij}$. To see this, write $p = x \otimes x$ for some $x \in \mathbf{C}^n$. If $e_1, \ldots, e_n$ is the canonical orthonormal basis of $\mathbf{C}^n$, then $x = \sum_{i=1}^n \lambda_i e_i$ for some scalars $\lambda_1, \ldots, \lambda_n$. Since $e_i \otimes e_j$ is the matrix with all entries zero except for the (i, j)–entry, which is 1, and since $p = \sum_{i,j=1}^n \lambda_i \bar{\lambda}_j e_i \otimes e_j$, we have $p = (\lambda_i \bar{\lambda}_j)_{ij}$.

6.4.5. Theorem. Let τ, ρ be positive linear functionals on C*-algebras A, B, respectively. Then the linear functional $\tau \otimes \rho$ on $A \otimes B$ is positive.

Proof. If $c \in A \otimes B$, then we have to show $(\tau \otimes \rho)(c^*c) \geq 0$. We write $c = \sum_{j=1}^n a_j \otimes b_j$, where $a_1, \ldots, a_n \in A$ and $b_1, \ldots, b_n \in B$. Then

$$(\tau \otimes \rho)(c^*c) = (\tau \otimes \rho)\Big(\sum_{i,j=1}^n a_i^* a_j \otimes b_i^* b_j\Big) = \sum_{i,j=1}^n \tau(a_i^* a_j)\rho(b_i^* b_j).$$

Now if $\lambda_1, \ldots, \lambda_n \in \mathbf{C}$, then

$$\sum_{i,j=1}^{n} \rho(b_i^* b_j)\bar{\lambda}_i \lambda_j = \rho((\sum_{i=1}^{n} \lambda_i b_i)^*(\sum_{j=1}^{n} \lambda_j b_j)) \geq 0,$$

since ρ is positive. Hence, the matrix $u = (\rho(b_i^* b_j))_{i,j}$ is a positive element of $M_n(\mathbf{C})$, so it can be diagonalised, and therefore, it can be written in the form $u = \sum_{j=1}^{n} t_j p_j$ with $t_1, \ldots, t_n \in \mathbf{R}^+$ and $p_1, \ldots, p_n$ rank-one projections in $M_n(\mathbf{C})$. Thus, to show that $(\tau \otimes \rho)(c^* c) \geq 0$, it is sufficient to show that $\sum_{i,j=1}^{n} \tau(a_i^* a_j)p_{ij} \geq 0$ for each rank-one projection $p = (p_{ij})_{ij}$ in $M_n(\mathbf{C})$. By Remark 6.4.3 any such projection p is of the form $p = (\bar{\lambda}_i \lambda_j)_{ij}$ for some scalars $\lambda_1, \ldots, \lambda_n \in \mathbf{C}$. Hence,

$$\sum_{i,j=1}^{n} \tau(a_i^* a_j)p_{ij} = \sum_{i,j=1}^{n} \tau(a_i^* a_j)\bar{\lambda}_i \lambda_j$$

$$= \tau((\sum_{i=1}^{n} \lambda_i a_i)^*(\sum_{j=1}^{n} \lambda_j a_j))$$

$$\geq 0,$$

since τ is positive.   □

6.4.6. Theorem. Let A and B be C*-algebras and suppose that γ is a C*-norm on $A \otimes B$. If τ, ρ are states on A, B, respectively, and $\tau \otimes \rho$ is continuous with respect to γ, then $\tau \otimes \rho$ extends uniquely to a state ω on $A \otimes_\gamma B$.

Proof. Since $A \otimes B$ is a dense vector subspace of $A \otimes_\gamma B$, it is clear that $\tau \otimes \rho$ extends uniquely to a continuous linear functional ω on $A \otimes_\gamma B$—the point of the theorem is that ω is positive and of norm 1.

If $c \in A \otimes_\gamma B$, then there is a sequence (c_n) of elements of $A \otimes B$ converging in the norm γ to c. Hence, $c^* c = \lim_{n\to\infty} c_n^* c_n$, so $\omega(c^* c) = \lim_{n\to\infty}(\tau \otimes \rho)(c_n^* c_n)$. Since $(\tau \otimes \rho)(c_n^* c_n) \geq 0$ for all n by Theorem 6.4.5, we have $\omega(c^* c) \geq 0$. Thus, ω is positive.

By Lemma 6.4.1 we may choose an approximate unit for $A \otimes_\gamma B$ of the form $(u_\lambda \otimes v_\lambda)_{\lambda \in \Lambda}$, where $(u_\lambda)_{\lambda \in \Lambda}$ and $(v_\lambda)_{\lambda \in \Lambda}$ are approximate units for A and B, respectively. Applying Theorem 3.3.3, $\|\omega\| = \lim_\lambda \omega(u_\lambda \otimes v_\lambda) = \lim_\lambda \tau(u_\lambda)\rho(v_\lambda) = 1$, since $\lim_\lambda \tau(u_\lambda) = \|\tau\| = 1$ and $\lim_\lambda \rho(v_\lambda) = \|\rho\| = 1$. Therefore, ω is a state of $A \otimes_\gamma B$.   □

We denote the state ω in Theorem 6.4.6 by $\tau \otimes_\gamma \rho$.

6.4.7. Theorem. Let A, B be C*-algebras and suppose γ is a C*-norm on $A \otimes B$. Let τ, ρ be states on A, B, respectively, such that $\tau \otimes \rho$ is continuous with respect to γ. Then there exists a unitary $u \colon H_\tau \,\hat\otimes\, H_\rho \to H_{\tau\otimes_\gamma\rho}$ such that for all $c \in A \otimes B$,

$$\varphi_{\tau\otimes_\gamma\rho}(c) = u(\varphi_\tau \,\hat\otimes\, \varphi_\rho)(c)u^*.$$

Proof. Let $\omega = \tau \otimes_\gamma \rho$ and $\pi = \varphi_\tau \,\hat\otimes\, \varphi_\rho$, and let y be the unit vector $x_\tau \otimes x_\rho$ in $H_\tau \,\hat\otimes\, H_\rho$. We claim that for all $c \in A \otimes B$ we have

$$\langle \varphi_\omega(c)(x_\omega), x_\omega \rangle = \langle \pi(c)(y), y \rangle. \tag{2}$$

To show this, we may suppose that $c = a \otimes b$, where $a \in A$ and $b \in B$. Then

$$
\begin{aligned}
\langle \varphi_\omega(c)(x_\omega), x_\omega \rangle &= \omega(a \otimes b) \\
&= \tau(a)\rho(b) \\
&= \langle \varphi_\tau(a)(x_\tau), x_\tau \rangle \langle \varphi_\rho(b)(x_\rho), x_\rho \rangle \\
&= \langle (\varphi_\tau(a) \,\hat\otimes\, \varphi_\rho(b))(x_\tau \otimes x_\rho), x_\tau \otimes x_\rho \rangle \\
&= \langle \pi(c)(y), y \rangle.
\end{aligned}
$$

Let $H_0 = \varphi_\omega(A \otimes B)x_\omega$ and $K_0 = \pi(A \otimes B)y$. The map

$$u_0 \colon K_0 \to H_0, \quad \pi(c)(y) \mapsto \varphi_\omega(c)x_\omega,$$

is well-defined (by Eq. (2)), linear, and isometric (again by Eq. (2)). Hence, by density of K_0 in $H_\tau \,\hat\otimes\, H_\rho$ and H_0 in H_ω, we can extend u_0 uniquely to a unitary $u \colon H_\tau \,\hat\otimes\, H_\rho \to H_\omega$. A routine verification shows that $\varphi_\omega(c) = u\pi(c)u^*$ for all $c \in A \otimes B$. $\qquad\square$

6.4.8. Theorem. Let A and B be non-zero C*-algebras, and suppose that $c \in A \otimes B$. Then

$$\|c\|_*^2 = \sup_{\substack{\tau \in S(A) \\ \rho \in S(B)}} \sup_{\substack{d \in A \otimes B \\ (\tau\otimes\rho)(d^*d) > 0}} \frac{(\tau \otimes \rho)(d^*c^*cd)}{(\tau \otimes \rho)(d^*d)}.$$

Proof. If ω is a state of $A \otimes_* B$, then

$$\|\varphi_\omega(c)\|^2 = \sup_{\substack{d \in A \otimes B \\ \omega(d^*d) > 0}} \frac{\omega(d^*c^*cd)}{\omega(d^*d)},$$

because $\|d + N_\omega\|^2 = \omega(d^*d)$ and $\varphi_\omega(A \otimes B)x_\omega$ is dense in H_ω. By Theorem 6.4.2 we have

$$\|c\|_*^2 = \sup_{\substack{\tau \in S(A) \\ \rho \in S(B)}} \|(\varphi_\tau \,\hat\otimes\, \varphi_\rho)(c^*c)\|.$$

Applying Corollary 6.4.3 and Theorems 6.4.6 and 6.4.7, we have

$$\|\varphi_{\tau \otimes_{\|\cdot\|_*} \rho}(d)\| = \|(\varphi_\tau \hat{\otimes} \varphi_\rho)(d)\|,$$

for all $\tau \in S(A)$, $\rho \in S(B)$, and $d \in A \otimes B$. Putting these equations together, we get

$$\|c\|_*^2 = \sup_{\substack{\tau \in S(A) \\ \rho \in S(B)}} \|\varphi_{\tau \otimes_{\|\cdot\|_*} \rho}(c^*c)\| = \sup_{\substack{\tau \in S(A) \\ \rho \in S(B)}} \sup_{d \in A \otimes B} \frac{(\tau \otimes \rho)(d^*c^*cd)}{(\tau \otimes \rho)(d^*d)}.$$

$\square$

6.4.9. Theorem. *If A and B are C*-algebras, the restriction to $A \otimes B$ of the spatial C*-norm on $A \otimes \tilde{B}$ is the spatial C*-norm on $A \otimes B$.*

Proof. Let γ be the restriction to $A \otimes B$ of the spatial C*-norm on $A \otimes \tilde{B}$. Applying Theorem 6.4.8, we get for $c \in A \otimes B$,

$$\gamma(c)^2 = \sup_{\substack{\tau \in S(A) \\ \rho \in S(\tilde{B})}} \sup_{\substack{d \in A \otimes \tilde{B} \\ (\tau \otimes \rho)(d^*d) > 0}} \frac{(\tau \otimes \rho)(d^*c^*cd)}{(\tau \otimes \rho)(d^*d)},$$

and also,

$$\|c\|_*^2 = \sup_{\substack{\tau \in S(A) \\ \rho \in S(B)}} \sup_{\substack{d \in A \otimes B \\ (\tau \otimes \rho)(d^*d) > 0}} \frac{(\tau \otimes \rho)(d^*c^*cd)}{(\tau \otimes \rho)(d^*d)}.$$

Using the fact that each $\rho \in S(B)$ has a unique extension $\tilde{\rho}$ in $S(\tilde{B})$ (Theorem 3.3.9), we therefore have $\gamma(c) \geq \|c\|_*$.

Now let (H, φ) and (K, ψ) be the universal representations of A and $\tilde{B}$, respectively. Let ψ_B denote the restriction of ψ to B. If $c \in A \otimes B$, then $(\varphi \hat{\otimes} \psi)(c) = (\varphi \hat{\otimes} \psi_B)(c)$. Since $\gamma(c) = \|(\varphi \hat{\otimes} \psi)(c)\|$ (by definition of the spatial C*-norm on $A \otimes \tilde{B}$), and since $\|(\varphi \hat{\otimes} \psi_B)(c)\| \leq \|c\|_*$ by Theorem 6.4.4, we have $\gamma(c) \leq \|c\|_*$. Therefore, $\gamma = \|\cdot\|_*$. $\square$

6.4.10. Theorem. *If A and B are C*-algebras, if B is non-unital, and if γ is a C*-norm on $A \otimes B$, then there is a C*-norm on $A \otimes \tilde{B}$ extending γ.*

Proof. Let (H, π) be a faithful non-degenerate representation of $A \otimes_\gamma B$. Since

$$\pi(a \otimes b) = \pi_A(a)\pi_B(b) = \pi_B(b)\pi_A(a) \tag{3}$$

for all a and b, it is clear that π_A and π_B are injective, because π is. Extend π_B to a unital *-homomorphism $\pi_B' : \tilde{B} \to B(H)$. We claim that π_B' is injective. To see this, let $b \in B$ and $\lambda \in \mathbb{C}$ and suppose that $\pi_B'(b + \lambda) = 0$. If $\lambda \neq 0$ this implies that $\pi_B(b') = 1$ for $b' = -b/\lambda \in B$. Since

π_B is injective, the element b' is therefore a unit for B, contradicting our assumption that B is non-unital. Hence, $\lambda = 0$ and therefore, $b = 0$. Thus, π'_B is injective as claimed.

By Eq. (3), the elements of $\pi_A(A)$ commute with the elements of $\pi'_B(\tilde{B})$. It follows from Remark 6.3.2 that we have a $*$-homomorphism $\pi' : A \otimes \tilde{B} \to B(H)$ extending π. Since $\gamma(c) = \|\pi(c)\|$ for all $c \in A \otimes_\gamma B$ by injectivity of π, to prove the theorem we have only to show that π' is injective (since in this case $c \mapsto \|\pi'(c)\|$ is a C*-norm on $A \otimes \tilde{B}$ extending γ).

Suppose that $d \in \ker(\pi')$. If $c \in A \otimes B$, then $dc \in A \otimes B$ and $\pi(dc) = 0$, so $dc = 0$, since π is injective. Let $\theta = \pi_A \,\hat{\otimes}\, \pi'_B$. Then $\theta(d)\theta(c) = 0$. This shows that $\theta(d)$ is equal to zero on $K_0 = \theta(A \otimes B)(H \,\hat{\otimes}\, H)$. But it is easily verified that K_0 is dense in $H \,\hat{\otimes}\, H$ (use the non-degeneracy of π_A and π_B). Hence, $\theta(d) = 0$. Since π_A and π'_B are injective, so is θ by Theorem 6.3.3. Therefore, $d = 0$, and π' is injective. $\square$

6.4.11. Lemma. *Let A and B be C*-algebras, and suppose that u and v are unitaries in $\tilde{A}$ and $\tilde{B}$, respectively. Then the unique $*$-isomorphism π on $A \otimes B$, such that $\pi(a \otimes b) = uau^* \otimes vbv^*$ for all $a \in A$ and $b \in B$, is an isometry for any C*-norm γ on $A \otimes B$.*

Proof. Since π has inverse the unique $*$-isomorphism π' on $A \otimes B$ such that $\pi'(a \otimes b) = u^*au \otimes v^*bv$ for all a and b, it suffices by symmetry to show that π is norm-decreasing. Applying Lemma 6.4.1, we may choose an approximate unit for $A \otimes_\gamma B$ of the form $(u_\lambda \otimes v_\lambda)_{\lambda \in \Lambda}$, where $(u_\lambda)_{\lambda \in \Lambda}$ and $(v_\lambda)_{\lambda \in \Lambda}$ are approximate units for A and B, respectively. Let $w = u \otimes v$ and $w_\lambda = u_\lambda \otimes v_\lambda$, and observe that $\pi(c) = wcw^*$ for all $c \in A \otimes B$ (we are regarding $A \otimes B$ as a $*$-subalgebra of $\tilde{A} \otimes \tilde{B}$). If $a \in A$ and $b \in B$, then $uau^* = \lim_\lambda uu_\lambda au_\lambda u^*$ in A and $vbv^* = \lim_\lambda vv_\lambda bv_\lambda v^*$ in B, so by continuity of the bilinear map $A \times B \to A \otimes_\gamma B$, $(a', b') \mapsto a' \otimes b'$, (this is given by Corollary 6.3.6), we have

$$
\begin{aligned}
w(a \otimes b)w^* &= uau^* \otimes vbv^* \\
&= \lim_\lambda uu_\lambda au_\lambda u^* \otimes vv_\lambda bv_\lambda v^* \\
&= \lim_\lambda ww_\lambda(a \otimes b)w_\lambda w^*
\end{aligned}
$$

in $A \otimes_\gamma B$. It follows that $wcw^* = \lim_\lambda ww_\lambda cw_\lambda w^*$ for all $c \in A \otimes B$; that is, $\pi(c) = \lim_\lambda \pi(w_\lambda cw_\lambda)$. Hence,

$$
\begin{aligned}
\gamma(\pi(c)) &= \lim_\lambda \gamma(\pi(w_\lambda cw_\lambda)) \\
&= \lim_\lambda \gamma(ww_\lambda cw_\lambda w^*) \\
&\leq \sup_{\lambda \in \Lambda} \gamma(ww_\lambda)\gamma(c)\gamma(w_\lambda w^*) \\
&\leq \sup_{\lambda \in \Lambda} \|uu_\lambda\| \|vv_\lambda\| \gamma(c) \|u_\lambda u^*\| \|v_\lambda v^*\| \\
&\leq \gamma(c).
\end{aligned}
$$

This proves the lemma.   □

If A, B are C*-algebras and γ is a C*-norm on $A \otimes B$, we denote by S_γ the set of all pairs $(\tau, \rho) \in \mathrm{PS}(A) \times \mathrm{PS}(B)$ such that $\tau \otimes \rho$ is continuous on $A \otimes B$ with respect to γ. The set S_γ plays a fundamental role in the proof that abelian C*-algebras are nuclear and that the spatial C*-norm is minimal.

6.4.12. Theorem. *Let A, B be C*-algebras and let γ be a C*-norm on $A \otimes B$. Then S_γ is closed in $\mathrm{PS}(A) \times \mathrm{PS}(B)$ (where the sets $\mathrm{PS}(A), \mathrm{PS}(B)$ are endowed with the weak* topologies). Moreover, if u, v are unitaries in $\tilde{A}, \tilde{B}$, respectively, and $(\tau, \rho) \in S_\gamma$, then $(\tau^u, \rho^v) \in S_\gamma$.*

Proof. If $\pi \colon A \otimes B \to A \otimes B$ is the unique *-isomorphism such that $\pi(a \otimes b) = uau^* \otimes vbv^*$ for all a and b, then $\tau^u \otimes \rho^v = (\tau \otimes \rho)\pi$, so continuity of $\tau^u \otimes \rho^v$ with respect to γ follows from Lemma 6.4.11 and the continuity of $\tau \otimes \rho$ with respect to γ.

The proof that S_γ is closed is a routine argument (use Theorem 6.4.6 to show that $|(\tau \otimes \rho)(c)| \leq \gamma(c)$ $(c \in A \otimes B)$ if $(\tau, \rho) \in S_\gamma$).   □

Let A and B be C*-algebras and γ a C*-norm on $A \otimes B$. If ω is a state of $A \otimes_\gamma B$ and $\pi = \varphi_\omega$, we define states ω_A and ω_B on A and B, respectively, by setting

$$\omega_A(a) = \langle \pi_A(a)(x_\omega), x_\omega \rangle \quad \text{and} \quad \omega_B(b) = \langle \pi_B(b)(x_\omega), x_\omega \rangle.$$

If $(u_\lambda)_{\lambda \in \Lambda}$ is an approximate unit for A, then $x_\omega = \lim_\lambda \pi_A(u_\lambda)(x_\omega)$ (by non-degeneracy of (H, π_A)), so for all $b \in B$,

$$\omega_B(b) = \lim_\lambda \omega(u_\lambda \otimes b),$$

since

$$\begin{aligned}
\omega_B(b) &= \langle \pi_B(b)(x_\omega), x_\omega \rangle \\
&= \lim_\lambda \langle \pi_B(b)\pi_A(u_\lambda)(x_\omega), x_\omega \rangle \\
&= \lim_\lambda \langle \pi(u_\lambda \otimes b)(x_\omega), x_\omega \rangle \\
&= \lim_\lambda \omega(u_\lambda \otimes b).
\end{aligned}$$

Similarly, if $(v_\mu)_{\mu \in M}$ is an approximate unit for B, then for all $a \in A$,

$$\omega_A(a) = \lim_\mu \omega(a \otimes v_\mu).$$

(In particular, if B is unital, then $\omega_A(a) = \omega(a \otimes 1)$.)

If $(\tau, \rho) \in S_\gamma$ and $\omega = \tau \otimes_\gamma \rho$, then $\tau = \omega_A$ and $\rho = \omega_B$. We prove this for τ (the proof for ρ is similar): $\omega_A(a) = \lim_\mu \omega(a \otimes v_\mu) = \lim_\mu \tau(a)\rho(v_\mu) = \tau(a)$, because $\lim_\mu \rho(v_\mu) = 1$, since ρ is a state on B.

6.4.13. Theorem. *Let A and B be C*-algebras and suppose that A or B is abelian. Suppose that γ is a C*-norm on $A \otimes B$ and let $(\tau, \rho) \in S_\gamma$. Then $\tau \otimes_\gamma \rho$ is a pure state of $A \otimes_\gamma B$.*

Proof. We show this in the case that A is abelian. Let $\omega = \tau \otimes_\gamma \rho$, and $(H, \pi) = (H_\omega, \varphi_\omega)$. Let $K = [\pi_A(A)x_\omega]$. Then K is a closed vector subspace of H invariant for $\pi_A(A)$, the map

$$\psi \colon A \to B(K), \quad a \mapsto \pi_A(a)_K,$$

is a *-homomorphism, and the vector x_ω is a unit cyclic vector for the representation (K, ψ) of A. Since $\langle \psi(a)(x_\omega), x_\omega \rangle = \tau(a) = \langle \varphi_\tau(a)(x_\tau), x_\tau \rangle$ for all $a \in A$, the representations (K, ψ) and (H_τ, φ_τ) are unitarily equivalent, by Theorem 5.1.4. Because τ is a pure state, (H_τ, φ_τ) is an irreducible representation, and therefore (K, ψ) is also irreducible. Hence, $\psi(A)' = \mathbb{C}1$ (Theorem 5.1.5). Since A is abelian, $\psi(A) \subseteq \psi(A)'$, so if $a \in A$, there is a scalar λ such that $\psi(a) = \lambda 1$. Hence, $\tau(a) = \langle \psi(a)(x_\omega), x_\omega \rangle = \langle \lambda x_\omega, x_\omega \rangle = \lambda$. Therefore, $\psi(a) = \tau(a)1$.

We claim now that $\pi_A(a)\pi_B(b) = \tau(a)\pi_B(b)$, for all $a \in A$ and $b \in B$. To see this it suffices to show that

$$\pi_A(a)\pi_B(b)(x) = \tau(a)\pi_B(b)(x), \tag{4}$$

for all $x \in H$ of the form $x = \pi_A(a')\pi_B(b')(x_\omega)$ (since the set of such elements has dense linear span H). However,

$$\begin{aligned}
\pi_A(a)\pi_B(b)\pi_A(a')\pi_B(b')(x_\omega) &= \pi_B(bb')\psi(a)\pi_A(a')(x_\omega) \\
&= \pi_B(bb')\tau(a)\pi_A(a')(x_\omega) \\
&= \tau(a)\pi_B(b)\pi_A(a')\pi_B(b')(x_\omega),
\end{aligned}$$

so Eq. (4) holds, and the claim is proved. It follows directly that $\pi(A \otimes_\gamma B) = \pi_B(B)$. Hence, x_ω is a cyclic vector for (H, π_B). Since $\langle \pi_B(b)(x_\omega), x_\omega \rangle = \rho(b) = \langle \varphi_\rho(b)(x_\rho), x_\rho \rangle$ for all $b \in B$, it follows from Theorem 5.1.4 that (H, π_B) and (H_ρ, φ_ρ) are unitarily equivalent representations of B. Since ρ is pure, (H_ρ, φ_ρ) is irreducible, so (H, π_B) is irreducible. Hence, $\pi(A \otimes_\gamma B)' = \pi_B(B)' = \mathbb{C}1$, by Theorem 5.1.5, so (H, π) is an irreducible representation of $A \otimes_\gamma B$. Therefore $\omega = \tau \otimes_\gamma \rho$ is a pure state of $A \otimes_\gamma B$. $\square$

6.4.14. Lemma. *Let A and B be C*-algebras and suppose that γ is a C*-norm on $A \otimes B$, that ω is a pure state of $A \otimes_\gamma B$, and that ω_A is a pure state of A. Then $(\omega_A, \omega_B) \in S_\gamma$ and $\omega = \omega_A \otimes_\gamma \omega_B$.*

Proof. Let $(H, \pi) = (H_\omega, \varphi_\omega)$ and $\tau = \omega_A$, $\rho = \omega_B$. Let $K = [\pi_A(A)x_\omega]$ and let ψ denote the *-homomorphism

$$A \to B(K), \quad a \mapsto \pi_A(a)_K.$$

The vector x_ω is cyclic for the representation (K, ψ). Since $\langle \psi(a)(x_\omega), x_\omega \rangle = \tau(a)$ for all $a \in A$, the representations (K, ψ) and (H_τ, φ_τ) are unitarily equivalent, and since τ is pure by hypothesis, (H_τ, φ_τ) is irreducible, and therefore so is (K, ψ). Let p be the projection of H onto K. Then $p \in \pi_A(A)'$ as K is invariant for $\pi_A(A)$. If q is a projection in $p\pi_A(A)'p$, then $q(H)$ is a closed vector subspace of K invariant for (K, ψ), so $q(H) = 0$ or K by irreducibility of (K, ψ). Hence, $q = 0$ or p. Thus, the von Neumann algebra $p\pi_A(A)'p$ contains only scalar projections, and since a von Neumann algebra is the closed linear span of its projections, it follows that $p\pi_A(A)'p = \mathbf{C}p$. Now $\pi_B(B) \subseteq \pi_A(A)'$, so if $b \in B$, then there exists a scalar λ such that $p\pi_B(b)p = \lambda p$. Hence,

$$\begin{aligned}
\rho(b) &= \langle \pi_B(b)(x_\omega), x_\omega \rangle \\
&= \langle \pi_B(b)p(x_\omega), p(x_\omega) \rangle \\
&= \langle p\pi_B(b)p(x_\omega), x_\omega \rangle \\
&= \langle \lambda x_\omega, x_\omega \rangle \\
&= \lambda.
\end{aligned}$$

Thus, $p\pi_B(b)p = \rho(b)p$. If $a \in A$, then

$$\begin{aligned}
\omega(a \otimes b) &= \langle \pi(a \otimes b)(x_\omega), x_\omega \rangle \\
&= \langle \pi_A(a)\pi_B(b)p(x_\omega), p(x_\omega) \rangle \\
&= \langle \pi_A(a)p\pi_B(b)p(x_\omega), x_\omega \rangle \\
&= \langle \pi_A(a)\rho(b)x_\omega, x_\omega \rangle \\
&= \rho(b)\langle \pi_A(a)(x_\omega), x_\omega \rangle \\
&= \rho(b)\tau(a) \\
&= (\tau \otimes \rho)(a \otimes b).
\end{aligned}$$

Hence, ω extends $\tau \otimes \rho$ to $A \otimes_\gamma B$, so $\omega = \tau \otimes_\gamma \rho$.

By Theorem 6.4.7 there is a unitary $u \colon H_\tau \hat{\otimes} H_\rho \to H$ such that $\pi(c) = u(\varphi_\tau \hat{\otimes} \varphi_\rho)(c)u^*$ for all $c \in A \otimes B$. Suppose that ρ is not pure (and we shall get a contradiction). In this case there exists a non-trivial closed vector subspace L of H_ρ invariant for $\varphi_\rho(B)$. Set $L' = H_\tau \hat{\otimes} L$, so L' is a non-trivial closed vector subspace of $H_\tau \hat{\otimes} H_\rho$ invariant for $(\varphi_\tau \hat{\otimes} \varphi_\rho)(c)$ for all $c \in A \otimes B$. Hence, $L'' = u(L')$ is a non-trivial closed vector subspace of H invariant for $\pi(c)$ for all $c \in A \otimes B$, and therefore for all $c \in A \otimes_\gamma B$. This is impossible, because (H, π) is irreducible (since ω is a pure state). Thus, to avoid contradiction we conclude that ρ is pure. Hence, $(\omega_A, \omega_B) \in S_\gamma$. $\square$

6.4.15. Theorem (Takesaki). *Every abelian C*-algebra is nuclear.*

Proof. Let A, B be C*-algebras where A is abelian, and suppose that γ is a C*-norm on $A \otimes B$. Let $\omega \in \mathrm{PS}(A \otimes_\gamma B)$ and set $(H, \pi) = (H_\omega, \varphi_\omega)$ and

$\tau = \omega_A$, $\rho = \omega_B$. Since A is abelian and (H, π) is an irreducible representation of $A \otimes_\gamma B$, we have $\pi_A(A) \subseteq \pi(A \otimes_\gamma B)' = \mathbf{C}1$. Hence, if $a \in A$ there is a scalar λ such that $\pi_A(a) = \lambda 1$. Consequently, $\tau(a) = \langle \pi_A(a)(x_\omega), x_\omega \rangle = \langle \lambda x_\omega, x_\omega \rangle = \lambda$, so $\pi_A(a) = \tau(a)1$. Therefore, τ is a multiplicative state on A, so, by Theorem 5.1.6, a pure state. By Lemma 6.4.14 ρ is a pure state of B and $(\tau, \rho) \in S_\gamma$ and $\omega = \tau \otimes_\gamma \rho$. By Theorem 6.4.7, if $c \in A \otimes B$, we have $\|\pi(c)\| = \|(\varphi_\tau \hat{\otimes} \varphi_\rho)(c)\|$. Since $\gamma(c) = \sup_{\omega \in \mathrm{PS}(A \otimes_\gamma B)} \|\varphi_\omega(c)\|$ (this is got by combining Theorems 5.1.7 and 5.1.12), therefore

$$\gamma(c) = \sup_{(\tau, \rho) \in S_\gamma} \|(\varphi_\tau \hat{\otimes} \varphi_\rho)(c)\|.$$

Thus, if we show that $S_\gamma = \mathrm{PS}(A) \times \mathrm{PS}(B)$, we shall have

$$\gamma(c) = \sup_{\substack{\tau \in \mathrm{PS}(A) \\ \rho \in \mathrm{PS}(B)}} \|(\varphi_\tau \hat{\otimes} \varphi_\rho)(c)\|,$$

and since the right-hand side of this equation is completely independent of the norm γ, we shall have shown that $A \otimes B$ has a unique C*-norm.

Suppose then that $S_\gamma \neq \mathrm{PS}(A) \times \mathrm{PS}(B)$. We shall derive a contradiction and thus prove the theorem. Since S_γ is relatively closed in $\mathrm{PS}(A) \times \mathrm{PS}(B)$ (Theorem 6.4.12), there exist a pair of non-empty, relatively weak* open sets U and V in $\mathrm{PS}(A)$ and $\mathrm{PS}(B)$, respectively, such that $S_\gamma \cap (U \times V) = \emptyset$. Using Theorem 6.4.12 again, we may and do suppose that U and V are unitarily invariant (if for each unitary $u \in \tilde{A}$ we set $U^u = \{\tau^u \mid \tau \in U\}$ and similarly define V^v for each unitary $v \in \tilde{B}$, then $U' = \cup_u U^u$ and $V' = \cup_v V^v$ are relatively weak* open unitarily invariant non-empty sets in $\mathrm{PS}(A)$ and $\mathrm{PS}(B)$, respectively, such that $S_\gamma \cap (U' \times V')$ is empty—thus, we may replace U, V by U', V' if necessary). The sets $S_A = \mathrm{PS}(A) \setminus U$ and $S_B = \mathrm{PS}(B) \setminus V$ are relatively weak* closed unitarily invariant sets in $\mathrm{PS}(A)$ and $\mathrm{PS}(B)$, respectively, and since $S_A \neq \mathrm{PS}(A)$ and $S_B \neq \mathrm{PS}(B)$, it follows from Theorem 5.4.10 that the closed ideals $S_A^\perp$ and $S_B^\perp$ in A and B, respectively, are non-zero, and therefore contain non-zero positive elements a and b, respectively. If $(\tau, \rho) \in S_\gamma$, then $(\tau \otimes_\gamma \rho)(a \otimes b) = \tau(a)\rho(b) = 0$, since either $\tau \notin U$ or $\rho \notin V$. However, by Theorem 5.1.11, there is a pure state $\omega \in \mathrm{PS}(A \otimes_\gamma B)$ such that $\gamma(a \otimes b) = \omega(a \otimes b)$, and by the first part of the proof of this theorem, $\omega = \tau \otimes_\gamma \rho$ for some $(\tau, \rho) \in S_\gamma$, so $\gamma(a \otimes b) = 0$, and therefore $a \otimes b = 0$. Hence, either a or b is zero, a contradiction. $\square$

We shall need the following elementary topological fact: Suppose that Ω is a compact Hausdorff space, and $U_1, \ldots, U_n$ are open sets such that $\Omega = U_1 \cup \ldots \cup U_n$. Then there exist continuous functions $h_1, \ldots, h_n$ from Ω to $[0,1]$ such that U_j contains the support of h_j for all j and $h_1 + \cdots + h_n = 1$ [Rud 1, Theorem 2.13].

If Ω is a locally compact Hausdorff space and X a Banach space, $C_0(\Omega, X)$ denotes the Banach space of all continuous functions g from Ω to X that vanish at infinity (this means that the function $\omega \mapsto \|g(\omega)\|$ vanishes at infinity). The operations on $C_0(\Omega, X)$ are the pointwise-defined ones and the norm is the supremum norm.

(We are particularly interested in $C_0(\Omega, X)$ when X is a C*-algebra, in which case $C_0(\Omega, X)$ is a C*-algebra also, with the pointwise-defined multiplication and involution.)

If $f \in C_0(\Omega)$, and $x \in X$, denote by fx the element of $C_0(\Omega, X)$ defined by setting $fx(\omega) = f(\omega)x$.

6.4.16. Lemma. *Let Ω be a locally compact Hausdorff space and X a Banach space. Then $C_0(\Omega, X)$ is the closed linear span of the functions fx ($f \in C_0(\Omega)$, $x \in X$).*

Proof. Let $g \in C_0(\Omega, X)$. Define an extension $\tilde{g}$ of g to the one-point compactification $\tilde{\Omega}$ of Ω, by setting $\tilde{g}(\infty) = 0$ where ∞ is the point at infinity. Since g is continuous and vanishes at infinity, the function $\tilde{g}$ is continuous.

Let $\varepsilon > 0$. The set $\tilde{g}(\tilde{\Omega})$ is compact and therefore totally bounded, so there exist elements $x_1, \ldots, x_n \in \tilde{g}(\tilde{\Omega})$ such that if

$$U_j = \{\omega \in \tilde{\Omega} \mid \|\tilde{g}(\omega) - x_j\| < \varepsilon\},$$

then $\tilde{\Omega} = U_1 \cup \ldots \cup U_n$. The sets $U_1, \ldots, U_n$ are open in $\tilde{\Omega}$, so, by the elementary topological fact quoted before this lemma, there exist continuous functions $h_1, \ldots, h_n$ from $\tilde{\Omega}$ to $[0, 1]$ such that the support of h_j is contained in U_j for $j = 1, \ldots, n$ and $h_1 + \cdots + h_n = 1$. Hence,

$$\|\tilde{g}(\omega) - \sum_{j=1}^{n} h_j(\omega)x_j\| = \|\sum_{j=1}^{n} h_j(\omega)(\tilde{g}(\omega) - x_j)\|$$

$$\leq \sum_{j=1}^{n} h_j(\omega)\|\tilde{g}(\omega) - x_j\|$$

$$\leq \sum_{j=1}^{n} h_j(\omega)\varepsilon$$

$$= \varepsilon.$$

In particular, $\|\sum_{j=1}^{n} h_j(\infty)x_j\| \leq \varepsilon$, since $\tilde{g}(\infty) = 0$. Let f_j be the restriction to Ω of $h_j - h_j(\infty)$. Then $f_j \in C_0(\Omega)$ and

$$\|g - \sum_{j=1}^{n} f_j x_j\|_\infty \leq \|\tilde{g} - \sum_{j=1}^{n} h_j x_j\|_\infty + \|\sum_{j=1}^{n} h_j(\infty)x_j\| \leq 2\varepsilon.$$

This proves the lemma.   □

Let Ω be a locally compact Hausdorff space and A a C*-algebra. Since the map

$$C_0(\Omega) \times A \to C_0(\Omega, A), \quad (f, a) \mapsto fa,$$

is bilinear, it induces a unique linear map $\pi \colon C_0(\Omega) \otimes A \to C_0(\Omega, A)$ such that $\pi(f \otimes a) = fa$ for all $f \in C_0(\Omega)$ and $a \in A$. We call π the *canonical map* from $C_0(\Omega) \otimes A$ to $C_0(\Omega, A)$.

6.4.17. Theorem. *If Ω is a locally compact Hausdorff space and A a C*-algebra, then the canonical map from $C_0(\Omega) \otimes A$ to $C_0(\Omega, A)$ extends uniquely to a $*$-isomorphism from $C_0(\Omega) \otimes_* A$ to $C_0(\Omega, A)$.*

Proof. Let π be the canonical map. It is readily verified that π is a $*$-homomorphism. If $c \in \ker(\pi)$, write $c = \sum_{j=1}^n f_j \otimes a_j$, where $f_1, \ldots, f_n \in C_0(\Omega)$ and $a_1, \ldots, a_n$ are linearly independent elements of A. Then $\pi(c) = 0$ implies that $\sum_{j=1}^n f_j(\omega) a_j = 0$ for all ω in Ω. By linear independence of $a_1, \ldots, a_n$, therefore, $f_1(\omega) = \cdots = f_n(\omega) = 0$. Hence, $f_1 = \cdots = f_n = 0$, so $c = 0$. Therefore, π is injective.

The function

$$C_0(\Omega) \otimes A \to \mathbf{R}^+, \quad c \mapsto \|\pi(c)\|,$$

is a C*-norm on $C_0(\Omega) \otimes A$, and by Theorem 6.4.15 it is the only C*-norm on this algebra, so $\|\pi(c)\| = \|c\|_*$ for all $c \in C_0(\Omega) \otimes A$. Hence, π extends uniquely to an isometric $*$-homomorphism $\pi' \colon C_0(\Omega) \otimes_* A \to C_0(\Omega, A)$. Since the range of π' contains the elements fa for all $f \in C_0(\Omega)$ and $a \in A$, it follows from Lemma 6.4.16 that π' is surjective.   □

If A and B are $*$-algebras, then the unique linear map $\theta \colon A \otimes B \to B \otimes A$, such that $\theta(a \otimes b) = b \otimes a$ for all $a \in A$ and $b \in B$, is a $*$-isomorphism. Hence, if $A \otimes B$ admits a unique C*-norm, so does $B \otimes A$. This simple observation is used in the proof of the following theorem.

6.4.18. Theorem. *For any C*-algebras A and B, the spatial C*-norm is the least C*-norm on $A \otimes B$.*

Proof. Let γ be a C*-norm on $A \otimes B$. If B is non-unital, then we can extend γ to a C*-norm on $A \otimes \tilde{B}$ by Theorem 6.4.10, and the spatial C*-norm on $A \otimes \tilde{B}$ extends the spatial C*-norm of $A \otimes B$ by Theorem 6.4.9. Thus, it suffices to prove the theorem in the case that B is unital, and therefore we assume B is unital.

We show first that $S_\gamma = \mathrm{PS}(A) \times \mathrm{PS}(B)$. Suppose the contrary (and we shall get a contradiction). As in the proof of Theorem 6.4.15, we can get relatively weak* closed unitarily invariant proper subsets S_A and S_B of $\mathrm{PS}(A)$ and $\mathrm{PS}(B)$, respectively, such that $S_\gamma \subseteq S_A \times \mathrm{PS}(B) \cup \mathrm{PS}(A) \times S_B$

and the ideals $S_A^\perp$ and $S_B^\perp$ contain non-zero positive elements a_0 and b_0, respectively. Thus, for all $(\tau, \rho) \in S_\gamma$,

$$(\tau \otimes_\gamma \rho)(a_0 \otimes b_0) = \tau(a_0)\rho(b_0) = 0. \tag{5}$$

Now let C be the C*-subalgebra of B generated by b_0 and 1. This is abelian, and therefore nuclear, by Theorem 6.4.15, so $A \otimes C$ has a unique C*-norm, and therefore $\gamma = \|.\|_*$ on $A \otimes C$. Thus, we may regard $A \otimes_* C$ as a C*-subalgebra of $A \otimes_\gamma B$. Choose pure states τ on A and ρ on C such that $\tau(a_0) = \|a_0\| > 0$ and $\rho(b_0) = \|b_0\| > 0$ (this is possible by Theorem 5.1.11). Then $\tau \otimes \rho$ extends to a pure state ω' on $A \otimes_* C$, by Corollary 6.4.3 and Theorem 6.4.13. It follows from Theorem 5.1.13 that ω' can be extended in turn to a pure state ω on $A \otimes_\gamma B$. For each $a \in A$, $\omega_A(a) = \omega(a \otimes 1)$, so $\omega_A(a) = \tau(a)\rho(1) = \tau(a)$, and therefore $\omega_A = \tau$ is a pure state on A. It follows from Lemma 6.4.14 that $(\omega_A, \omega_B) \in S_\gamma$ and $\omega = \omega_A \otimes_\gamma \omega_B$. Hence, $\omega(a_0 \otimes b_0) = 0$ by Eq. (5), and yet $\omega(a_0 \otimes b_0) = (\tau \otimes \rho)(a_0 \otimes b_0) = \tau(a_0)\rho(b_0) > 0$. This contradiction shows that $S_\gamma = \mathrm{PS}(A) \times \mathrm{PS}(B)$.

The states of a C*-algebra are weak* limits of nets of convex combinations of the zero functional and the pure states (by Theorems 5.1.8 and A.14). Let the positive functionals τ, ρ on A, B, respectively, be convex combinations of the zero functional and pure states. Hence, there exist $\tau_1, \dots, \tau_n \in \{0\} \cup \mathrm{PS}(A)$ and $\rho_1, \dots, \rho_m \in \{0\} \cup \mathrm{PS}(B)$ and non-negative numbers $t_1, \dots, t_n, s_1, \dots, s_m$ such that $\sum_{i=1}^n t_i = 1$, $\sum_{j=1}^m s_j = 1$, and $\tau = \sum_{i=1}^n t_i \tau_i$ and $\rho = \sum_{j=1}^m s_j \rho_j$. Therefore, the functional

$$\tau \otimes \rho = \sum_{i=1}^n \sum_{j=1}^m t_i s_j \tau_i \otimes \rho_j$$

is continuous with respect to γ, because this is the case for each $\tau_i \otimes \rho_j$ (since $S_\gamma = \mathrm{PS}(A) \times \mathrm{PS}(B)$). Hence, if we suppose now that τ, ρ are arbitrary states of A, B, respectively, then there exist nets $(\tau_\lambda)_{\lambda \in \Lambda}$ and $(\rho_\mu)_{\mu \in M}$ of positive linear functionals on A and B, respectively, converging weak* to τ and ρ, respectively, such that $\|\tau_\lambda\|, \|\rho_\mu\| \leq 1$ and $\tau_\lambda \otimes \rho_\mu$ is continuous with respect to γ for all $\lambda \in \Lambda$ and $\mu \in M$. Reasoning as in the proof of Theorem 6.4.6, $\tau_\lambda \otimes \rho_\mu$ has a unique extension to $A \otimes_\gamma B$ which is a positive linear functional of norm $\|\tau_\lambda\|\|\rho_\mu\|$. Therefore, for all $c \in A \otimes B$,

$$|(\tau_\lambda \otimes \rho_\mu)(c)| \leq \gamma(c).$$

Since $(\tau \otimes \rho)(c) = \lim_{\lambda,\mu}(\tau_\lambda \otimes \rho_\mu)(c)$, therefore $|(\tau \otimes \rho)(c)| \leq \gamma(c)$ for all $c \in A \otimes B$. Hence, $\tau \otimes \rho$ is continuous with respect to γ.

Let D be the unitisation of $A \otimes_\gamma B$, let τ, ρ be states of A, B, respectively, and let ω be the unique state on D extending $\tau \otimes_\gamma \rho$. If $d \in D$, then the linear functional

$$\omega^d: D \to \mathbf{C}, \quad c \mapsto \omega(d^*cd),$$

is obviously positive. Since $\gamma(c^*c)1-c^*c \geq 0$, we have $\omega^d(\gamma(c^*c)1-c^*c) \geq 0$; that is, $\gamma(c^*c)\omega^d(1) \geq \omega^d(c^*c)$. Hence, if $\omega(d^*d) > 0$, we have $\gamma(c)^2 \geq \omega(d^*c^*cd)/\omega(d^*d)$, and therefore, by Theorem 6.4.8, for all $c \in A \otimes B$ we have $\|c\|_*^2 \leq \gamma(c)^2$. Thus, $\|.\|_*$ is the least C*-norm on $A \otimes B$. $\square$

6.4.4. Remark. Let A and B be C*-algebras and let γ be a C*-norm on $A \otimes B$. Then $\gamma(a \otimes b) = \|a\|\|b\|$. This is true, since $\|a\|\|b\| = \|a \otimes b\|_* \leq \gamma(a \otimes b)$ (by Theorem 6.4.18) $\leq \|a\|\|b\|$ (by Corollary 6.3.6).

6.4.19. Theorem. *If (H,φ) and (K,ψ) are faithful representations of C*-algebras A and B, respectively, then*

$$\|(\varphi \hat{\otimes} \psi)(c)\| = \|c\|_* \qquad (c \in A \otimes B).$$

Proof. The function

$$\gamma: A \otimes B \to \mathbf{R}^+, \quad c \mapsto \|(\varphi \hat{\otimes} \psi)(c)\|,$$

is a C*-norm, since $\varphi \hat{\otimes} \psi: A \otimes B \to B(H \hat{\otimes} K)$ is injective by Theorem 6.3.3. Hence, $\gamma(c) \geq \|c\|_*$ for all $c \in A \otimes B$ by Theorem 6.4.18. However, by Theorem 6.4.4, the reverse inequality also holds, so $\gamma = \|.\|_*$. $\square$

6.5. Nuclear C*-Algebras and Short Exact Sequences

We continue our investigation of nuclear C*-algebras. The principal result of this section is Theorem 6.5.3, which asserts that extensions of nuclear C*-algebras by nuclear C*-algebras are themselves nuclear.

6.5.1. Theorem. *Let A, B, A', B' be C*-algebras and let $\varphi: A \to A'$ and $\psi: B \to B'$ be *-homomorphisms. Then there is a unique *-homomorphism $\pi: A \otimes_* B \to A' \otimes_* B'$ such that*

$$\pi(a \otimes b) = \varphi(a) \otimes \psi(b) \qquad (a \in A, \ b \in B).$$

Moreover, if φ and ψ are injective, so is π.

Proof. Let (H',φ') and (K',ψ') be faithful representations of A' and B', respectively. Then $\varphi' \hat{\otimes} \psi'$ is isometric on $A' \otimes B'$ for the spatial C*-norm by Theorem 6.4.19. Let $\pi = \varphi \otimes \psi$. Then $(\varphi' \hat{\otimes} \psi')\pi = \varphi'\varphi \hat{\otimes} \psi'\psi$, so by Theorem 6.4.4, $\|(\varphi' \hat{\otimes} \psi')\pi(c)\| \leq \|c\|_*$. Hence, $\|\pi(c)\|_* = \|(\varphi' \hat{\otimes} \psi')\pi(c)\| \leq \|c\|_*$ for all $c \in A \otimes B$.

If the *-homomorphisms φ and ψ are injective, so are $\varphi'\varphi$ and $\psi'\psi$. Therefore, by Theorem 6.4.19 $\|(\varphi'\varphi \hat{\otimes} \psi'\psi)(c)\| = \|c\|_*$ for all $c \in A \otimes B$, so π is isometric for the spatial C*-norms. The theorem now follows, since we can extend π to a *-homomorphism from $A \otimes_* B$ to $A' \otimes_* B'$. $\square$

We denote the *-homomorphism π in Theorem 6.5.1 by $\varphi \otimes_* \psi$.

6.5.1. *Remark.* If A and B are C*-subalgebras of the C*-algebras A' and B', respectively, and i and j are the inclusion *-homomorphisms, then by Theorem 6.5.1 the *-homomorphism $i \otimes_* j \colon A \otimes_* B \to A' \otimes_* B'$ is injective. Thus, we may regard $A \otimes_* B$ as a C*-subalgebra of $A' \otimes_* B'$.

Let J, A, and B be C*-algebras. Suppose that $j \colon J \to A$ is an injective *-homomorphism and $\pi \colon A \to B$ is a surjective *-homomorphism, and that $\mathrm{im}(j) = \ker(\pi)$. Then we say that the sequence

$$0 \to J \xrightarrow{j} A \xrightarrow{\pi} B \to 0$$

is a *short exact sequence* of C*-algebras. In this case we also say that A is an *extension* of B by J.

If $\pi \colon A \to B$ is a surjective *-homomorphism of C*-algebras, $J = \ker(\pi)$, and $j \colon J \to A$ is the inclusion map, then

$$0 \to J \xrightarrow{j} A \xrightarrow{\pi} B \to 0$$

is a short exact sequence of C*-algebras.

6.5.2. Theorem. *Let J, A, B, and D be C*-algebras and suppose that*

$$0 \to J \xrightarrow{j} A \xrightarrow{\pi} B \to 0$$

is a short exact sequence of C-algebras. Suppose also that $B \otimes D$ has a unique C*-norm (this is the case if B or D is nuclear). Then*

$$0 \to J \otimes_* D \xrightarrow{j \, \otimes_* \, \mathrm{id}} A \otimes_* D \xrightarrow{\pi \, \otimes_* \, \mathrm{id}} B \otimes_* D \to 0$$

is a short exact sequence of C-algebras.*

Proof. Let $\bar{j} = j \otimes_* \mathrm{id}_D$ and $\bar{\pi} = \pi \otimes_* \mathrm{id}_D$. That $\bar{j}$ is injective follows from Theorem 6.5.1. The map $\bar{\pi}$ is surjective, since $\mathrm{im}(\bar{\pi})$ contains $\bar{\pi}(A \otimes D) = \pi(A) \otimes D = B \otimes D$. The C*-subalgebra $\mathrm{im}(\bar{j}) = \mathrm{im}(j) \otimes_* D$ of $A \otimes_* D$ is an ideal, since $\mathrm{im}(j)$ is a closed ideal of A. Let Q be the quotient C*-algebra $(A \otimes_* D)/\mathrm{im}(\bar{j})$, and ψ the quotient map from $A \otimes_* D$ to Q. Clearly, $\bar{\pi}(\mathrm{im}(\bar{j})) = 0$, so there exists a unique *-homomorphism $\pi' \colon Q \to B \otimes_* D$ such that $\pi'\psi = \bar{\pi}$. We shall show that π' is a *-isomorphism, and this will imply that $\ker(\bar{\pi}) = \ker(\psi) = \mathrm{im}(\bar{j})$, thus proving the theorem.

That π' is surjective is immediate from the surjectivity of $\bar{\pi}$. We show injectivity of π' by constructing a left inverse. The map

$$B \times D \to Q, \quad (\pi(a), d) \mapsto a \otimes d + \mathrm{im}(\bar{j}),$$

is well-defined, and it is readily verified that it is bilinear, so it induces a unique linear map $\varphi \colon B \otimes D \to Q$ such that $\varphi(\pi(a) \otimes d) = a \otimes d + \operatorname{im}(\bar{j})$ for all $a \in A$ and $d \in D$. It is easily checked that φ is a *-homomorphism. The function

$$B \otimes D \to \mathbf{R}^+, \quad c \mapsto \max(\|\varphi(c)\|, \|c\|_*),$$

is a C*-norm, so, by the assumption that $B \otimes D$ has a unique C*-norm, we have $\max(\|\varphi(c)\|, \|c\|_*) = \|c\|_*$, and therefore, $\|\varphi(c)\| \leq \|c\|_*$, for all $c \in B \otimes D$. Hence, φ extends to a *-homomorphism from $B \otimes_* D$ to Q which we shall also denote by φ. Now $\varphi\pi' = \operatorname{id}_Q$, since for all $a \in A$ and $d \in D$ we have $\varphi\pi'(a \otimes d + \operatorname{im}(\bar{j})) = \varphi\bar{\pi}(a \otimes d) = \varphi(\pi(a) \otimes d) = a \otimes d + \operatorname{im}(\bar{j})$. Therefore, π' is injective, and the theorem is proved. $\square$

6.5.3. Theorem. *An extension of a nuclear C*-algebra by a nuclear C*-algebra is itself nuclear.*

Proof. Let J, A, and B be C*-algebras, and suppose that

$$0 \to J \xrightarrow{j} A \xrightarrow{\pi} B \to 0$$

is a short exact sequence of C*-algebras, and that J and B are nuclear. We prove that A also is nuclear.

Let D be an arbitrary C*-algebra. Since $B \otimes D$ has a unique C*-norm, it follows from Theorem 6.5.2 that the following is a short exact sequence:

$$0 \to J \otimes_* D \xrightarrow{\bar{j}} A \otimes_* D \xrightarrow{\bar{\pi}} B \otimes_* D \to 0.$$

We are using $\bar{j}$ and $\bar{\pi}$ to denote $j \otimes_* \operatorname{id}$ and $\pi \otimes_* \operatorname{id}$, respectively. The identity map on $A \otimes D$ extends to a *-homomorphism $\varphi \colon A \otimes_{max} D \to A \otimes_* D$, since $\|.\|_* \leq \|.\|_{max}$. We shall have proved the theorem if we show that $\|.\|_{max} = \|.\|_*$ on $A \otimes D$, since any C*-norm must lie between $\|.\|_*$ and $\|.\|_{max}$ (Theorem 6.4.18). Therefore, we need only show that φ is injective.

Let j' denote the unique *-homomorphism from $J \otimes D$ to $A \otimes_{max} D$ such that $j'(a \otimes d) = j(a) \otimes d$ for all $a \in J$ and $d \in D$. By nuclearity of J, the C*-norm

$$J \otimes D \to \mathbf{R}^+, \quad c \mapsto \max(\|j'(c)\|_{max}, \|c\|_*),$$

is the same as the spatial C*-norm $\|.\|_*$ on $J \otimes D$. Hence, j' is norm-decreasing for $\|.\|_*$ and so extends to a *-homomorphism from $J \otimes_* D$ to $A \otimes_{max} D$ which we shall also denote by j'. Clearly, $\bar{j} = \varphi j'$.

There is a unique *-homomorphism $\pi' \colon A \otimes D \to B \otimes_* D$ such that $\pi'(a \otimes d) = \pi(a) \otimes d$ for all $a \in A$ and $d \in D$. The function

$$A \otimes D \to \mathbf{R}^+, \quad c \mapsto \max(\|\pi'(c)\|_*, \|c\|_{max}),$$

is a C*-norm and, therefore, it is dominated by the maximal C*-norm $\|.\|_{max}$. Hence, π' is norm-decreasing for $\|.\|_{max}$, so π' can be extended to a *-homomorphism from $A \otimes_{max} D$ to $B \otimes_* D$ which we shall also denote by π'.

Let Q be the quotient algebra of $A \otimes_{max} D$ by the closed ideal $\mathrm{im}(j')$, and let $\psi \colon A \otimes_{max} D \to Q$ be the quotient map. By a construction similar to that carried out in the proof of Theorem 6.5.2, there is a unique *-homomorphism $\theta \colon B \otimes_* D \to Q$ such that $\theta(\pi(a) \otimes d) = a \otimes d + \mathrm{im}(j')$ for all $a \in A$ and $d \in D$ (this uses nuclearity of B). We therefore get a commutative diagram:

$$
\begin{array}{ccccc}
J \otimes_* D & \xrightarrow{j \otimes_* \mathrm{id}} & A \otimes_* D & \xrightarrow{\pi \otimes_* \mathrm{id}} & B \otimes_* D \\
& \searrow{j'} & \uparrow{\varphi} & \nearrow{\pi'} & \downarrow{\theta} \\
& & A \otimes_{max} D & \xrightarrow{\psi} & Q.
\end{array}
$$

Now suppose that $c \in \ker(\varphi)$. Then $0 = \bar\pi\varphi(c) = \pi'(c)$, so $0 = \theta\pi'(c) = \psi(c)$. Hence, $c = j'(c_0)$ for some element $c_0 \in J \otimes_* D$, and therefore $\bar\jmath(c_0) = \varphi j'(c_0) = \varphi(c) = 0$. Since $\bar\jmath$ is injective by Theorem 6.5.1, we have $c_0 = 0$ and therefore $c = j'(c_0) = 0$. Thus, φ is injective and the theorem is proved. $\square$

6.5.1. Example. Let $\mathbf{A}$ denote the Toeplitz algebra (the C*-algebra generated by all Toeplitz operators on the Hardy space H^2 having continuous symbol). This algebra was investigated in Section 3.5, where it was shown that its commutator ideal is $K(H^2)$ (Theorem 3.5.10). The algebras $K(H^2)$ and $\mathbf{A}/K(H^2)$ are nuclear (by Example 6.3.2 and Theorem 6.4.15, respectively), so by Theorem 6.5.3, $\mathbf{A}$ is nuclear.

6. Exercises

1. Let $(A_n, \varphi_n)_{n=1}^\infty$ and $(B_n, \psi_n)_{n=1}^\infty$ be direct sequences of C*-algebras with direct limits A and B, respectively. Let $\varphi^n \colon A_n \to A$ and $\psi^n \colon B_n \to B$ be the natural maps. Suppose there are *-homomorphisms $\pi_n \colon A_n \to B_n$ such that for each n the following diagram commutes:

$$
\begin{array}{ccc}
A_n & \xrightarrow{\varphi_n} & A_{n+1} \\
\downarrow{\pi_n} & & \downarrow{\pi_{n+1}} \\
B_n & \xrightarrow{\psi_n} & B_{n+1}.
\end{array}
$$

Show that there exists a unique *-homomorphism $\pi: A \to B$ such that for each n the following diagram commutes:

$$
\begin{array}{ccc}
A_n & \xrightarrow{\varphi^n} & A \\
\downarrow \pi_n & & \downarrow \pi \\
B_n & \xrightarrow{\psi^n} & B.
\end{array}
$$

Show that if all the π_n are *-isomorphisms, then π is a *-isomorphism.

2. Show that every non-zero finite-dimensional C*-algebra admits a faithful tracial state. Give an example of a unital simple C*-algebra not having a tracial state.

3. Let A be a C*-algebra. A *trace* on A is a function $\tau: A^+ \to [0, +\infty]$ such that

$$
\begin{aligned}
\tau(a + b) &= \tau(a) + \tau(b) \\
\tau(ta) &= t\tau(a) \\
\tau(c^*c) &= \tau(cc^*)
\end{aligned}
$$

for all $a, b \in A^+$, $c \in A$, and all $t \in \mathbf{R}^+$. We use the convention that $0.(+\infty) = 0$.

The motivating example is the usual trace function on $B(H)$. Another example is got on $C_0(\mathbf{R})$ by setting $\tau(f) = \int f \, dm$ where $f \in C_0(\mathbf{R})^+$ and m is ordinary Lebesgue measure on $\mathbf{R}$.

Traces (and their generalisation, *weights*) play a fundamental role, especially in von Neumann algebra theory ([Ped], [Tak]).

Let

$$
A_\tau^2 = \{a \in A \mid \tau(a^*a) < \infty\}.
$$

Show that

$$
(a + b)^*(a + b) \leq 2a^*a + 2b^*b
$$

and

$$
(ab)^*ab \leq \|a\|^2 b^*b,
$$

and deduce that A_τ^2 is a self-adjoint ideal of A.

Let A_τ be the linear span of all products ab, where $a, b \in A_\tau^2$. Show that A_τ is a self-adjoint ideal of A.

Show that for arbitrary $a, b \in A$,

$$
a^*b = \tfrac{1}{4} \sum_{k=0}^{3} i^k (b + i^k a)^*(b + i^k a),
$$

and if a^*b is self-adjoint,

$$a^*b = \tfrac{1}{4}[(b+a)^*(b+a) - (b-a)^*(b-a)].$$

Let

$$A_\tau^+ = \{a \in A^+ \mid \tau(a) < \infty\}.$$

Show that A_τ is the linear span of A_τ^+ and $A_\tau^+ = A_\tau \cap A^+$.

Show that there is a unique positive linear extension (also denoted τ) of τ to A_τ. Show that

$$\tau(ab) = \tau(ba)$$

for all $a, b \in A_\tau^2$, and deduce that this equation also holds for all $a \in A$ and $b \in A_\tau$.

4. Show that an AF-algebra admits a sequential approximate unit consisting of projections.

5. Let u be a normal operator on a Hilbert space H. Show that there is a commuting sequence of projections on H such that the C*-algebra that they generate contains u.

Use this to construct an example of a C*-subalgebra of an AF-algebra which is not an AF-algebra.

6. If A is an AF-algebra, show that $M_n(A)$ is one also.

7. Show that if A and B are AF-algebras, then $A \otimes_* B$ is an AF-algebra.

8. Show that if $A, B,$ and C are *-algebras, then the unique linear map $\varphi \colon (A \otimes B) \otimes C \to B \otimes (A \otimes C)$, such that $\varphi((a \otimes b) \otimes c) = b \otimes (a \otimes c)$ for all $a \in A$, $b \in B$, and $c \in C$, is a *-isomorphism.

Deduce that if A is a nuclear C*-algebra, so is $M_n(A)$.

9. If $H_1, H_2,$ and H_3 are Hilbert spaces, show that there exists a unique unitary $u \colon (H_1 \,\hat{\otimes}\, H_2) \,\hat{\otimes}\, H_3 \to H_1 \,\hat{\otimes}\, (H_2 \,\hat{\otimes}\, H_3)$ such that

$$u((x_1 \otimes x_2) \otimes x_3) = x_1 \otimes (x_2 \otimes x_3) \qquad (x_j \in H_j,\ j = 1, 2, 3).$$

Show that

$$u((v_1 \,\hat{\otimes}\, v_2) \,\hat{\otimes}\, v_3)u^* = v_1 \,\hat{\otimes}\, (v_2 \,\hat{\otimes}\, v_3) \qquad (v_j \in B(H_j),\ j = 1, 2, 3).$$

Deduce that if $A_1, A_2,$ and A_3 are C*-algebras, then there exists a unique *-isomorphism $\theta \colon (A_1 \otimes_* A_2) \otimes_* A_3 \to A_1 \otimes_* (A_2 \otimes_* A_3)$ such that

$$\theta((a_1 \otimes a_2) \otimes a_3) = a_1 \otimes (a_2 \otimes a_3) \qquad (a_j \in A_j,\ j = 1, 2, 3).$$

10. If A, B are C*-algebras, show that there exists a unique *-isomorphism $\theta \colon A \otimes_* B \to B \otimes_* A$ such that $\theta(a \otimes b) = b \otimes a$ ($a \in A$, $b \in B$).

6. Addenda

The hyperfinite factor exhibited in Theorem 6.2.5 is of Type II$_1$ if the UHF algebra A is infinite-dimensional.

If A is a separable C*-algebra, then A is an AF-algebra if and only if for any $a_1, \ldots, a_n$ in A and $\varepsilon > 0$ there is a finite-dimensional C*-subalgebra B of A and there exist $b_1, \ldots, b_n$ in B such that $\|a_j - b_j\| < \varepsilon$ for $1 \leq j \leq n$.

A hereditary C*-subalgebra of an AF-algebra is an AF-algebra.

If I is a closed ideal in a C*-algebra A such that I and A/I are AF-algebras, then A is an AF-algebra. This was first proved by L. Brown using K-theory.

Reference: [Eff].

If A and B are simple C*-algebras, then $A \otimes_* B$ is simple.

Postliminal C*-algebras are nuclear.

If I is a closed ideal in a nuclear C*-algebra A, then I and A/I are nuclear. It follows from this and Theorem 6.3.10 that the direct limit of a sequence of nuclear C*-algebras is nuclear.

A hereditary C*-subalgebra of a nuclear C*-algebra is nuclear. An arbitrary C*-subalgebra of a nuclear C*-algebra need not be nuclear.

If H is an infinite-dimensional Hilbert space, then $B(H)$ is non-nuclear.

References: [Lan], [Sak], [Tak].

CHAPTER 7

K-Theory of C*-Algebras

One of the most important recent developments in C*-algebra theory has been the introduction of homological algebraic methods. Specifically, the theory we investigate in this chapter, the K-theory of C*-algebras, has had some spectacular successes in solving long-open problems. The basic idea of this theory is to associate with each C*-algebra A two abelian groups $K_0(A)$ and $K_1(A)$, which reflect some of the properties of A. In the first section of this chapter, we present some elementary results concerning $K_0(A)$, and in the second we use $K_0(A)$ to show how AF-algebras can be classified. In Sections 3, 4, and 5 we establish the basic properties of K-theory, including Bott periodicity.

7.1. Elements of K-Theory

Let A be a $*$-algebra. If $a = (a_{ij})$ and $b = (b_{jk})$ are, respectively, an $m \times n$ matrix and $n \times p$ matrix with entries in A, then the product $c = ab$ is an $m \times p$ matrix with (i, k)-entry given by $c_{ik} = \sum_{j=1}^{n} a_{ij} b_{jk}$. Also, the *adjoint* a^* is the $n \times m$ matrix with (j, i)-entry a_{ij}^*.

We shall have frequent need to use block matrices, so we shall state here a few elementary results concerning them. Let $r = (r_1, \ldots, r_m)$ and $c = (c_1, \ldots, c_n)$ be tuples of positive integers, and suppose that for each integer i, such that $1 \leq i \leq m$, and j, such that $1 \leq j \leq n$, we have an $r_i \times c_j$ matrix A_{ij} with entries in A. The $r \times c$ *block* matrix

$$
a = \begin{pmatrix}
A_{11} & A_{12} & \ldots & A_{1n} \\
A_{21} & A_{22} & \ldots & A_{2n} \\
\vdots & \vdots & \ddots & \vdots \\
A_{m1} & A_{m2} & \ldots & A_{mn}
\end{pmatrix}
\tag{1}
$$

217

is regarded in an obvious fashion as an $(r_1 + \cdots + r_m) \times (c_1 + \cdots + c_n)$ matrix with entries in A. The matrix a^* is the $c \times r$ block matrix

$$a^* = \begin{pmatrix} A_{11}^* & A_{21}^* & \cdots & A_{m1}^* \\ A_{12}^* & A_{22}^* & \cdots & A_{m2}^* \\ \vdots & \vdots & \ddots & \vdots \\ A_{1n}^* & A_{2n}^* & \cdots & A_{mn}^* \end{pmatrix}.$$

If b is a $c \times d$ block matrix, where $d = (d_1, \ldots, d_p)$, and b is given in block form by

$$b = \begin{pmatrix} B_{11} & B_{12} & \cdots & B_{1p} \\ B_{21} & B_{22} & \cdots & B_{2p} \\ \vdots & \vdots & \ddots & \vdots \\ B_{n1} & B_{n2} & \cdots & B_{np} \end{pmatrix},$$

then the product ab is the $r \times d$ block matrix

$$ab = \begin{pmatrix} C_{11} & C_{12} & \cdots & C_{1p} \\ C_{21} & C_{22} & \cdots & C_{2p} \\ \vdots & \vdots & \ddots & \vdots \\ C_{m1} & C_{m2} & \cdots & C_{mp} \end{pmatrix},$$

where $C_{ik} = \sum_{j=1}^{n} A_{ij} B_{jk}$. In words, to multiply two block matrices, perform the usual matrix multiplication on the corresponding blocks. The proofs of these results are elementary exercises.

If the block matrix a in Eq. (1) has $m = n$ and zero off-diagonal entries—that is, A_{ij} is the zero matrix for $i \neq j$—we shall denote a by $A_{11} \oplus A_{22} \oplus \cdots \oplus A_{mm}$.

We denote by 0_n the $n \times n$ matrix all of whose entries are 0, and if A is unital, we denote by 1_n the $n \times n$ matrix all of whose entries are zero, except for those on the main diagonal, all of which equal 1.

Let A be an arbitrary $*$-algebra, and set

$$P[A] = \cup_{n=1}^{\infty} \{ p \in M_n(A) \mid p \text{ is a projection} \}.$$

If $p, q \in P[A]$ we say p and q are *equivalent*, and write $p \sim q$, if there is a rectangular matrix u with entries in A such that $p = u^*u$ and $q = uu^*$. If this is the case, we may suppose that $u = uu^*u$, by replacing u by qup if necessary. It is a straightforward exercise to check that $\sim$ is an equivalence relation on $P[A]$. If p and q are contained in the same algebra $M_n(A)$, then $p \sim q$ if and only if p and q are Murray–von Neumann equivalent in the sense that we defined this in Section 4.1.

7.1.1. Theorem. *Let A be a $*$-algebra, and suppose that p, q, p', q' are projections in $P[A]$.*

(1) *If $p \sim p'$ and $q \sim q'$, then $p \oplus q \sim p' \oplus q'$.*
(2) *$p \oplus q \sim q \oplus p$.*
(3) *If $p, q \in M_n(A)$ and $pq = 0$, then $p + q \sim p \oplus q$.*

Proof. Suppose that $p \sim p'$ and $q \sim q'$. If $p = u^*u$, $p' = uu^*$, $q = v^*v$, and $q' = vv^*$, then $p \oplus q = w^*w$ and $p' \oplus q' = ww^*$, where $w = u \oplus v$. Therefore, $p \oplus q \sim p' \oplus q'$.

To see that $p \oplus q \sim q \oplus p$, set

$$u = \begin{pmatrix} 0 & q \\ p & 0 \end{pmatrix}.$$

Then $p \oplus q = u^*u$ and $q \oplus p = uu^*$.

Observe that if $p \in M_n(A)$, then $p \sim p \oplus 0_m$, for if $u = (p, 0_{nm})$ where 0_{nm} is the $n \times m$ matrix all of whose entries are zero, then $u^*u = p \oplus 0_m$ and $uu^* = p$.

Suppose now that $p, q \in M_n(A)$ and $pq = 0$. Set

$$u = \begin{pmatrix} p & q \\ 0_n & 0_n \end{pmatrix}.$$

Then $u^*u = p \oplus q$ and $uu^* = (p + q) \oplus 0_n$. Hence, we have $p + q \sim (p + q) \oplus 0_n \sim p \oplus q$. $\qquad\square$

Let A be a unital $*$-algebra. We say elements p, q of $P[A]$ are *stably equivalent*, and we write $p \approx q$, if there is a positive integer n such that $1_n \oplus p \sim 1_n \oplus q$. It is easy to check that $\approx$ is an equivalence relation on $P[A]$. Observe that if $p \approx p'$ and $q \approx q'$, then $p \oplus q \approx p' \oplus q'$. For $p \in P[A]$, let $[p]$ denote its stable equivalence class, and denote by $K_0(A)^+$ the set of all these equivalence classes. For $[p], [q] \in K_0(A)^+$, define $[p] + [q] = [p \oplus q]$.

If there is a possibility of ambiguity, we write $[p]_A$ for the stable equivalence class relative to the algebra A.

7.1.2. Theorem. *If A is a unital $*$-algebra, then $K_0(A)^+$ is a cancellative abelian semigroup with zero element $[0]$.*

Proof. Associativity and commutativity are immediate. It is also clear that $[0_n]$ is the zero element of $K_0(A)^+$.

To show that $K_0(A)^+$ is cancellative, suppose that $[p] + [q] = [p] + [r]$, and we shall show that $[q] = [r]$. For some integer m, we have $p \oplus q \oplus 1_m \sim p \oplus r \oplus 1_m$. If we suppose that $p \in M_n(A)$, then $(1_n - p) \oplus p \oplus q \oplus 1_m \sim (1_n - p) \oplus p \oplus r \oplus 1_m$. But $(1_n - p) \oplus p \sim 1_n$, by Theorem 7.1.1, Condition (3), so $1_n \oplus q \oplus 1_m \sim 1_n \oplus r \oplus 1_m$, and therefore $1_{n+m} \oplus q \sim 1_{n+m} \oplus r$. Hence, $[q] = [r]$. $\qquad\square$

Let N be a cancellative abelian semigroup with a zero element. We define an equivalence relation $\sim$ on $N \times N$ by setting $(x, y) \sim (z, t)$ if $x + t = y + z$. Denote by $[x, y]$ the equivalence class of (x, y). The set $G(N)$ of equivalence classes is an abelian group under the well-defined operation

$$[x, y] + [z, t] = [x + z, y + t].$$

(The inverse of $[x, y]$ is $[y, x]$.) We call $G(N)$ the *enveloping* or *Grothendieck* group of N. The map

$$\varphi: N \to G(N), \quad x \mapsto [x, 0]$$

is a homomorphism (that is, $\varphi(x + y) = \varphi(x) + \varphi(y)$ for all $x, y \in N$), and since it is also injective, we can and do identify N as a subsemigroup of $G(N)$ by identifying x with $[x, 0]$. Hence, $G(N) = \{x - y \mid x, y \in N\}$.

If $\psi: N \to G$ is a homomorphism, where G is an abelian group, then there is a unique homomorphism $\tilde{\psi}: G(N) \to G$ extending ψ.

The elementary proofs of these results are left as exercises.

If A is a unital $*$-algebra, we define $K_0(A)$ to be the Grothendieck group of $K_0(A)^+$.

If $\varphi: A \to B$ is a $*$-homomorphism of $*$-algebras and $a = (a_{ij})$ is an $m \times n$ matrix with entries in A, set $\varphi(a) = (\varphi(a_{ij}))$, so $\varphi(a)$ is an $m \times n$ matrix with entries in B. If b is an $n \times p$ matrix with entries in A, then $\varphi(ab) = \varphi(a)\varphi(b)$. Observe also that $\varphi(a^*) = (\varphi(a))^*$. If $p \sim q$, then $\varphi(p) \sim \varphi(q)$ and if $\varphi: A \to B$ is a unital $*$-homomorphism of unital $*$-algebras, then $p \approx q \Rightarrow \varphi(p) \approx \varphi(q)$. Hence, there is a well-defined map $\varphi_*: K_0(A)^+ \to K_0(B)^+$ given by setting $\varphi_*[p] = [\varphi(p)]$. Since $\varphi(p \oplus q) = \varphi(p) \oplus \varphi(q)$, we have $\varphi_*([p] + [q]) = \varphi_*([p]) + \varphi_*([q])$; that is, φ_* is a homomorphism. Hence, there is a unique homomorphism $\varphi_*: K_0(A) \to K_0(B)$ such that $\varphi_*([p]) = [\varphi(p)]$.

If $\varphi: A \to B$ and $\psi: B \to C$ are unital $*$-homomorphisms of unital $*$-algebras, then $(\psi\varphi)_* = \psi_*\varphi_*$. Also, $(\mathrm{id}_A)_* = \mathrm{id}_{K_0(A)}$. Thus, we have a covariant functor

$$A \mapsto K_0(A), \qquad \varphi \mapsto \varphi_*,$$

from the category of all unital $*$-algebras to the category of abelian groups.

7.1.1. Example. It is easy to check that projections $p, q \in P[\mathbf{C}]$ are equivalent if and only if they have the same rank. Also, for any $p, q \in P[\mathbf{C}]$, we have $\mathrm{rank}(p \oplus q) = \mathrm{rank}(p) + \mathrm{rank}(q)$. Thus, we may define a homomorphism $\mathrm{rank}: K_0(\mathbf{C})^+ \to \mathbf{Z}$ by setting $\mathrm{rank}([p]) = \mathrm{rank}(p)$. Hence, we can extend uniquely to get a homomorphism $\mathrm{rank}: K_0(\mathbf{C}) \to \mathbf{Z}$. This function is an isomorphism: It is surjective, since $1 = \mathrm{rank}([1_1])$, and it is injective, since if $x \in \ker(\mathrm{rank})$ we can write $x = [p] - [q]$, where $p, q \in P[\mathbf{C}]$

are projections of the same rank, and therefore equivalent, from which $[p] = [q]$, and $x = 0$.

Motivated by this example, one should think of $K_0(A)$ as a "dimension" group, and think of $[p]$ as the "generalised dimension" of p.

7.1.2. Example. A non-trivial algebra may have trivial K_0-group. For instance, if H is a separable infinite-dimensional Hilbert space, then for any pair of infinite-rank projections p, q on H we have $p \sim q$ (cf. Remark 4.1.5). Since $M_n(B(H)) = B(H^{(n)})$, it follows that for any $p \in P[B(H)]$, we have $1_1 \oplus p \sim 1_1$, so $[p] = 0$. Hence, $K_0(B(H)) = 0$.

7.1.1. Remark. It is easy to verify that for any positive integer n and any $p, q \in P[M_n(\mathbf{C})]$, we have $p \approx q$ if and only if $p \sim q$. From this it follows easily that for any finite-dimensional C*-algebra A and $p, q \in P[A]$, p and q are stably equivalent if and only if they are equivalent.

7.2. The K-Theory of AF-Algebras

We show in this section that the K_0-group of a unital AF-algebra A, endowed with some additional structure, is a complete isomorphism invariant of A (Elliott's theorem). The additional structure on $K_0(A)$ is a naturally defined partial ordering (together with the "base point" $[1_1]$). The C*-algebraic concept that enables us to get the ordering is stable finiteness:

A unital C*-algebra A is *stably finite* if, for every positive integer n and $u \in M_n(A)$ such that $u^*u = 1$, we have $uu^* = 1$.

7.2.1. Theorem. *If A is a unital AF-algebra, then it is stably finite.*

Proof. First observe that if $A = M_n(\mathbf{C})$ and u is an element of A having a left inverse, that is, there is an element $v \in A$ such that $vu = 1$, then by elementary linear algebra $uv = 1$. Since a finite-dimensional C*-algebra A is a direct sum of a finite number of such matrix algebras $M_n(\mathbf{C})$, it follows in this case that an element of A that is left invertible is invertible.

Now suppose that A is a unital AF-algebra. To prove the theorem, we have to show that if $u \in M_n(A)$ and $u^*u = 1$, then $uu^* = 1$. Since $M_n(A)$ is a unital AF-algebra, it suffices to show the result in the case $n = 1$. Suppose then $u \in A$ and $u^*u = 1$. There is a sequence (u_n) in A converging to u such that each u_n is contained in a finite-dimensional C*-subalgebra A_n of A containing the unit of A. Since $1 = u^*u = \lim_{n \to \infty} u_n^* u_n$, we may suppose that $\|1 - u_n^* u_n\| < 1$ (by going to a subsequence if necessary) and therefore $u_n^* u_n$ is invertible in A_n. It follows that u_n is left invertible in A_n, and therefore (since A_n is finite-dimensional), u_n is invertible in A_n. If v_n is the inverse of u_n^*, then $u = \lim_{n \to \infty} v_n u_n^* u_n$, so $u = \lim_{n \to \infty} v_n$. Hence, $uu^* = \lim_{n \to \infty} v_n u_n^*$, and since $v_n u_n^* = 1$, therefore $uu^* = 1$. $\square$

A *partially ordered group* is a pair $(G, \leq)$ consisting of an abelian group G and a partial order $\leq$ on G such that if

$$G^+ = \{x \in G \mid 0 \leq x\},$$

then $G = G^+ - G^+$, and if $x \leq y$, then $x + z \leq y + z$, for all $x, y, z \in G$.

If G is an abelian group and N is a subset such that $N + N \subseteq N$, $G = N - N$, and $N \cap (-N) = \{0\}$, we call N a *cone* on G. If G is a partially ordered group, then G^+ is a cone on G.

If G is an abelian group and N is a cone on G, then we define a partial order $\leq$ on G by setting $x \leq y$ if $y - x \in N$. Clearly, $(G, \leq)$ is a partially ordered group with $G^+ = N$. We say $\leq$ is the partial order *induced* by N.

Partially ordered groups exist in great abundance. For instance, every subgroup of the additive group $\mathbf{R}$ is a partially ordered group with the order induced from $\mathbf{R}$. An important example is given by $\mathbf{Z}^k$. This is a partially ordered group with the partial order induced by the cone $\mathbf{N}^k$.

7.2.2. Theorem. *If A is a stably finite unital C*-algebra, then $K_0(A)^+$ is a cone on $K_0(A)$.*

Proof. The only thing not obvious is that $x \in K_0(A)^+ \cap (-K_0(A)^+) \Rightarrow x = 0$. Suppose that $x = [p] = -[q]$ for some $p, q \in P[A]$. Then $[p \oplus q] = 0$, so if $r = p \oplus q$, and $r \in M_n(A)$, then $[1_n] = [1_n - r]$, and therefore, for some positive integer m, we have $1_m \oplus (1_n - r) \sim 1_m \oplus 1_n = 1_{m+n}$. Thus, there exists $u \in M_{n+m}(A)$ such that $u^*u = 1_{m+n}$ and $uu^* = 1_m \oplus (1_n - r)$. Since A is stably finite, $uu^* = 1_{m+n}$, so $r = 0$. Hence, p and q are zero projections, and therefore, $x = 0$. $\square$

If A is as in Theorem 7.2.2, then $K_0(A)$ is a partially ordered group with partial order induced by $K_0(A)^+$.

7.2.3. Lemma. *Let A be a C*-algebra and $p_1, \ldots, p_n$ projections in $P[A]$, and q a projection in A such that $q \sim p_1 \oplus \cdots \oplus p_n$. Then there exist pairwise orthogonal projections $q_1, \ldots, q_n$ in A such that $q_i \sim p_i$ $(i = 1, \ldots, n)$, and $q = q_1 + \cdots + q_n$.*

Proof. There exists a rectangular matrix w with entries in A such that $w^*w = q$ and $ww^* = p_1 \oplus \cdots \oplus p_n$. Write w as a block matrix,

$$w = \begin{pmatrix} w_1 \\ \vdots \\ w_n \end{pmatrix},$$

where each w_i is an $m_i \times 1$ matrix. We have $q = w^*w = \sum_{i=1}^n w_i^* w_i$, and

$$ww^* = \begin{pmatrix} w_1 \\ \vdots \\ w_n \end{pmatrix} (w_1^*, \ldots, w_n^*) = \begin{pmatrix} w_1 w_1^* & \cdots & w_1 w_n^* \\ w_2 w_1^* & \cdots & w_2 w_n^* \\ \vdots & \ddots & \vdots \\ w_n w_1^* & \cdots & w_n w_n^* \end{pmatrix},$$

so $w_i w_i^* = p_i$ $(i = 1, \ldots, n)$ and $w_i w_j^* = 0$ for $i \neq j$ $(i, j = 1, \ldots, n)$.

Set $q_i = w_i^* w_i$. Each element q_i is a self-adjoint element of A and $q_i^3 = q_i^2$, so by the functional calculus, q_i is a projection. Also, $q_1 + \cdots + q_n = q$ and $q_i \sim p_i$. The q_i are pairwise orthogonal, since $q_i q_j = w_i^*(w_i w_j^*)w_j = w_i^* 0 w_j = 0$ if $i \neq j$. $\square$

If $1 \leq i, j \leq n$, define the element $e_{ij} \in M_n(\mathbf{C})$ to be the matrix with all its entries zero except for the (i, j) entry, which is 1. The matrices e_{ij} $(i, j = 1, \ldots, n)$ form a linear basis for $M_n(\mathbf{C})$, called the *canonical* basis. We shall make frequent use of the following elementary facts:

$$e_{ij} e_{kl} = \delta_{jk} e_{il} \quad \text{and} \quad e_{ij}^* = e_{ji}.$$

Since $M_n(\mathbf{C}) = B(\mathbf{C}^n)$, and $M_m(M_n(\mathbf{C})) = M_m(B(\mathbf{C}^n)) = B(\mathbf{C}^{mn})$, it is clear that every projection in $M_m(M_n(\mathbf{C}))$ is unitarily equivalent to a diagonal matrix with only zeros and ones on the diagonal, so if $p \in P[M_n(\mathbf{C})]$, then $[p] = k[e_{11}]$ for some integer k.

Now suppose that A is the C*-algebra $M_{n_1}(\mathbf{C}) \oplus \cdots \oplus M_{n_k}(\mathbf{C})$. We regard $M_{n_l}(\mathbf{C})$ as a C*-subalgebra of A in an obvious way. We denote by e_{ij}^l $(i, j = 1, \ldots, n_l)$ the canonical basis of $M_{n_l}(\mathbf{C})$, and call the elements e_{ij}^l $(l = 1, \ldots, k; i, j = 1, \ldots, n_l)$ the *canonical* basis of A.

The homomorphism

$$\tau \colon \mathbf{Z}^k \to K_0(A), \quad (m_1, \ldots, m_k) \mapsto \sum_{l=1}^{k} m_l [e_{11}^l],$$

is the *canonical* map from $\mathbf{Z}^k$ to $K_0(A)$.

If $\varphi \colon G_1 \to G_2$ is a group homomorphism between partially ordered groups G_1 and G_2, we say φ is *positive* if $\varphi(G_1^+) \subseteq G_2^+$. If, in addition, φ is bijective and φ^{-1} is also positive, we call φ an *order isomorphism*, and we say G_1 and G_2 are *order isomorphic* if such an order isomorphism exists.

7.2.4. Theorem. If $A = M_{n_1}(\mathbf{C}) \oplus \cdots \oplus M_{n_k}(\mathbf{C})$, then the canonical map $\tau \colon \mathbf{Z}^k \to K_0(A)$ is an order isomorphism.

Proof. If p is a projection in $P[A]$, then $p = (p_1, \ldots, p_k) = p_1 + \cdots + p_k$, where each p_l is a projection in $P[M_{n_l}(\mathbf{C})]$. Each p_l is equivalent to a direct sum $e_{11}^l \oplus \cdots \oplus e_{11}^l$ in $P[M_{n_l}(\mathbf{C})]$, and therefore in $P[A]$. This shows that the elements $[e_{11}^1], \ldots, [e_{11}^k]$ generate the group $K_0(A)$, so τ is surjective, and this also shows that $\tau((\mathbf{Z}^k)^+) = K_0(A)^+$.

Let $\pi_l \colon A \to M_{n_l}(\mathbf{C})$ be the projection *-homomorphism. Suppose that $\tau(m_1, \ldots, m_k) = 0$; that is, $\sum_{l=1}^{k} m_l[e_{11}^l] = 0$. Then, for each l', we have $(\pi_{l'})_*(\sum_{l=1}^{k} m_l[e_{11}^l]) = m_{l'}[e_{11}^{l'}] = 0$, so $e_{11}^{l'} \oplus \cdots \oplus e_{11}^{l'}$ ($|m_{l'}|$-summands) is equivalent to zero. Hence, $m_{l'} = 0$. Thus, τ is injective. $\square$

7.2.5. Corollary. *If A is a non-zero finite-dimensional C*-algebra, then $K_0(A)$ is a free abelian group with a basis $x_1, \ldots, x_k$ such that*

$$K_0(A)^+ = \mathbf{N}x_1 + \cdots + \mathbf{N}x_k.$$

Proof. This follows from Theorems 6.3.8 and 7.2.4.   □

If A and B are unital C*-algebras and $\tau \colon K_0(A) \to K_0(B)$, we say τ is *unital* if $\tau([1_1]) = [1_1]$.

7.2.6. Theorem. *Let A and B be non-zero finite-dimensional C*-algebras.*

(1) *Suppose that $\tau \colon K_0(A) \to K_0(B)$ is a unital positive homomorphism. Then there is a unital *-homomorphism $\varphi \colon A \to B$ such that $\varphi_* = \tau$.*
(2) *If $\varphi, \psi \colon A \to B$ are unital *-homomorphisms, then $\varphi_* = \psi_*$ if and only if $\psi = (\mathrm{Ad}\, u)\varphi$ for some unitary $u \in B$.*

Proof. Since there exists a *-isomorphism π from A to a direct sum of matrix algebras $M_n(\mathbf{C})$, and this induces the isomorphism π_* between the corresponding K_0-groups, we may suppose $A = M_{n_1}(\mathbf{C}) \oplus \cdots \oplus M_{n_k}(\mathbf{C})$ for some positive integers $n_1, \ldots, n_k$. For $l = 1, \ldots, k$ and $(i, j = 1, \ldots, n_l)$, let e^l_{ij} denote the canonical basis elements of A. Let e_l be the unit of $M_{n_l}(\mathbf{C})$ and let $1_A, 1_B$ be the units of A, B, respectively. Recall that, over a finite-dimensional C*-algebra, stable equivalence and equivalence are the same (Remark 7.1.1).

We have $\tau[e_l] = [p_l]$ for some projection $p_l \in P[B]$, since τ is positive. Hence,

$$[p_1 \oplus \cdots \oplus p_k] = \tau\Big(\sum_{l=1}^{k}[e_l]\Big) = \tau[1_A] = [1_B],$$

so $p_1 \oplus \cdots \oplus p_k \sim 1_B$. Therefore, by Lemma 7.2.3, there exists pairwise orthogonal projections $q_1, \ldots, q_k \in B$ such that $q_1 + \cdots + q_k = 1_B$ and $q_l \sim p_l$ for $l = 1, \ldots, k$. Note that $\tau[e_l] = [q_l]$.

Now $\tau[e^l_{11}] = [p^l_{11}]$ for some projection p^l_{11} in $P[B]$, so

$$n_l[p^l_{11}] = \tau(n_l[e^l_{11}]) = \tau[e_l] = [q_l].$$

Hence,

$$p^l_{11} \oplus \cdots \oplus p^l_{11} \ (n_l \text{ summands}) \sim q_l.$$

Since q_l is a projection in B, it follows from Lemma 7.2.3 that there exist pairwise orthogonal projections $q^l_{11}, q^l_{22}, \ldots, q^l_{n_l, n_l}$ in B such that $\sum_{j=1}^{n_l} q^l_{jj} = q_l$, and $q^l_{jj} \sim p^l_{11}$ for $j = 1, \ldots, n_l$. Note that $\tau[e^l_{11}] = [q^l_{jj}]$.

Since $q^l_{jj} \sim q^l_{11}$ for all j, there exist partial isometries $u^l_j \in B$ such that $q^l_{jj} = u^l_j(u^l_j)^*$ and $q^l_{11} = (u^l_j)^* u^l_j$. Set $q^l_{ij} = u^l_i(u^l_j)^*$ for $i, j = 1, \ldots, n_l$, and note that this is consistent with our previous use of the symbols q^l_{jj}.

Elementary computations show that $(q_{ij}^l)^* = q_{ji}^l$ and $q_{ij}^l q_{mn}^l = \delta_{jm} q_{in}^l$, where $i, j, m, n = 1, \ldots, n_l$. It is straightforward to show from this that the unique linear map $\varphi: A \to B$, such that $\varphi(e_{ij}^l) = q_{ij}^l$ for $l = 1, \ldots, k$ and $i, j = 1, \ldots, n_l$, is a unital $*$-homomorphism. Since the elements $[e_{11}^1], \ldots, [e_{11}^k]$ generate the group $K_0(A)$ (Theorem 7.2.4) and $\varphi_*[e_{11}^l] = [q_{11}^l] = \tau[e_{11}^l]$ for $l = 1, \ldots, k$, we have $\varphi_* = \tau$. This proves Condition (1).

Suppose now that φ, ψ are arbitrary unital $*$-homomorphisms from A to B. It is easily checked that if u is a unitary of B, then $(\operatorname{Ad} u)_* = \operatorname{id}$. Therefore, if $\psi = (\operatorname{Ad} u)\varphi$, then $\psi_* = \varphi_*$.

Suppose conversely that $\psi_* = \varphi_*$. Set $p_{ij}^l = \varphi(e_{ij}^l)$ and $q_{ij}^l = \psi(e_{ij}^l)$ for all $l = 1, \ldots, k$ and $i, j = 1 \ldots, n_l$. Then $[p_{ij}^l] = \varphi_*[e_{ij}^l] = \psi_*[e_{ij}^l] = [q_{ij}^l]$, so $p_{ij}^l \sim q_{ij}^l$. Hence, there exist partial isometries $u_l \in B$ such that $p_{11}^l = u_l^* u_l$ and $q_{11}^l = u_l u_l^*$. Set

$$ u = \sum_{l=1}^k \sum_{i=1}^{n_l} q_{i1}^l u_l p_{1i}^l. $$

A direct computation shows that u is a unitary in B and that $u p_{ij}^l = q_{ij}^l u$ for all l, i, j. Hence, $\psi(e_{ij}^l) = (\operatorname{Ad} u)\varphi(e_{ij}^l)$ for all l, i, j, so $\psi = (\operatorname{Ad} u)\varphi$. Therefore, Condition (2) holds.   □

7.2.7. Lemma. *Let p, q be projections in a C*-algebra A and suppose that there is an element $u \in A$ such that $\|p - u^* u\|$ and $\|q - u u^*\|$ are less than one and $u = qup$. Then $p \sim q$.*

Proof. The inequality $\|p - u^* u\| < 1$ implies that $u^* u$ is invertible in the C*-algebra pAp and, similarly, the inequality $\|q - u u^*\| < 1$ implies that $u u^*$ is invertible in qAq. Let z be the inverse of $|u|$ in pAp, and put $w = uz$. Then $w^* w = z u^* u z = z |u|^2 z = p$. Also $u u^* w w^* = u u^* u z^2 u^* = u |u|^2 z^2 u^* = u u^*$, so $w w^* = q$ by invertibility of $u u^*$ in qAq. Thus, $p \sim q$.□

7.2.8. Lemma. *Suppose that A is a unital C*-algebra and $(A_n)_{n=1}^\infty$ is an increasing sequence of C*-subalgebras of A containing the unit of A. Suppose also that $\cup_{n=1}^\infty A_n$ is dense in A.*

(1) *If $p \in P[A]$, then there exists $q \in P[A_k]$ for some integer k such that $[p]_A = [q]_A$.*

(2) *If $p, q \in P[A_k]$ for some k and $[p]_A = [q]_A$, then there exists an integer $m > k$ such that $[p]_{A_m} = [q]_{A_m}$.*

Proof. For each integer l, the sequence of C*-algebras $(M_l(A_n))_{n=1}^\infty$ is increasing, and the union $\cup_{n=1}^\infty M_l(A_n)$ is dense in $M_l(A)$. Moreover, each $M_l(A_n)$ contains the unit of $M_l(A)$. Thus, to prove the theorem, it suffices to show that if p is a projection in A, then it is equivalent to a projection in some A_k, and to show that if p, q are projections in some A_k equivalent in A, then they are equivalent in A_m for some $m > k$.

Suppose first that p is a projection in A. Then there is a sequence (u_n) of elements in $\cup_n A_n$ converging to p, and by replacing u_n by $\mathrm{Re}(u_n)$ if necessary, we may suppose that the u_n are self-adjoint. Since $(u_n^2)_n$ also converges to p, there exists n such that $\|p - u_n\| < 1/2$ and $\|u_n - u_n^2\| < 1/4$. Hence, by Lemma 6.2.2, there exists a projection q in the C*-algebra generated by u_n such that $\|u_n - q\| < 1/2$. Consequently, q belongs to some A_k, and since $\|p - q\| < 1$, there is a unitary u in A such that $q = upu^*$ (Lemma 6.2.1), and therefore, $q \sim p$.

Now suppose that p, q are projections in some A_k equivalent in A. There exists $u \in A$ such that $p = u^*u$ and $q = uu^*$ and $u = qup$. Hence, there is a sequence (u_n) in $\cup_n A_n$ converging to u, and we may suppose that $u_n = qu_np$ for all n (replace u_n by qu_np if necessary). For sufficiently large n, we have $\|p - u_n^*u_n\| < 1$ and $\|q - u_nu_n^*\| < 1$, and we may choose such an n so that u_n belongs to some A_m with $m > k$. By Lemma 7.2.7, p and q are equivalent in A_m. $\qquad\square$

The following elementary lemma will be used in the proof of Elliott's theorem.

7.2.9. Lemma. *Let A, B, and C be unital stably finite C*-algebras. Suppose that A is finite-dimensional and that $\tau \colon K_0(A) \to K_0(C)$ and $\rho \colon K_0(B) \to K_0(C)$ are positive homomorphisms such that $\tau(K_0(A)^+) \subseteq \rho(K_0(B)^+)$. Then there is a positive homomorphism $\tau' \colon K_0(A) \to K_0(B)$ such that $\rho\tau' = \tau$.*

Proof. By Corollary 7.2.5, there is a basis $x_1, \ldots, x_k$ of $K_0(A)$ as a free abelian group such that $K_0(A)^+ = \mathbf{N}x_1 + \cdots + \mathbf{N}x_k$. The assumption that $\tau(K_0(A)^+) \subseteq \rho(K_0(B)^+)$ implies that there are elements $y_1, \ldots, y_k$ in $K_0(B)^+$ such that $\tau(x_j) = \rho(y_j)$ for $1 \leq j \leq k$. Let τ' be the unique homomorphism from $K_0(A)$ to $K_0(B)$ such that $\tau'(x_j) = y_j$ $(1 \leq j \leq k)$. Clearly, $\rho\tau' = \tau$. Moreover, $\tau'(K_0(A)^+) = \mathbf{N}y_1 + \cdots + \mathbf{N}y_k \subseteq K_0(B)^+$, so τ' is positive. $\qquad\square$

7.2.10. Theorem (Elliott). *Let A and B be unital AF-algebras and τ a unital order isomorphism from $K_0(A)$ to $K_0(B)$. Then there is a *-isomorphism φ from A to B such that $\varphi_* = \tau$.*

Proof. There exist increasing sequences $(A_n)_{n=1}^\infty$ and $(B_n)_{n=1}^\infty$ of finite-dimensional C*-subalgebras of A and B, respectively, such that $(\cup_n A_n)^- = A$ and $(\cup_n B_n)^- = B$, and we may suppose that each A_n and B_n contains the unit of A and B, respectively. Denote by $\varphi^n \colon A_n \to A$ and $\psi^n \colon B_n \to B$ the inclusion *-homomorphisms.

Let ρ be the inverse of τ, so ρ is also a unital order isomorphism.

It is immediate from Lemma 7.2.8 that $K_0(A)^+$ is the union of the increasing sequence $(\varphi_*^n((K_0(A_n))^+))_{n=1}^\infty$, so $K_0(A) = \cup_n\varphi_*^n(K_0(A_n))$. Similarly, $K_0(B)^+$ is the union of the increasing sequence $(\psi_*^n((K_0(B_n))^+))_{n=1}^\infty$, and $K_0(B) = \cup_n \psi_*^n(K_0(B_n))$.

Set $n_1 = 1$. By Corollary 7.2.5, there is a basis $x_1, \ldots, x_k$ for the free abelian group $K_0(A_{n_1})$ such that $K_0(A_{n_1})^+ = \mathbf{N}x_1 + \cdots + \mathbf{N}x_k$. Hence, $\tau\varphi_*^{n_1}(K_0(A_{n_1})^+) = \mathbf{N}\tau\varphi_*^{n_1}(x_1) + \cdots + \mathbf{N}\tau\varphi_*^{n_1}(x_k)$. Since $K_0(B)^+$ is the increasing union of the sets $\psi_*^m(K_0(B_m)^+)$ $(m = 1, 2, \ldots)$, it follows that the elements $\tau\varphi_*^{n_1}(x_1), \ldots, \tau\varphi_*^{n_1}(x_k)$ belong to $\psi_*^m(K_0(B_m)^+)$ for some $m > n_1$. Hence, $\tau\varphi_*^{n_1}(K_0(A_{n_1})^+) \subseteq \psi_*^m(K_0(B_m)^+)$, so by Lemma 7.2.9 there is a positive homomorphism $\tilde{\tau} \colon K_0(A_{n_1}) \to K_0(B_m)$ such that the following diagram commutes:

$$
\begin{array}{ccc}
K_0(A_{n_1}) & \xrightarrow{\varphi_*^{n_1}} & K_0(A) \\
\downarrow{\tilde{\tau}} & & \downarrow{\tau} \\
K_0(B_m) & \xrightarrow{\psi_*^m} & K_0(B).
\end{array}
$$

Now $\tilde{\tau}[1_A] = [e]_{B_m}$ say, and therefore $[e]_B = \psi_*^m[e]_{B_m} = \tau\varphi_*^{n_1}[1_A] = [1_B]_B$. By Lemma 7.2.8, Condition (2), there exists $m_1 > m$ such that $[e]_{B_{m_1}} = [1_B]_{B_{m_1}}$. Let $\tilde{\psi} \colon B_m \to B_{m_1}$ be the inclusion, and set $\tau^1 = \tilde{\psi}_* \tilde{\tau}$. Then τ^1 is a unital positive homomorphism and the diagram

$$
\begin{array}{ccc}
K_0(A_{n_1}) & \xrightarrow{\varphi_*^{n_1}} & K_0(A) \\
\downarrow{\tau^1} & & \downarrow{\tau} \\
K_0(B_{m_1}) & \xrightarrow{\psi_*^{m_1}} & K_0(B)
\end{array}
$$

commutes. By a similar argument applied to $\rho\psi_*^{m_1}$, there exists an integer $n > m_1$ and a positive homomorphism $\tilde{\rho} \colon K_0(B_{m_1}) \to K_0(A_n)$ such that the diagram

$$
\begin{array}{ccc}
K_0(B_{m_1}) & \xrightarrow{\psi_*^{m_1}} & K_0(B) \\
\downarrow{\tilde{\rho}} & & \downarrow{\rho} \\
K_0(A_n) & \xrightarrow{\varphi_*^n} & K_0(A)
\end{array}
$$

commutes.

We can write $x_j = [p_j]_{A_{n_1}}$ for projections $p_1, \ldots, p_k \in P[A_{n_1}]$, and similarly, $\tilde{\rho}\tau^1(x_j) = [q_j]_{A_n}$ for projections $q_1, \ldots, q_k \in P[A_n]$. We have $[p_j]_A = [q_j]_A$, since $\varphi_*^n \tilde{\rho}\tau^1 = \rho\psi_*^{m_1}\tau^1 = \rho\tau\varphi_*^{n_1} = \varphi_*^{n_1}$. Applying Lemma 7.2.8, Condition (2), there exists $n_2 > n$ such that $P[A_{n_2}]$ contains $p_1, \ldots, p_k$ and $q_1, \ldots, q_k$, and

$$
[p_j]_{A_{n_2}} = [q_j]_{A_{n_2}} \qquad (j = 1, \ldots, k). \tag{1}
$$

Set $\rho^1 = \tilde{\varphi}_* \tilde{\rho}$, where $\tilde{\varphi} \colon A_n \to A_{n_2}$ is the inclusion. Then the following diagram commutes:

$$
\begin{array}{ccc}
K_0(B_{m_1}) & \xrightarrow{\psi^{m_1}_*} & K_0(B) \\
\downarrow \rho^1 & & \downarrow \rho \\
K_0(A_{n_2}) & \xrightarrow{\varphi^{n_2}_*} & K_0(A).
\end{array}
$$

Moreover, $\rho^1 \tau^1 = \varphi_{1*}$, where $\varphi_1 \colon A_{n_1} \to A_{n_2}$ is the inclusion, since for each j, we have $\rho^1 \tau^1(x_j) = \tilde{\varphi}_* \tilde{\rho} \tau^1(x_j) = \tilde{\varphi}_*[q_j]_{A_n} = [q_j]_{A_{n_2}} = [p_j]_{A_{n_2}}$ (by Eq. (1)) $= \varphi_{1*}[p_j]_{A_{n_1}} = \varphi_{1*}(x_j)$.

Continuing in the above fashion, we inductively construct two sequences of integers such that $n_1 < m_1 < n_2 < m_2 < \ldots$, and positive homomorphisms $\tau^k \colon K_0(A_{n_k}) \to K_0(B_{m_k})$ and $\rho^k \colon K_0(B_{m_k}) \to K_0(A_{n_{k+1}})$ such that the diagrams

$$
\begin{array}{ccc}
K_0(A_{n_k}) & \xrightarrow{\varphi^{n_k}_*} & K_0(A) \\
\downarrow \tau^k & & \downarrow \tau \\
K_0(B_{m_k}) & \xrightarrow{\psi^{m_k}_*} & K_0(B)
\end{array}
\tag{2}
$$

and

$$
\begin{array}{ccc}
K_0(B_{m_k}) & \xrightarrow{\psi^{m_k}_*} & K_0(B) \\
\downarrow \rho^k & & \downarrow \rho \\
K_0(A_{n_{k+1}}) & \xrightarrow{\varphi^{n_{k+1}}_*} & K_0(A)
\end{array}
\tag{3}
$$

commute, and $\rho^k \tau^k = \varphi_{k*}$ and $\tau^{k+1} \rho^k = \psi_{k*}$, where $\varphi_k \colon A_{n_k} \to A_{n_{k+1}}$ and $\psi_k \colon B_{m_k} \to B_{m_{k+1}}$ are the inclusions. Note that τ^k and ρ^k are necessarily unital (because τ^1 is unital and $\rho^k \tau^k = \varphi_{k*}$, and $\tau^{k+1} \rho^k = \psi_{k*}$ for all k, so by induction τ^k and ρ^k are unital for all k).

By Theorem 7.2.6, there are unital $*$-homomorphisms $\alpha^1 \colon A_{n_1} \to B_{m_1}$ and $\beta^1 \colon B_{m_1} \to A_{n_2}$ such that $\alpha^1_* = \tau^1$ and $\beta^1_* = \rho^1$. We have $(\beta^1 \alpha^1)_* = \rho^1 \tau^1 = \varphi_{1*}$, so by Theorem 7.2.6 again, there is a unitary u in A_{n_2} such that $(\mathrm{Ad}\, u)\beta^1 \alpha^1 = \varphi_1$. Since $(\mathrm{Ad}\, u)_* = \mathrm{id}$, we may suppose that $\beta^1 \alpha^1 = \varphi_1$ (replacing β^1 by $(\mathrm{Ad}\, u)\beta^1$ if necessary). Continuing in this fashion, we construct by induction unital $*$-homomorphisms $\alpha^k \colon A_{n_k} \to B_{m_k}$ and $\beta^k \colon B_{m_k} \to A_{n_{k+1}}$ such that for all k we have $\alpha^k_* = \tau^k$, $\beta^k_* = \rho^k$, $\beta^k \alpha^k = \varphi_k$, and $\alpha^{k+1} \beta^k = \psi_k$.

If $a \in A_{n_k}$, then $\alpha^k(a) = \alpha^{k+1}(a)$, because $\alpha^k(a) = \psi_k \alpha^k(a) = \alpha^{k+1} \beta^k \alpha^k(a) = \alpha^{k+1} \varphi_k(a) = \alpha^{k+1}(a)$.

Hence, if $A' = \cup_k A_{n_k}$, we may well-define a map $\varphi \colon A' \to B$ by setting $\varphi(a) = \alpha^k(a)$ if $a \in A_{n_k}$. Since the maps α^k are norm-decreasing

*-homomorphisms, so is φ. Hence, it extends from the dense *-subalgebra A' of A to a *-homomorphism on A, again denoted by φ. In like manner we get a *-homomorphism $\psi: B \to A$ such that if $b \in B_{m_k}$, then $\psi(b) = \beta^k(b)$. If $a \in A_{n_k}$, then $\psi\varphi(a) = \beta^k\alpha^k(a) = \varphi_k(a) = a$. This shows that $\psi\varphi = \mathrm{id}$, and similarly, one shows that $\varphi\psi = \mathrm{id}$. Thus, φ is a *-isomorphism.

Now suppose that $p \in P[A_{n_k}]$. Then $\tau([p]_A) = \tau\varphi_*^{n_k}([p]_{A_{n_k}}) = \psi_*^{m_k}\alpha_*^k([p]_{A_{n_k}})$ (the last equality follows from commutativity of Diagram (2) and the fact that $\tau^k = \alpha_*^k$). Hence, $\tau([p]_A) = \psi_*^{m_k}([\alpha^k(p)]_{B_{m_k}}) = [\varphi(p)]_B = \varphi_*([p]_A)$. This shows that $\tau = \varphi_*$ on the sets $\varphi_*^{n_k}(K_0(A_{n_k})^+)$ for all k, and since these sets have union $K_0(A)^+$, and this generates $K_0(A)$, it follows that $\tau = \varphi_*$.   □

7.2.11. Corollary. *Two unital AF-algebras are *-isomorphic if and only if there is a unital order isomorphism between their K_0-groups.*

Proof. If φ is a *-isomorphism of unital AF-algebras, then φ_* is a unital order isomorphism. This gives the forward implication. The reverse implication is given by Theorem 7.2.10.   □

7.3. Three Fundamental Results in K-Theory

The three results referred to in the title of this section are weak exactness, homotopy invariance, and continuity of the functor K_0, or more precisely, of $\tilde{K}_0$. The latter is the extension of K_0 to the class of all C*-algebras. For technical reasons $\tilde{K}_0$ is defined differently than K_0, but if an algebra A is unital, then $K_0(A)$ and $\tilde{K}_0(A)$ are isomorphic groups.

Let A be a C*-algebra, which may be unital or non-unital. If $\tau: \tilde{A} \to \mathbf{C}$ is the canonical *-homomorphism, we set $\tilde{K}_0(A) = \ker(\tau_*)$, so $\tilde{K}_0(A)$ is a subgroup of $K_0(\tilde{A})$. If $\varphi: A \to B$ is a *-homomorphism of C*-algebras and $\tilde{\varphi}: \tilde{A} \to \tilde{B}$ is the unique unital *-homomorphism extending φ, then $\tilde{\varphi}_*(\tilde{K}_0(A)) \subseteq \tilde{K}_0(B)$. Hence, we get a homomorphism $\varphi_*: \tilde{K}_0(A) \to \tilde{K}_0(B)$ by restricting $\tilde{\varphi}_*$. It is straightforward to show that these constructions give a covariant functor

$$A \to \tilde{K}_0(A), \qquad \varphi \mapsto \varphi_*$$

from the category of all C*-algebras and *-homomorphisms to the category of all abelian groups and homomorphisms.

Suppose now A is a unital C*-algebra, and e denotes its unit. If $p \in P[A]$, then $\tau(p) = 0$, so $[p]_{\tilde{A}} \in \tilde{K}_0(A)$. Moreover, if $p, q \in P[A]$ and $p \approx q$ relative to A, then for some integer n we have $e_n \oplus p \sim e_n \oplus q$ relative to A, and therefore relative to $\tilde{A}$. Hence, $[e_n]_{\tilde{A}} + [p]_{\tilde{A}} = [e_n \oplus p]_{\tilde{A}} = [e_n \oplus q]_{\tilde{A}} = [e_n]_{\tilde{A}} + [q]_{\tilde{A}}$, so $[p]_{\tilde{A}} = [q]_{\tilde{A}}$. Thus, we get a well-defined map $K_0(A)^+ \to \tilde{K}_0(A)$, $[p]_A \mapsto [p]_{\tilde{A}}$, which is clearly a homomorphism, and which therefore extends uniquely to a homomorphism $j_A\colon K_0(A) \to \tilde{K}_0(A)$. We call j_A the *natural* homomorphism from $K_0(A)$ to $\tilde{K}_0(A)$. In the language of category theory, the following theorem asserts that j_A implements a natural isomorphism between the two functors K_0 and $\tilde{K}_0$ on the category of all unital C*-algebras and unital *-homomorphisms.

7.3.1. Theorem.   *If A is a unital C*-algebra, then the natural map $j_A\colon K_0(A) \to \tilde{K}_0(A)$ is an isomorphism. Moreover, if $\varphi\colon A \to B$ is a unital *-homomorphism into a unital C*-algebra B, then the diagram*

$$
\begin{array}{ccc}
K_0(A) & \overset{\varphi_*}{\longrightarrow} & K_0(B) \\
\downarrow j_A & & \downarrow j_B \\
\tilde{K}_0(A) & \overset{\varphi_*}{\longrightarrow} & \tilde{K}_0(B)
\end{array}
$$

commutes.

Proof.   If e is the unit of A, then the map

$$\psi\colon \tilde{A} \to A, \quad a \mapsto eae,$$

is a unital *-homomorphism. If $j\colon \tilde{K}_0(A) \to K_0(\tilde{A})$ is the inclusion, then it is easily checked that $\psi_* j j_A = \mathrm{id}$ and $j_A \psi_* j = \mathrm{id}$. Thus, j_A is a bijection. Commutativity of the diagram is trivial. $\square$

  We now need some elementary results on the unitary group of a unital C*-algebra. These results will be used in connection with weak exactness, which we shall be looking at presently.

  If A is a unital C*-algebra, we denote by $U_n(A)$ the group of unitaries of $M_n(A)$. We let $U_n^0(A)$ denote the connected component of the unit in $U_n(A)$. We shall also write $U(A), U^0(A)$ for $U_1(A), U_1^0(A)$, respectively.

7.3.2. Theorem.   *Let A be a unital C*-algebra. Then $U^0(A)$ is a normal subgroup of $U(A)$. If $u \in A$, then $u \in U^0(A)$ if and only if there exist elements $a_1, \dots, a_n \in A_{sa}$ such that $u = e^{ia_1} \cdots e^{ia_n}$.*

Proof.   Let V be the set of all elements u of A which can be written in the form $u = e^{ia_1} \cdots e^{ia_n}$ for some n and some $a_1, \dots, a_n \in A_{sa}$. Any such element u belongs to $U^0(A)$, since the function

$$[0,1] \to U(A), \quad t \mapsto e^{ita_1} \cdots e^{ita_n},$$

is a continuous path in $U(A)$ from 1 to u. Obviously, V is a subgroup of $U(A)$. By Theorem 2.1.12, V is a neighbourhood of 1 in $U(A)$, and therefore (since V is a subgroup) V is a neighbourhood of all its points; that is, V is open in $U(A)$. Since this implies that the cosets of V are also open, and since the complement of V in $U(A)$ is a union of cosets, it follows that V is closed in $U(A)$. Thus, V is a non-empty clopen subset of the connected set $U^0(A)$, and therefore $V = U^0(A)$. If $u \in U(A)$, then $uU^0(A)u^{-1}$ is a connected set of unitaries containing 1, so $uU^0(A)u^{-1} \subseteq U^0(A)$. Hence, $U^0(A)$ is normal in $U(A)$. $\qquad\square$

7.3.3. Corollary. *Suppose that φ is a unital surjective $*$-homomorphism from a unital C*-algebra A to a unital C*-algebra B. If $v \in U^0(B)$, then there exists $u \in U^0(A)$ such that $v = \varphi(u)$.*

Proof. If $b \in B_{sa}$, then $b = \varphi(a)$ for some $a \in A$, and therefore $b = \varphi(\mathrm{Re}(a))$; that is, we may suppose that $a \in A_{sa}$. If $v \in U^0(B)$, then $v = e^{ib_1} \cdots e^{ib_n}$ for some $b_j \in B_{sa}$. Hence, there exists $a_1, \ldots, a_n \in A_{sa}$ such that $\varphi(a_j) = b_j$ for all j, so $u = e^{ia_1} \cdots e^{ia_n} \in U^0(A)$ and $\varphi(u) = v$. $\square$

7.3.1. Remark. If u is a *symmetry*, that is, a self-adjoint unitary, in a C*-algebra A, then $u \in U^0(A)$. This follows from the computation $e^{i\pi u/2} = \sum_{n \text{ even}} (i\pi/2)^n/n! + u \sum_{n \text{ odd}} (i\pi/2)^n/n! = \cos(\pi/2) + iu \sin(\pi/2) = iu$. Thus, $u = e^{i\pi a}$, where $a = (u-1)/2$.

7.3.4. Lemma. *Let p, q be equivalent projections in a unital C*-algebra A. Then there exists $u \in U_2^0(A)$ such that $u(p \oplus 0)u^* = q \oplus 0$ in $M_2(A)$.*

Proof. Let v be a partial isometry in A such that $p = v^*v$ and $q = vv^*$, and set

$$u = \begin{pmatrix} v & 1 - vv^* \\ 1 - v^*v & v^* \end{pmatrix}.$$

Simple computations show that u is a unitary, and that if

$$w = \begin{pmatrix} 0 & 1 \\ 1 & 0 \end{pmatrix},$$

then w and wu are symmetries. Hence, u is the product $u = w(wu)$ of symmetries, and therefore $u \in U_2^0(A)$ (*cf.* Remark 7.3.1). That we have

$$\begin{pmatrix} q & 0 \\ 0 & 0 \end{pmatrix} = u \begin{pmatrix} p & 0 \\ 0 & 0 \end{pmatrix} u^*$$

again follows by direct computation. $\qquad\square$

If $G \xrightarrow{\tau} G' \xrightarrow{\rho} G''$ is a sequence of homomorphisms of abelian groups, it is *exact* if $\mathrm{im}(\tau) = \ker(\rho)$. The sequence

$$0 \to G \xrightarrow{\tau} G' \xrightarrow{\rho} G'' \to 0 \qquad (1)$$

is a *short exact sequence of groups* if τ is injective, ρ is surjective, and the sequence $G \xrightarrow{\tau} G' \xrightarrow{\rho} G''$ is exact. The short exact sequence (1) is said to *split* if there is a homomorphism $\rho': G'' \to G'$ such that $\rho\rho' = \mathrm{id}$. In this case there is a unique homomorphism $\tau': G' \to G$ such that $\tau'\tau = \mathrm{id}$ and $\tau\tau' + \rho'\rho = \mathrm{id}_{G'}$. Hence, the map

$$G' \to G \oplus G'', \quad x \mapsto (\tau'(x), \rho(x)),$$

is an isomorphism. The proofs of these observations are elementary exercises.

A sequence of homomorphisms between groups

$$\ldots G_n \xrightarrow{\varphi_n} G_{n+1} \xrightarrow{\varphi_{n+1}} G_{n+2} \ldots$$

(where the sequence may be finite or extend infinitely in either direction) is said to be *exact* if $\mathrm{im}(\varphi_n) = \ker(\varphi_{n+1})$ for all relevant n.

7.3.2. Remark. Let A be a C*-algebra and let $x \in \tilde{K}_0(A)$. If 1 is the unit of $\tilde{A}$, then there exists an integer n and a projection $p \in P[\tilde{A}]$ such that $x = [p] - [1_n]$. To see this, observe that $x = [r] - [q]$ for some projections r, q, both of which we may suppose to be elements of $M_n(\tilde{A})$ for some n. Then $x = [r] + [1_n - q] - [1_n] = [p] - [1_n]$, where $p = r \oplus (1_n - q)$.

The property of the functor $\tilde{K}_0$ asserted in the following theorem is referred to as *weak exactness*.

7.3.5. Theorem. *Let*

$$0 \to J \xrightarrow{j} A \xrightarrow{\varphi} B \to 0$$

be a short exact sequence of C-algebras and *-homomorphisms. Then the sequence*

$$\tilde{K}_0(J) \xrightarrow{j_*} \tilde{K}_0(A) \xrightarrow{\varphi_*} \tilde{K}_0(B)$$

is exact.

Proof. We may assume that J is an ideal in A and that j is the inclusion map. Since $\varphi j = 0$, we have $\varphi_* j_* = 0_* = 0$, so $\mathrm{im}(j_*) \subseteq \ker(\varphi_*)$.

To show the reverse inclusion, let $x \in \ker(\varphi_*)$. Then we can write $x = [p] - [1_n]$ for p a projection in some $M_m(\tilde{A})$ and for some integer $n < m$. (We may suppose the unit of $\tilde{J}$ and that of $\tilde{A}$ are the same and we shall use 1 to denote the unit for $\tilde{J}, \tilde{A},$ and $\tilde{B}$.) Now $[\varphi(p)] = [1_n]$ in $K_0(\tilde{B})$, so there is an integer k such that $1_k \oplus \varphi(p) \sim 1_k \oplus 1_n \oplus 0_{m-n} = 1_{k+n} \oplus 0_{m-n}$ in $M_{k+m}(\tilde{B})$. It follows from Lemma 7.3.4 that there exists $v \in U_{2k+2m}^0(\tilde{B})$ such that

$$1_{k+n} \oplus 0_{k+2m-n} = v(1_k \oplus \varphi(p) \oplus 0_{k+m})v^*.$$

By Corollary 7.3.3, there exists $u \in U_{2k+2m}^0(\tilde{A})$ such that $\varphi(u) = v$. Let $r = u(1_k \oplus p \oplus 0_{k+m})u^*$, so r is a projection in $M_{2k+2m}(\tilde{A})$ equivalent to $1_k \oplus p \oplus 0_{k+m}$. Since $\varphi(r) = v(1_k \oplus \varphi(p) \oplus 0_{k+m})v^* = 1_{k+n} \oplus 0_{k+2m-n}$, it follows that $r \in M_{2k+2m}(\tilde{J})$. It is easily checked that the element $[r] - [1_{k+n}]$ of $K_0(\tilde{J})$ actually lies in $\tilde{K}_0(J)$. Finally, $j_*([r] - [1_{k+n}]) = [1_k \oplus p] - [1_{k+n}] = [p] - [1_n] = x$, so $x \in \operatorname{im}(j_*)$. Hence, $\ker(\varphi_*) = \operatorname{im}(j_*)$.   □

7.3.6. Theorem. *If A_1 and A_2 are C*-algebras, then $\tilde{K}_0(A_1 \oplus A_2)$ is isomorphic to $\tilde{K}_0(A_1) \oplus \tilde{K}_0(A_2)$.*

Proof. Let $\varphi_i: A_i \to A_1 \oplus A_2$ and $\pi_i: A_1 \oplus A_2 \to A_i$ be the inclusion and projection *-homomorphisms, respectively. The sequence

$$0 \to A_1 \xrightarrow{\varphi_1} A_1 \oplus A_2 \xrightarrow{\pi_2} A_2 \to 0$$

is a short exact sequence of C*-algebras, so by Theorem 7.3.5 the sequence

$$\tilde{K}_0(A_1) \xrightarrow{\varphi_{1*}} \tilde{K}_0(A_1 \oplus A_2) \xrightarrow{\pi_{2*}} \tilde{K}_0(A_2)$$

is exact. Since $\pi_i \varphi_i = \operatorname{id}$, and, therefore, $\pi_{i*}\varphi_{i*} = \operatorname{id}$ ($i = 1, 2$), the homomorphism φ_{1*} is injective, the homomorphism π_{2*} is surjective, and

$$0 \to \tilde{K}_0(A_1) \xrightarrow{\varphi_{1*}} \tilde{K}_0(A_1 \oplus A_2) \xrightarrow{\pi_{2*}} \tilde{K}_0(A_2) \to 0$$

is a split short exact sequence. Hence, $\tilde{K}_0(A_1 \oplus A_2)$ is isomorphic to $\tilde{K}_0(A_1) \oplus \tilde{K}_0(A_2)$.   □

If $\varphi, \psi: A \to B$ are *-homomorphisms of C*-algebras A, B, we say φ and ψ are *homotopic*, and write $\varphi \approx \psi$, if for each $t \in [0, 1]$ there is a *-homomorphism $\varphi_t: A \to B$, where $\varphi_0 = \varphi$ and $\varphi_1 = \psi$, and for each $a \in A$ we have continuity of the map

$$[0, 1] \to B, \quad t \mapsto \varphi_t(a).$$

We then call $(\varphi_t)_{0 \leq t \leq 1}$ a *homotopy* from φ to ψ. The relation $\varphi \approx \psi$ is an equivalence relation on the *-homomorphisms from A to B.

7.3.1. Example. Let J be a closed ideal in a unital C*-algebra A, and let $a \in A_{sa}$. Set $u = e^{ia}$, and let $\varphi: J \to J$ be the restriction to J of $\operatorname{Ad} u$. Then φ is homotopic to the identity map id_J of J. A homotopy $(\varphi_t)_t$ from id_J to φ is got by letting φ_t be the restriction to J of $\operatorname{Ad} e^{ita}$ for all $t \in [0, 1]$.

7.3.2. Example. If H is a Hilbert space and the map $\varphi: K(H) \to K(H)$ is a *-isomorphism, then φ is homotopic to $\operatorname{id}_{K(H)}$. This is immediate from Theorems 2.4.8 and 2.5.8 and from Example 7.3.1.

7.3.7. Theorem. *If* $\varphi, \psi: A \to B$ *are homotopic* *-*homomorphisms between C*-algebras* A, B, *then* $\varphi_* = \psi_*: \tilde{K}_0(A) \to \tilde{K}_0(B)$.

Proof. If $(\varphi_t)_t$ is a homotopy from φ to ψ, then it is easily checked that $(\tilde{\varphi}_t)_t$ is a homotopy from $\tilde{\varphi}$ to $\tilde{\psi}$ in which all $\tilde{\varphi}_t$ are unital. We may therefore suppose that A, B are unital and that there is a homotopy $(\varphi_t)_t$ from φ to ψ such that all φ_t are unital, and show that $\varphi_* = \psi_*: K_0(A) \to K_0(B)$. In this case, if $p \in M_n(A)$ is a projection and $p_t = \varphi_t(p)$, then the map

$$[0,1] \to M_n(B), \quad t \mapsto p_t,$$

is uniformly continuous, so there is a partition $0 = t_0 < t_1 < \ldots < t_m = 1$ of $[0,1]$ such that $\|p_{t_j} - p_{t_{j+1}}\| < 1$ $(0 \leq j < m)$. Hence, p_{t_j} and $p_{t_{j+1}}$ are unitarily equivalent by Lemma 6.2.1, so $p_{t_j} \sim p_{t_{j+1}}$, and therefore, $\varphi(p) \sim \psi(p)$. Consequently, $\varphi_*([p]) = \psi_*([p])$, and since p was an arbitrary element of $P[A]$, we have $\varphi_* = \psi_*$. $\square$

The following lemma says if $A = \varinjlim A_n$, then $M_k(A) = \varinjlim M_k(A_n)$. It is of independent interest, but for us its importance is its application to proving "continuity" of the K_0-functor (Theorem 7.3.10).

7.3.8. Lemma. *Let* A *be the direct limit of the sequence of C*-algebras* $(A_n, \varphi_n)_{n=1}^{\infty}$, *and let* B *be the direct limit of the corresponding sequence* $(M_k(A_n), \varphi_n)_{n=1}^{\infty}$, *where* k *is a fixed integer. Denote by* $\varphi^n: A_n \to A$ *and* $\psi^n: M_k(A_n) \to B$ *the natural maps for each integer* n. *Then there is a unique* *-*isomorphism* $\pi: B \to M_k(A)$ *such that for each* n *the diagram*

$$
\begin{array}{ccc}
M_k(A_n) & \xrightarrow{\psi^n} & B \\
& {\scriptstyle \varphi^n}\searrow & \downarrow {\scriptstyle \pi} \\
& & M_k(A)
\end{array}
$$

commutes.

Proof. Since the diagram

$$
\begin{array}{ccc}
M_k(A_n) & \xrightarrow{\varphi_n} & M_k(A_{n+1}) \\
& {\scriptstyle \varphi^n}\searrow & \downarrow {\scriptstyle \varphi^{n+1}} \\
& & M_k(A)
\end{array}
$$

commutes for each n, it follows from Theorem 6.1.2 that there is a unique *-homomorphism $\pi: B \to M_k(A)$ such that the diagram

$$
\begin{array}{ccc}
M_k(A_n) & \xrightarrow{\psi^n} & B \\
& {\scriptstyle \varphi^n}\searrow & \downarrow {\scriptstyle \pi} \\
& & M_k(A)
\end{array}
$$

commutes for each n. As $\cup_n \varphi^n(A_n)$ is dense in A, so $\cup_n \varphi^n(M_k(A_n))$ is dense in $M_k(A)$, and it follows that π is surjective.

To show that π is injective, it suffices to show that it is injective when restricted to the C*-subalgebras $\psi^n(M_k(A_n))$, since it is then isometric on these algebras, and therefore, by continuity of π and density of $\cup_n \psi^n(M_k(A_n))$ in B, it follows that π is isometric on B. Suppose then that $\pi(\psi^n(a)) = 0$, where $a \in M_k(A_n)$. Let $\varepsilon > 0$. If $b \in A_n$ and $\varphi^n(b) = 0$, then there exists $m \geq n$ such that $\|\varphi_{nm}(b)\| < \varepsilon$ (cf. Remark 6.1.2). Applying this to the entries a_{ij} of the matrix a, since $\varphi^n(a) = 0$, there exists $m \geq n$ such that $\|\varphi_{nm}(a_{ij})\| < \varepsilon$ ($1 \leq i, j \leq k$). Hence, $\|\varphi_{nm}(a)\| \leq \sum_{i,j=1}^{k} \|\varphi_{nm}(a_{ij})\| < k^2\varepsilon$ (cf. Remark 3.4.1). Consequently, $\|\psi^n(a)\| = \|\psi^m \varphi_{nm}(a)\| \leq \|\varphi_{nm}(a)\| < k^2\varepsilon$. Letting $\varepsilon \to 0$ we get $\|\psi^n(a)\| = 0$, so π is injective on $\psi^n(M_k(A_n))$ as required. This proves the theorem.   □

If $(A_n, \varphi_n)_{n=1}^{\infty}$ is a direct sequence of C*-algebras, we say it is *unital* if the algebras A_n and the *-homomorphisms φ_n are unital. In this case the algebra $A = \varinjlim A_n$ is unital, as are the natural *-homomorphisms $\varphi^n \colon A_n \to A$.

7.3.9. Lemma. *Let A be the direct limit of a unital sequence $(A_n, \varphi_n)_{n=1}^{\infty}$ of C*-algebras, and for each n let $\varphi^n \colon A_n \to A$ be the natural map.*

(1) *If p is a projection in A, then there is an integer n and a projection $q \in A_n$ such that p is unitarily equivalent to $\varphi^n(q)$ in A.*

(2) *If n is given and p, q are projections in A_n such that $\varphi^n(p) \sim \varphi^n(q)$ in A, then there is an integer $m \geq n$ such that $\varphi_{nm}(p) \sim \varphi_{nm}(q)$ in A_m.*

Proof. Let p be a projection in A. Since $A = (\cup_{n=1}^{\infty} \varphi^n(A_n))^-$ there is a sequence $(\varphi^{n_k}(a_k))_{k=1}^{\infty}$ in $\cup_n \varphi^n(A_n)$ converging to p. As $p = p^*$ we may suppose that each a_k is self-adjoint (replace a_k by $\mathrm{Re}(a_k)$ if necessary). Since $p = p^2$, the sequence $(\varphi^{n_k}(a_k^2))_{k=1}^{\infty}$ also converges to p, and therefore $(\varphi^{n_k}(a_k - a_k^2))_{k=1}^{\infty}$ converges to 0. Hence, there exists an integer m and a self-adjoint element $a \in A_m$ such that $\|p - \varphi^m(a)\| < 1/2$ and $\|\varphi^m(a - a^2)\| < 1/4$. It follows that there exists $n \geq m$ such that $\|\varphi_{mn}(a - a^2)\| < 1/4$. Set $b = \varphi_{mn}(a)$. Then b is a self-adjoint element of A_n such that $\|b - b^2\| < 1/4$, and therefore, by Lemma 6.2.2, there is a projection $q \in A_n$ such that $\|b - q\| < 1/2$. Using the equality $\varphi^m(a) = \varphi^n(b)$, we have

$$\begin{aligned}
\|p - \varphi^n(q)\| &\leq \|p - \varphi^m(a)\| + \|\varphi^n(b) - \varphi^n(q)\| \\
&\leq \|p - \varphi^m(a)\| + \|b - q\| \\
&< 1/2 + 1/2 = 1,
\end{aligned}$$

so by Lemma 6.2.1 the projections p and $\varphi^n(q)$ are unitarily equivalent in A. This proves Condition (1).

Now suppose that n is a given integer, and that p, q are projections in A_n such that $\varphi^n(p) \sim \varphi^n(q)$ in A. Then there is a partial isometry u in A such that $\varphi^n(p) = u^*u$ and $\varphi^n(q) = uu^*$. Now u is the limit of a sequence $(\varphi^{n_k}(v_k))_{k=1}^{\infty}$, where $v_k \in A_{n_k}$ and $n_k \geq n$, and since $u = uu^*u = \varphi^n(q)u = u\varphi^n(p)$, we may suppose that $v_k = \varphi_{nn_k}(q)v_k\varphi_{nn_k}(p)$ (replace v_k by $\varphi_{nn_k}(q)v_k\varphi_{nn_k}(p)$ if necessary). Clearly, $\varphi^n(p) = \lim_{k\to\infty} \varphi^{n_k}(v_k^*v_k)$ and $\varphi^n(q) = \lim_{k\to\infty} \varphi^{n_k}(v_kv_k^*)$. Hence, there exists an integer $k \geq n$ and $v \in A_k$ such that $v = \varphi_{nk}(q)v\varphi_{nk}(p)$ and $\|\varphi^k(\varphi_{nk}(p) - v^*v)\| < 1$ and $\|\varphi^k(\varphi_{nk}(q) - vv^*)\| < 1$. It follows that there is an integer $m \geq k$ such that $\|\varphi_{km}(\varphi_{nk}(p) - v^*v)\| < 1$ and $\|\varphi_{km}(\varphi_{nk}(q) - vv^*)\| < 1$. Therefore, if $w = \varphi_{km}(v)$, then $w \in A_m$ and we have $\|\varphi_{nm}(p) - w^*w\| < 1$ and $\|\varphi_{nm}(q) - ww^*\| < 1$ and $\varphi_{nm}(q)w\varphi_{nm}(p) = w$. Hence, by Lemma 7.2.7, the projections $\varphi_{nm}(p)$ and $\varphi_{nm}(q)$ are equivalent in A_m. This proves Condition (2). $\qquad\qquad\qquad\qquad\qquad\qquad\qquad\qquad\qquad\qquad\qquad\qquad\square$

The content of the following theorem is the continuity of K_0. It says that if $A = \varinjlim A_n$, where $(A_n, \varphi_n)_n$ is a unital sequence of C*-algebras, then $K_0(A) = \varinjlim K_0(A_n)$.

7.3.10. Theorem. *Let A be the direct limit of a unital sequence of C*-algebras $(A_n, \varphi_n)_{n=1}^{\infty}$, and let G be the direct limit of the corresponding sequence of abelian groups $(K_0(A_n), \varphi_{n*})_{n=1}^{\infty}$. Denote by $\varphi^n \colon A_n \to A$ and $\tau^n \colon K_0(A_n) \to G$ the natural maps. Then there is a unique isomorphism $\tau \colon G \to K_0(A)$ such that the diagram*

$$
\begin{array}{ccc}
K_0(A_n) & \xrightarrow{\ \tau^n\ } & G \\
& {\scriptstyle \varphi_*^n}\searrow & \downarrow{\scriptstyle \tau} \\
& & K_0(A)
\end{array}
$$

commutes for each integer n.

Proof. For each integer n the diagram

$$
\begin{array}{ccc}
K_0(A_n) & \xrightarrow{\ \varphi_{n*}\ } & K_0(A_{n+1}) \\
& {\scriptstyle \varphi_*^n}\searrow & \downarrow{\scriptstyle \varphi_*^{n+1}} \\
& & K_0(A)
\end{array}
$$

commutes, so there is a unique homomorphism $\tau \colon G \to K_0(A)$ such that for each n the diagram

$$
\begin{array}{ccc}
K_0(A_n) & \xrightarrow{\ \tau^n\ } & G \\
& {\scriptstyle \varphi_*^n}\searrow & \downarrow{\scriptstyle \tau} \\
& & K_0(A)
\end{array}
$$

commutes.

For each integer k, let B_k be the direct limit of the direct sequence $(M_k(A_n), \varphi_n)_{n=1}^{\infty}$, and for each n, let $\varphi_k^n : M_k(A_n) \to B_k$ be the natural map. By Lemma 7.3.8, there is a unique $*$-isomorphism $\pi_k : B_k \to M_k(A)$ such that for each n the diagram

$$
\begin{array}{ccc}
M_k(A_n) & \xrightarrow{\varphi_k^n} & B_k \\
& \varphi^n \searrow & \downarrow \pi_k \\
& & M_k(A)
\end{array}
$$

commutes.

We show first that τ is surjective. Let $p \in P[A]$. Then $p \in M_k(A)$ for some integer k. Hence, $p = \pi_k(q)$ for some projection $q \in B_k$. By Lemma 7.3.9, Condition (1), there is a projection $r \in M_k(A_n)$ for some n such that $q \sim \varphi_k^n(r)$ in B_k. Hence, $p \sim \varphi^n(r)$ in $M_k(A)$, since $\pi_k \varphi_k^n = \varphi^n$. Consequently, $[p]_A = \varphi_*^n([r]_{A_n}) = \tau \tau^n([r]_{A_n})$, since $\varphi_*^n = \tau \tau^n$. This shows that $K_0(A)^+ \subseteq \text{im}(\tau)$, and, since $K_0(A)^+$ generates $K_0(A)$, we have $K_0(A) = \text{im}(\tau)$; that is, τ is surjective.

Now we show that τ is injective. Suppose that $x \in \ker(\tau)$. Since $G = \cup_{n=1}^{\infty} \text{im}(\tau^n)$, we can write $x = \tau^n([p]_{A_n} - [q]_{A_n})$ for some projections p, q in $P[A_n]$. We may suppose that $p, q \in M_k(A_n)$. Since $\tau(x) = 0$, we have $[\varphi^n(p)]_A = [\varphi^n(q)]_A$, as $\tau \tau^n = \varphi_*^n$. Thus, $\varphi^n(p) \approx \varphi^n(q)$ relative to A, so there is an integer l such that $1_l \oplus \varphi^n(p) \sim 1_l \oplus \varphi^n(q)$; that is, $\varphi^n(1_l \oplus p) \sim \varphi^n(1_l \oplus q)$ in the C*-algebra $M_{l+k}(A)$. Applying the $*$-isomorphism π_{l+k}^{-1}, we get $\varphi_{l+k}^n(1_l \oplus p) \sim \varphi_{l+k}^n(1_l \oplus q)$ in B_{l+k}. Hence, by Lemma 7.3.9, Condition (2), there is an integer $m \geq n$ such that $\varphi_{nm}(1_l \oplus p) \sim \varphi_{nm}(1_l \oplus q)$ in $M_{l+k}(A_m)$. Therefore, $\varphi_{nm*}([p]_{A_n}) = \varphi_{nm*}([q]_{A_n})$, and if we apply τ^m to both sides and observe that $\tau^m \varphi_{nm*} = \tau^n$, we get $\tau^n([p]_{A_n}) = \tau^n([q]_{A_n})$. Hence, $x = \tau^n([p]_{A_n} - [q]_{A_n}) = 0$. This shows that τ is injective and completes the proof. $\qquad\qquad\square$

7.3.3. *Example.* Let $s : \mathbf{N} \setminus \{0\} \to \mathbf{N} \setminus \{0\}$, and define $s!$, ε_s, and M_s as in Section 6.2. If $s' : \mathbf{N} \setminus \{0\} \to \mathbf{N} \setminus \{0\}$, we saw in Theorem 6.2.3 that M_s and $M_{s'}$ are $*$-isomorphic implies $\varepsilon_s = \varepsilon_{s'}$. We shall now use K-theory to prove that M_s and $M_{s'}$ are $*$-isomorphic if $\varepsilon_s = \varepsilon_{s'}$. It is convenient to "normalise" the sequences s, s': If $\tilde{s}$ is the sequence $(1, s_1, s_2, \dots,)$, then $M_{\tilde{s}}$ and M_s are $*$-isomorphic and $\varepsilon_{\tilde{s}} = \varepsilon_s$. Thus, to prove our result we may confine ourselves to sequences s such that $s_1 = 1$.

Let $A_n = M_{s!(n)}(\mathbf{C})$ and let the map $\varphi_n : A_n \to A_{n+1}$ be the canonical $*$-homomorphism.

Denote by $\mathbf{Z}(s)$ the additive group of rational numbers r which can be written in the form $r = m/s!(n)$ for $m \in \mathbf{Z}$ and $n \geq 1$. We make $\mathbf{Z}(s)$ into an ordered group by endowing it with the usual order from $\mathbf{R}$.

If e^n is the matrix in A_n such that all its entries are zero except for the $(1,1)$-entry, which is 1, then $K_0(A_n)^+ = \mathbf{N}[e^n]_{A_n}$, and $K_0(A_n) = \mathbf{Z}[e^n]_{A_n}$ (Theorem 7.2.4). Let $\rho^n \colon K_0(A_n) \to \mathbf{Z}(s)$ be the unique positive homomorphism such that $\rho^n([e^n]_{A_n}) = 1/s!(n)$. Since $\rho^n = \rho^{n+1}\varphi_{n*}$ for all n, there is a unique homomorphism ρ from $\varinjlim K_0(A_n)$ to $\mathbf{Z}(s)$ such that $\rho^n = \rho\tau^n$ for all n, where the $\tau^n \colon K_0(A_n) \to \varinjlim K_0(A_n)$ are the natural maps. A routine verification shows that ρ is an isomorphism. By Theorem 7.3.10, there is a unique isomorphism $\sigma \colon \varinjlim K_0(A_n) \to K_0(M_s)$ such that $\sigma\tau^n = \varphi_*^n$ for all n, where $\varphi^n \colon A_n \to M_s$ are the natural maps. Hence, $\tau = \rho\sigma^{-1}$ is an isomorphism from $K_0(M_s)$ to $\mathbf{Z}(s)$. An application of Lemmas 7.3.8 and 7.3.9 shows that $K_0(M_s)^+ = \cup_{n=1}^\infty \varphi_*^n(K_0(A_n)^+)$. Using this, it is easily checked that τ is in fact an order isomorphism. Also, $\tau([1_1]_{M_s}) = 1$ (use the fact that $\rho^1([e^1]_{A_1}) = 1$).

Now suppose that s' is another function from $\mathbf{N} \setminus \{0\}$ to itself such that $s_1' = 1$. By Corollary 7.2.11, the AF-algebras M_s and $M_{s'}$ are *-isomorphic if and only if there is a unital order isomorphism from $K_0(M_s)$ to $K_0(M_{s'})$, and by our computations above this is equivalent to saying there is an order isomorphism $\tau \colon \mathbf{Z}(s) \to \mathbf{Z}(s')$ such that $\tau(1) = 1$. It is elementary that the latter condition is equivalent to $\mathbf{Z}(s) = \mathbf{Z}(s')$, which is in turn easily seen to be equivalent to $\varepsilon_s = \varepsilon_{s'}$.

We want to extend our continuity result to $\tilde{K}_0$. First some technical details on unitisations and direct limits are needed.

7.3.3. Remark. Let A be the direct limit of the sequence of C*-algebras $(A_n, \varphi_n)_{n=1}^\infty$. Then $\tilde{A} = \varinjlim \tilde{A}_n$. We formulate this more precisely: Let B be the direct limit of the unital sequence $(\tilde{A}_n, \tilde{\varphi}_n)_{n=1}^\infty$, and let $\varphi^n \colon A_n \to A$ and $\psi^n \colon \tilde{A}_n \to B$ be the natural maps. Then there is a unique *-isomorphism $\varphi \colon B \to \tilde{A}$ such that for each n the diagram

$$\begin{array}{ccc} \tilde{A}_n & \xrightarrow{\psi^n} & B \\ & \tilde{\varphi}^n \searrow & \downarrow \varphi \\ & & \tilde{A} \end{array}$$

commutes. We show only that φ is injective, as the rest is straightforward. It suffices to show for each n that φ is injective on $\psi^n(\tilde{A}_n)$. Let $a \in A_n$ and $\lambda \in \mathbf{C}$, and suppose that $\varphi(\psi^n(a + \lambda)) = 0$. Let $\varepsilon > 0$. Now $\tilde{\varphi}^n = \varphi\psi^n$, so $\varphi^n(a) = -\lambda$, and therefore $\lambda = 0$. Hence, there exists $m \geq n$ such that $\|\varphi_{nm}(a)\| < \varepsilon$, and therefore $\|\psi^n(a+\lambda)\| = \|\psi^m\tilde{\varphi}_{nm}(a+\lambda)\| \leq \|\varphi_{nm}(a)\| < \varepsilon$. Letting $\varepsilon \to 0$, this gives $\psi^n(a + \lambda) = 0$, so φ is injective on $\psi^n(\tilde{A}_n)$ as required.

7.3.11. Lemma. Let A be the direct limit of the sequence of C*-algebras $(A_n, \varphi_n)_{n=1}^\infty$, and suppose G is the direct limit of the sequence of abelian groups $(K_0(\tilde{A}_n), \tilde{\varphi}_{n*})_{n=1}^{\infty}$. Denote by $\varphi^n: A_n \to A$ and $\tau^n: K_0(\tilde{A}_n) \to G$ the natural maps. Then there is a unique isomorphism $\tau: G \to K_0(\tilde{A})$ such that the diagram

$$K_0(\tilde{A}_n) \quad \xrightarrow{\ \tau^n\ } \quad G$$
$$\tilde{\varphi}_*^n \searrow \quad \downarrow \tau$$
$$K_0(\tilde{A})$$

commutes for all n.

Proof. Uniqueness of τ is clear. To see existence let B be the direct limit of the sequence $(\tilde{A}_n, \tilde{\varphi}_n)_{n=1}^{\infty}$, and let $\psi^n: \tilde{A}_n \to B$ be the natural *-homomorphism. There is a unique *-isomorphism $\varphi: B \to \tilde{A}$ such that for each n the diagram

$$\tilde{A}_n \quad \xrightarrow{\ \psi^n\ } \quad B$$
$$\tilde{\varphi}^n \searrow \quad \downarrow \varphi$$
$$\tilde{A}$$

commutes (cf. Remark 7.3.3). Also, by Theorem 7.3.10, there is a unique isomorphism $\rho: G \to K_0(B)$ such that the diagram

$$K_0(\tilde{A}_n) \quad \xrightarrow{\ \tau^n\ } \quad G$$
$$\psi_*^n \searrow \quad \downarrow \rho$$
$$K_0(B)$$

commutes for each n. Set $\tau = \varphi_* \rho$. Then $\tau \tau^n = \varphi_* \rho \tau^n = \varphi_* \psi_*^n = \tilde{\varphi}_*^n$ for all n, and the lemma is proved. $\square$

7.3.12. Theorem. Let A be the direct limit of a sequence of C^*-algebras $(A_n, \varphi_n)_{n=1}^{\infty}$, and G the direct limit of the corresponding sequence of abelian groups $(\tilde{K}_0(A_n), \varphi_{n*})_{n=1}^{\infty}$. Denote by $\varphi^n: A_n \to A$ and $\tau^n: \tilde{K}_0(A_n) \to G$ the natural maps. Then there is a unique isomorphism $\tau: G \to \tilde{K}_0(A)$ such that the diagram

$$\tilde{K}_0(A_n) \quad \xrightarrow{\ \tau^n\ } \quad G$$
$$\varphi_*^n \searrow \quad \downarrow \tau$$
$$\tilde{K}_0(A)$$

commutes for each integer n.

Proof. The result follows from earlier results by a diagram chase, so we begin by setting up the diagrams:

Since the diagram

$$\tilde{K}_0(A_n) \xrightarrow{\varphi_{n*}} \tilde{K}_0(A_{n+1})$$

$$\varphi_*^n \searrow \quad \downarrow \varphi_*^{n+1}$$

$$\tilde{K}_0(A)$$

commutes for each n, there is a unique homomorphism $\tau : G \to \tilde{K}_0(A)$ such that the diagram

$$\tilde{K}_0(A_n) \xrightarrow{\tau^n} G$$

$$\varphi_*^n \searrow \quad \downarrow \tau \qquad\qquad (1)$$

$$\tilde{K}_0(A)$$

commutes for each n.

Let $\tilde{G}$ be the direct limit of the sequence $(K_0(\tilde{A}_n), \tilde{\varphi}_{n*})_{n=1}^{\infty}$, and for each n denote by $\tilde{\tau}^n : K_0(\tilde{A}_n) \to \tilde{G}$ the natural map. By Lemma 7.3.11, there is a unique isomorphism $\tilde{\tau} : \tilde{G} \to K_0(\tilde{A})$ such that the diagram

$$K_0(\tilde{A}_n) \xrightarrow{\tilde{\tau}^n} \tilde{G}$$

$$\tilde{\varphi}_*^n \searrow \quad \downarrow \tilde{\tau} \qquad\qquad (2)$$

$$K_0(\tilde{A})$$

commutes for all n.

With these diagrams in place, we can now show injectivity of τ. Suppose that $x \in \ker(\tau)$. For some integer n, and some $y \in \tilde{K}_0(A_n)$, we have $x = \tau^n(y)$, so $0 = \tau\tau^n(y) = \varphi_*^n(y)$ by commutativity of Diagram (1). Since $\tilde{\varphi}_*^n(y) = \varphi_*^n(y)$, as $y \in \tilde{K}_0(A_n)$, and $\tilde{\varphi}_*^n = \tilde{\tau}\tilde{\tau}^n$ by commutativity of Diagram (2), we have $\tilde{\tau}\tilde{\tau}^n(y) = 0$, and because $\tilde{\tau}$ is an isomorphism this implies that $\tilde{\tau}^n(y) = 0$. Hence, there exists $m \geq n$ such that $\varphi_{nm*}(y) = \tilde{\varphi}_{nm*}(y) = 0$. Therefore, $x = \tau^n(y) = \tau^m\varphi_{nm*}(y) = 0$. Thus, τ is injective.

Now we show surjectivity of τ. Let $\rho_n : \tilde{A}_n \to \mathbf{C}$ and $\rho : \tilde{A} \to \mathbf{C}$ be the canonical maps. Suppose that $z \in \tilde{K}_0(A)$. Then $z \in K_0(\tilde{A})$, so $z = \tilde{\tau}(y)$ for some $y \in \tilde{G}$, because $\tilde{\tau}$ is surjective. Since $\tilde{G} = \cup_n \text{im}(\tilde{\tau}^n)$, there exists an integer n, and $x \in K_0(\tilde{A}_n)$, such that $y = \tilde{\tau}^n(x)$. Hence, $z = \tilde{\tau}\tilde{\tau}^n(x) = \tilde{\varphi}_*^n(x)$ by commutativity of Diagram (2). However, $x \in \tilde{K}_0(A_n) = \ker(\rho_{n*})$, since $z \in \tilde{K}_0(A) = \ker(\rho_*)$ and $\rho_{n*}(x) = \rho_*\tilde{\varphi}_*^n(x) = \rho_*(z)$. Therefore, $z = \varphi_*^n(x) = \tau\tau^n(x)$ by commutativity of Diagram (1), so $z \in \text{im}(\tau)$. $\square$

7.4. Stability

The most important result of this section asserts that if H is a separable infinite-dimensional Hilbert space, then $\tilde{K}_0(A) = \tilde{K}_0(K(H) \otimes_* A)$. This is referred to as *stability* of the functor $\tilde{K}_0$. It is a fundamental result, and will be used in the next section to prove Bott periodicity. Our line of attack is to show first that $\tilde{K}_0(A) = \tilde{K}_0(M_2(A))$, and then derive stability using continuity of $\tilde{K}_0$.

If A is a C*-algebra, the map

$$\kappa: A \to M_2(A), \quad a \mapsto \begin{pmatrix} a & 0 \\ 0 & 0 \end{pmatrix},$$

is an injective *-homomorphism, which we shall call the *inclusion* of A in $M_2(A)$. In cases of ambiguity we shall write κ^A rather than κ.

7.4.1. Remark. If A is a unital C*-algebra, then every element $x \in \tilde{K}_0(A)$ can be written $x = [p]_{\tilde{A}} - [q]_{\tilde{A}}$, where p, q belong to $P[A]$. We may even suppose that $q = (1_A)_n$ for some integer n, where 1_A is the unit of A. This follows from the natural isomorphism of $\tilde{K}_0(A)$ with $K_0(A)$ (Theorem 7.3.1), and the same trick we used in Remark 7.3.2.

7.4.1. Theorem. *Suppose A is a unital C*-algebra and $\kappa: A \to M_2(A)$ is the inclusion of A in $M_2(A)$. Then the map $\kappa_*: \tilde{K}_0(A) \to \tilde{K}_0(M_2(A))$ is an isomorphism.*

Proof. Let n be a positive integer. If σ is a permutation of $\{1, \ldots, n\}$, let u_σ be the unitary in $M_n(A)$ defined by setting $(u_\sigma)_{ij} = \delta_{\sigma(i),j}$. If p is a projection in $M_n(A)$, a routine verification shows that

$$\begin{pmatrix} p & 0_n \\ 0_n & 0_n \end{pmatrix} = u_\sigma \begin{pmatrix} \kappa(p_{11}) & \cdots & \kappa(p_{1n}) \\ \vdots & \ddots & \vdots \\ \kappa(p_{n1}) & \cdots & \kappa(p_{nn}) \end{pmatrix} u_\sigma^*,$$

where σ is the permutation of $\{1, \ldots, 2n\}$ given by

$$\sigma = \begin{pmatrix} 1 & 2 & 3 & \cdots & n & n+1 & n+2 & \cdots & 2n \\ 1 & 3 & 5 & \cdots & 2n-1 & 2 & 4 & \cdots & 2n \end{pmatrix}.$$

Hence, $p \sim \kappa(p)$ relative to A.

Suppose now that p is a projection in $M_m(M_2(A))$. Then $p \sim \kappa(p)$ relative to $M_2(A)$, so $[p]_{M_2(A)^-} = [\kappa(p)]_{M_2(A)^-} = \kappa_*([p]_{\tilde{A}})$. This shows that κ_* is surjective.

Now we show injectivity of κ_*. Suppose that $x \in \ker(\kappa_*)$, so that for some positive integer n we have $x = [p]_{\tilde{A}} - [q]_{\tilde{A}}$ for projections p, q in $M_n(A)$. Now $[\kappa(p)]_{M_2(A)^-} = [\kappa(q)]_{M_2(A)^-}$, so $[\kappa(p)]_{M_2(A)} = [\kappa(q)]_{M_2(A)}$ (using the natural isomorphism of $\tilde{K}_0(M_2(A))$ and $K_0(M_2(A))$). Hence, $\kappa(p) \approx \kappa(q)$ relative to $M_2(A)$. Therefore, $\kappa(p) \approx \kappa(q)$ relative to A, so $p \sim \kappa(p) \approx \kappa(q) \sim q$ again relative to A, so $p \approx q$ relative to A. Hence, $[p]_{\tilde{A}} = [q]_{\tilde{A}}$, so $x = 0$. Thus, κ_* is injective. $\qquad\square$

7.4.2. Remark. If A is a C*-algebra and $i: A \to \tilde{A}$ is the inclusion and $\rho: \tilde{A} \to \mathbf{C}$ is the canonical map, then

$$0 \to \tilde{K}_0(A) \xrightarrow{i_*} \tilde{K}_0(\tilde{A}) \xrightarrow{\rho_*} \tilde{K}_0(\mathbf{C}) \to 0$$

is a short exact sequence. This is immediate from the natural isomorphism of $\tilde{K}_0$ and K_0 on unital C*-algebras (Theorem 7.3.1) and the fact that

$$0 \to \tilde{K}_0(A) \xrightarrow{j} K_0(\tilde{A}) \xrightarrow{\rho_*} K_0(\mathbf{C}) \to 0$$

is a short exact sequence (j denotes the inclusion).

7.4.2. Theorem. If A is a C*-algebra and $\kappa: A \to M_2(A)$ is the inclusion, then $\kappa_*: \tilde{K}_0(A) \to \tilde{K}_0(M_2(A))$ is an isomorphism.

Proof. If $i: A \to \tilde{A}$ is the inclusion and $\rho: \tilde{A} \to \mathbf{C}$ is the canonical map, then we have a commutative diagram

$$
\begin{array}{ccccc}
A & \xrightarrow{i} & \tilde{A} & \xrightarrow{\rho} & \mathbf{C} \\
\downarrow{\kappa^A} & & \downarrow{\kappa^{\tilde{A}}} & & \downarrow{\kappa^{\mathbf{C}}} \\
M_2(A) & \xrightarrow{i} & M_2(\tilde{A}) & \xrightarrow{\rho} & M_2(\mathbf{C})
\end{array}
$$

and therefore the corresponding diagram on the $\tilde{K}_0$-level commutes:

$$
\begin{array}{ccccc}
\tilde{K}_0(A) & \xrightarrow{i_*} & \tilde{K}_0(\tilde{A}) & \xrightarrow{\rho_*} & \tilde{K}_0(\mathbf{C}) \\
\downarrow{\kappa_*^A} & & \downarrow{\kappa_*^{\tilde{A}}} & & \downarrow{\kappa_*^{\mathbf{C}}} \\
\tilde{K}_0(M_2(A)) & \xrightarrow{j} & \tilde{K}_0(M_2(\tilde{A})) & \xrightarrow{\rho_*} & \tilde{K}_0(M_2(\mathbf{C})).
\end{array}
$$

To avoid ambiguity in the argument to follow, we are denoting the map $i_*: \tilde{K}_0(M_2(A)) \to \tilde{K}_0(M_2(\tilde{A}))$ by j. The top row in the second diagram is a short exact sequence (*cf.* Remark 7.4.2), so, in particular, i_* is injective. Since $\kappa_*^{\tilde{A}}$ is an isomorphism by Theorem 7.4.1, and since $j\kappa_*^A = \kappa_*^{\tilde{A}}i_*$, it is clear that κ_*^A is injective.

If we assume that the map j is injective, we can show that κ_*^A is surjective: If $x \in \tilde{K}_0(M_2(A))$, then $j(x) = \kappa_*^{\tilde{A}}(y)$ for some $y \in \tilde{K}_0(\tilde{A})$, by surjectivity of $\kappa^{\tilde{A}}$. Since $\kappa_*^{\mathbf{C}}\rho_*(y) = \rho_*\kappa_*^{\tilde{A}}(y) = \rho_*j(x) = 0$ and $\kappa_*^{\mathbf{C}}$ is injective by Theorem 7.4.1, we have $\rho_*(y) = 0$, and therefore $y \in i_*(\tilde{K}_0(A))$. Thus, $y = i_*(z)$ for some $z \in \tilde{K}_0(A)$, and $j(x) = \kappa_*^{\tilde{A}}i_*(z) = j\kappa_*^A(z)$. Because j is assumed injective, we have, therefore, $x = \kappa_*^A(z)$. Hence, κ_*^A is surjective.

Thus, to prove the theorem we need only show that j is injective. Let $B = M_2(A) + \mathbf{C}1_2$, and let $k \colon M_2(A) \to B$ and $\psi \colon B \to M_2(\tilde{A})$ be the inclusion $*$-homomorphisms. The diagram

$$M_2(A) \xrightarrow{\ i\ } M_2(\tilde{A})$$
$$k \searrow \quad \uparrow \psi$$
$$B$$

commutes, so the diagram

$$\tilde{K}_0(M_2(A)) \xrightarrow{\ j\ } \tilde{K}_0(M_2(\tilde{A}))$$
$$k_* \searrow \quad \uparrow \psi_*$$
$$\tilde{K}_0(B)$$

commutes. By Remark 7.4.2, the map k_* is injective (identify B with $M_2(A)\tilde{\ }$). Thus, to show that j is injective, it suffices to show that ψ_* is injective.

We need another map: Denote by τ the unique unital $*$-homomorphism from B to $\mathbf{C}1_2$ having kernel $M_2(A)$.

If $x \in \ker(\psi_*)$, then by Remark 7.4.1 there exist an integer k and projections $p, q \in M_k(B)$ such that $x = [p]_{\tilde{B}} - [q]_{\tilde{B}}$. We may even suppose q is of the form $q = (1_2)_n \oplus 0_r$. Hence, $\tau(q) = q$. Now $p \approx q$ relative to $M_2(\tilde{A})$, and to prove the theorem we need only show this implies that $p \approx q$ relative to B. There is an integer m such that $(1_2)_m \oplus p \sim (1_2)_m \oplus q$ in $M_{m+k}(M_2(\tilde{A}))$. Thus, replacing p and q by $(1_2)_m \oplus p$ and $(1_2)_m \oplus q$ if necessary, it suffices to show that if p, q are projections in $M_k(B)$ such that $p \sim q$ in $M_k(M_2(\tilde{A}))$ and $\tau(q) = q$, then $p \sim q$ in $M_k(B)$. Since ρ is a $*$-homomorphism, we have $\rho(p) \sim \rho(q)$ in $M_k(M_2(\mathbf{C}))$. Now $\rho(p) = \tau(p)$ and $\rho(q) = \tau(q)$ belong to the subalgebra $M_k(\mathbf{C}1_2)$, and to show that they are equivalent in this subalgebra we have only to show they have the same rank (cf. Example 7.1.1). But for projections in $M_k(\mathbf{C}1_2)$ the rank is the same as the trace, and the normalised trace on $M_k(\mathbf{C}1_2)$ is just the restriction of the normalised trace on $M_k(M_2(\mathbf{C}))$, so $\tau(p), \tau(q)$ have the same trace in $M_k(\mathbf{C}1_2)$ because they are equivalent in $M_k(M_2(\mathbf{C}))$. Hence, there is a partial isometry $w \in M_k(\mathbf{C}1_2)$ such that $\tau(p) = w^*w$ and $\tau(q) = ww^*$. Since $p \sim q$ in $M_k(M_2(\tilde{A}))$, there exists $v \in M_k(M_2(\tilde{A}))$ such that $p = v^*v$ and $q = vv^*$. Set $u = w\rho(v)^*v$. Then $u = u_1 + u_2$, where $u_1 = w\rho(v)^*(v - \rho(v))$ and $u_2 = w\rho(v)^*\rho(v)$. Because $M_k(M_2(A))$ is an ideal in $M_k(M_2(\tilde{A}))$, and $v - \rho(v) \in M_k(M_2(A))$, we have $u_1 \in M_k(M_2(A))$, and therefore $u_1 \in M_k(B)$. As $u_2 = w\rho(v^*v) = w\rho(p) = w\tau(p)$, so $u_2 \in M_k(\mathbf{C}1_2)$, and therefore, $u_2 \in M_k(B)$. Hence, $u \in M_k(B)$. Finally, $u^*u = v^*\rho(v)w^*w\rho(v)^*v = v^*\rho(v)\rho(p)\rho(v)^*v = v^*\rho(vv^*vv^*)v = v^*qv = p$ and

$uu^* = w\rho(v)^*vv^*\rho(v)w^* = w\rho(v)^*q\rho(v)w^* = w\rho(v^*vv^*v)w^* = w\tau(p)w^* = \tau(q) = q$, so $p \sim q$ in $M_k(B)$ as required. $\qquad\qquad\qquad\qquad\qquad\square$

If A is a C*-algebra, if H is a Hilbert space, and if p is a rank-one projection in $K(H)$, then the map

$$\varphi: A \to K(H) \otimes_* A, \quad a \mapsto p \otimes a,$$

is a *-homomorphism. Write $p_* = \varphi_*: \tilde{K}_0(A) \to \tilde{K}_0(K(H) \otimes_* A)$. If q is another rank-one projection in $K(H)$, then there is a unitary $u \in B(H)$ such that $q = upu^*$. Hence, by Theorem 2.5.8, there is a self-adjoint operator $v \in B(H)$ such that $u = e^{iv}$. For $t \in [0, 1]$, set $u_t = e^{itv}$, so u_t is a unitary in $B(H)$. If $\varphi_t = \psi_t\varphi: A \to K(H) \otimes_* A$, where $\psi_t = \operatorname{Ad} u_t \otimes_* \operatorname{id}_A: K(H) \otimes_* A \to K(H) \otimes_* A$, then it is easy to check that $(\varphi_t)_t$ is a homotopy. Hence, $p_* = \varphi_{0*} = \varphi_{1*} = q_*$.

The homomorphism p_* is called the *canonical* map from $\tilde{K}_0(A)$ to $\tilde{K}_0(K(H) \otimes_* A)$. The following is a key result of K-theory, and is referred to as *stability* of $\tilde{K}_0$:

7.4.3. Theorem. *If A is a C*-algebra and H is a separable infinite-dimensional Hilbert space the canonical map from $\tilde{K}_0(A)$ to $\tilde{K}_0(K(H) \otimes_* A)$ is an isomorphism.*

Proof. Set $K = K(H)$. Let $(e_n)_{n=1}^\infty$ be an orthonormal basis of H, and write e_{ij} for the operator in $B(H)$ given by $e_{ij}(x) = \langle x, e_j \rangle e_i$. Set $p_n = \sum_{j=1}^n e_{jj}$ and note that the map

$$\psi_n: M_n(A) \to p_n K p_n \otimes A, \quad (a_{ij}) \mapsto \sum_{i,j=1}^n e_{ij} \otimes a_{ij},$$

is a *-isomorphism.

Set $B_n = M_{2^{n-1}}(A)$ for $n \geq 1$. The map

$$\pi^n: B_n \to K \otimes_* A, \quad a \mapsto \psi_{2^{n-1}}(a),$$

is an isometric *-homomorphism, and $\pi^n = \pi^{n+1}\kappa_n$, where $\kappa_n: B_n \to B_{n+1}$ is the inclusion; that is,

$$\kappa_n(a) = \begin{pmatrix} a & 0 \\ 0 & 0 \end{pmatrix}.$$

Let B be the direct limit of the sequence $(B_n, \kappa_n)_{n=1}^\infty$ and for each n let $\kappa^n: B_n \to B$ be the natural map. Then there is a unique *-homomorphism $\pi: B \to K \otimes_* A$ such that for all n we have $\pi^n = \pi\kappa^n$. Since

$$K = (\cup_{n=1}^\infty p_{2^{n-1}} K p_{2^{n-1}})^-,$$

we have

$$K \otimes_* A = (\cup_{n=1}^{\infty} p_{2n-1} K p_{2n-1} \otimes A)^- = (\cup_{n=1}^{\infty} \pi^n(B_n))^-,$$

and therefore π is surjective. Moreover, π is isometric on each subalgebra $\kappa^n(B_n)$, and therefore, by density of $\cup_{n=1}^{\infty} \kappa^n(B_n)$ in B, it follows that π is isometric on B. Thus, π is a *-isomorphism and therefore π_* is an isomorphism.

If G is the direct limit of the sequence of abelian groups $(\tilde{K}_0(B_n), \kappa_{n*})$, and if for each n we denote by τ^n the natural map from $\tilde{K}_0(B_n)$ to G, then by Theorem 7.3.12, there is a unique isomorphism $\tau : G \to \tilde{K}_0(B)$ such that $\kappa_*^n = \tau\tau^n$ for all n. It follows from Theorem 7.4.2 that each map $\kappa_{n*} : B_n \to B_{n+1}$ is an isomorphism, and therefore each map τ^n is an isomorphism. Hence, $\kappa_*^n = \tau\tau^n$ is an isomorphism, and therefore $\pi_*^n = \pi_*\kappa_*^n$ is an isomorphism. Since π^1 is the map from $A = M_1(A)$ to $K \otimes_* A$ given by $\pi^1(a) = \psi_1(a) = e_{11} \otimes a$, the map π_*^1 is the canonical map from $\tilde{K}_0(A)$ to $\tilde{K}_0(K \otimes_* A)$. This proves the theorem. $\square$

7.4.3. Remark. If $\varphi, \psi : A \to B$ are *-homomorphisms of C*-algebras, we say they are *orthogonal* if $\varphi(a)\psi(a') = 0$ $(a, a' \in A)$. In this case $\varphi + \psi$ is a *-homomorphism. If $p \in P[A]$ then $\varphi(p)$ and $\psi(p)$ are orthogonal projections, so

$$(\varphi + \psi)_*[p]_{\tilde{A}} = [\varphi(p) + \psi(p)]_{\tilde{A}} = [\varphi(p)]_{\tilde{A}} + [\psi(p)]_{\tilde{A}} = (\varphi_* + \psi_*)[p]_{\tilde{A}}.$$

If A is unital, the elements of $\tilde{K}_0(A)$ are of the form $[p]_{\tilde{A}} - [q]_{\tilde{A}}$ $(p, q \in P[A])$, so clearly, $(\varphi + \psi)_* = \varphi_* + \psi_* : \tilde{K}_0(A) \to \tilde{K}_0(B)$.

7.5. Bott Periodicity

We shall find it convenient to adopt the following notation: If Ω is a locally compact Hausdorff space and A is a C*-algebra, we set $A\Omega = C_0(\Omega, A)$. If $\varphi : A \to B$ is a *-homomorphism between C*-algebras A and B, the map

$$\bar{\varphi} : A\Omega \to B\Omega, \quad f \mapsto \varphi \circ f,$$

is a *-homomorphism.

7.5.1. Theorem. Let A, B be C*-algebras and let Ω be a locally compact Hausdorff space. If $(\varphi_t)_t$ is a homotopy of *-homomorphisms from A to B, then $(\bar{\varphi}_t)_t$ is a homotopy of *-homomorphisms from $A\Omega$ to $B\Omega$.

Proof. It is easily checked that if C is the set of all $g \in A\Omega$ such that

$$\psi_g : [0, 1] \to B\Omega, \quad t \mapsto \bar{\varphi}_t(g),$$

is continuous, then C is a C*-subalgebra of $A\Omega$. If $f \in C_0(\Omega)$ and $a \in A$, then $\bar{\varphi}_t(fa) = f\varphi_t(a)$, so the map ψ_g is continuous in the case that g is of the form $g = fa$. Since the elements of $A\Omega$ of this form have closed linear span $A\Omega$ by Lemma 6.4.16, it follows that $C = A\Omega$ and therefore $(\bar{\varphi}_t)_t$ is a homotopy. $\square$

If A is a C*-algebra, then the C*-algebra

$$C(A) = \{f \in A[0,1] \mid f(1) = 0\}$$

is a closed ideal in $A[0,1]$, called the *cone* of A.

A C*-algebra A is said to be *contractible* if the identity map id: $A \to A$ is homotopic to the zero map. In this case $\tilde{K}_0(A) = 0$, by Theorem 7.3.7.

7.5.2. Theorem. *If A is a C*-algebra, then its cone $C(A)$ is contractible.*

Proof. Let $f \in C(A)$ and for $t \in [0,1]$ define $\varphi_t(f) \in C(A)$ by setting $\varphi_t(f)(s) = f(1 - t + st)$ $(0 \le s \le 1)$. It is easily checked that the map

$$\varphi_t: C(A) \to C(A), \quad f \mapsto \varphi_t(f),$$

is a *-homomorphism. Since the map

$$h: [0,1]^2 \to A, \quad (s,t) \mapsto f(1 - t + st),$$

is continuous, and therefore uniformly continuous, it follows that if $\varepsilon > 0$ there exists some $\delta > 0$ such that

$$\max(|s - s'|, |t - t'|) < \delta \Rightarrow \|h(s,t) - h(s',t')\| < \varepsilon/2.$$

Thus, if $|t - t'| < \delta$, then $\|\varphi_t(f) - \varphi_{t'}(f)\| < \varepsilon$. Hence, the map

$$[0,1] \to C(A), \quad t \mapsto \varphi_t(f),$$

is continuous for all $f \in C(A)$, and therefore $(\varphi_t)_t$ is a homotopy on $C(A)$. Since $\varphi_0 = 0$ and $\varphi_1 = $ id, the zero and identity maps on $C(A)$ are homotopic; that is, $C(A)$ is contractible. $\square$

7.5.3. Theorem. *If A is a contractible C*-algebra and Ω is a locally compact Hausdorff space, then $A\Omega$ is also contractible.*

Proof. If $(\varphi_t)_t$ is a homotopy from the zero map of A to the identity map of A, then $(\bar{\varphi}_t)_\tau$ is a homotopy (by Theorem 7.5.1) from the zero map of $A\Omega$ to the identity map of $A\Omega$. Thus, $A\Omega$ is contractible. $\square$

If A is a C*-algebra, we define its *suspension* to be the C*-algebra

$$S(A) = \{f \in A[0,1] \mid f(0) = f(1) = 0\}.$$

Thus, $S(A)$ is a closed ideal in $C(A)$. If $\varphi: A \to B$ is a *-homomorphism of C*-algebras, then $\bar{\varphi}$ maps $S(A)$ into $S(B)$, so if we denote its restriction by $S(\varphi): S(A) \to S(B)$, then $S(\varphi)$ is a *-homomorphism, and it is clear from Theorem 7.5.1 that if $(\varphi_t)_t$ is a homotopy of *-homomorphisms from A to B, then $(S(\varphi_t))_t$ is a homotopy of *-homomorphisms from $S(A)$ to $S(B)$. Hence, if A is contractible, so is its suspension $S(A)$.

If A is an arbitrary C*-algebra, set $\tilde{K}_1(A) = \tilde{K}_0(S(A))$. If $\varphi: A \to B$ is a *-homomorphism of C*-algebras, denote by φ_* the homomorphism $(S(\varphi))_*: \tilde{K}_1(A) \to \tilde{K}_1(B)$. If there is a possibility of ambiguity, we shall write $\tilde{K}_0(\varphi)$ and $\tilde{K}_1(\varphi)$ for the homomorphisms $\varphi_*: \tilde{K}_0(A) \to \tilde{K}_0(B)$ and $\varphi_*: \tilde{K}_1(A) \to \tilde{K}_1(B)$, respectively. It is straightforward to verify that

$$A \mapsto \tilde{K}_1(A), \qquad \varphi \mapsto \varphi_*,$$

gives a covariant functor from the category of C*-algebras to the category of abelian groups.

7.5.4. Lemma. *Let A be a unital C*-algebra and $\alpha < \beta$ real numbers. If $p: [\alpha, \beta] \to A$ is a continuous path of projections, then there is a continuous path of unitaries $u: [\alpha, \beta] \to A$ such that $p(t) = u(t)p(\alpha)u^*(t)$ for all t in $[\alpha, \beta]$ and $u(\alpha) = 1$.*

Proof. Suppose first that $\|p(t) - p(\alpha)\| < 1$ for all t. Set

$$v(t) = 1 - p(\alpha) - p(t) + 2p(t)p(\alpha).$$

It follows from Lemma 6.2.1 that $v(t)$ is invertible and $u(t) = v(t)|v(t)|^{-1}$ is a unitary such that $p(t) = u(t)p(\alpha)u^*(t)$ for all t, and $u(\alpha) = 1$. It is easily checked that the function $t \mapsto u(t)$ is continuous.

We reduce the general case to the preceding case by using the uniform continuity of p. There exists a partition $\alpha_0 < \alpha_1 < \cdots < \alpha_n$ of $[\alpha, \beta]$ such that $\|p(t) - p(s)\| < 1$ for all $t, s \in [\alpha_i, \alpha_{i+1}]$. Therefore, there is a continuous path of unitaries $u_i: [\alpha_i, \alpha_{i+1}] \to A$ such that $p(t) = u_i(t)p(\alpha_i)u_i^*(t)$ for $t \in [\alpha_i, \alpha_{i+1}]$, and $u_i(\alpha_i) = 1$. Set $u = u_0$ on $[\alpha_0, \alpha_1]$, and if $i > 0$ and $t \in [\alpha_i, \alpha_{i+1}]$, set $u(t) = u_i(t)u_{i-1}(\alpha_i) \cdots u_0(\alpha_1)$. This gives a well-defined continuous path $t \mapsto u(t)$ of unitaries such that $p(t) = u(t)p(\alpha)u^*(t)$ for all $t \in [\alpha, \beta]$ and $u(\alpha) = 1$. $\qquad \square$

7.5.1. Remark. Let Ω be a locally compact Hausdorff space and A a C*-algebra. If $f \in M_n(C_0(\Omega, A))$, define $g = \varphi(f) \in C_0(\Omega, M_n(A))$ by setting $g(\omega) = (f_{ij}(\omega))$ for all $\omega \in \Omega$. It is a straightforward exercise to show that this gives a *-isomorphism $f \mapsto \varphi(f)$ from $M_n(C_0(\Omega, A))$ onto $C_0(\Omega, M_n(A))$. We shall call this the *canonical* *-isomorphism.

7.5.5. Lemma. *Let A be a unital C^*-algebra such that $U_n(A) = U_n^0(A)$ for all $n \geq 1$. Then $\tilde{K}_1(A) = 0$.*

Proof. The inclusion $S(A) \to C([0,1], A)$ has a unique extension to an injective unital $*$-homomorphism $\varphi \colon S(A)^\sim \to C([0,1], A)$. The inflation $\varphi \colon M_n(S(A)^\sim) \to M_n(C([0,1], A))$ is therefore an injective $*$-homomorphism also. Composing this together with the canonical $*$-isomorphism θ from $M_n(C([0,1], A))$ to $C([0,1], M_n(A))$, we get a $*$-isomorphism $\psi_n \colon M_n(S(A)^\sim) \to \Omega_n(A)$, where $\Omega_n(A) = \theta\varphi M_n(S(A)^\sim) = \{f \in C([0,1], M_n(A)) \mid f(0) = f(1) \in M_n(\mathbf{C})\}$. Moreover, if $\tau \colon S(A)^\sim \to \mathbf{C}$ is the canonical map, and ε denotes the $*$-homomorphism

$$\Omega_n(A) \to M_n(\mathbf{C}), \quad f \mapsto f(0),$$

then the diagram

$$
\begin{array}{ccc}
M_n(S(A)^\sim) & \xrightarrow{\psi_n} & \Omega_n(A) \\
& \tau \searrow & \downarrow \varepsilon \\
& & M_n(\mathbf{C})
\end{array}
$$

commutes.

Now suppose that $x \in \tilde{K}_1(A) = \tilde{K}_0(S(A))$. Then we may write $x = [p] - [1_n]$ for some projection p in $M_{2n}(S(A)^\sim)$. Since $\tau(p) \sim \tau(1_n \oplus 0_n)$ in $M_{2n}(\mathbf{C})$, there is a unitary u' in $M_{2n}(\mathbf{C})$ such that $u'\tau(p)u'^* = 1_n \oplus 0_n$. Hence, replacing p by $u'pu'^*$ if necessary, we may suppose that $\tau(p) = 1_n \oplus 0_n$.

The element $q = \psi_{2n}(p) \in \Omega_{2n}(A)$ is a continuous path of projections in $M_{2n}(A)$, so by Lemma 7.5.4 there is a continuous path $u \colon [0,1] \to M_{2n}(A)$ of unitaries such that $q(t) = u(t)q(0)u^*(t)$ for all t, and $u(0) = 1_{2n}$. Now $q(1) = q(0) = 1_n \oplus 0_n$, so $u(1)(1_n \oplus 0_n)u^*(1) = 1_n \oplus 0_n$. This implies that $u(1)$ can be written in the form $u_1 \oplus u_2$, where $u_1, u_2 \in U_n(A)$. Since $U_n(A) = U_n^0(A)$ by hypothesis, there exist continuous paths $v_1, v_2 \colon [0,1] \to U_n(A)$ of unitaries such that $v_i(0) = 1_n$ and $v_i(1) = u_i^*$ for $i = 1, 2$. Set $w(t) = u(t)(v_1(t) \oplus v_2(t))$. Then

$$w \colon [0,1] \to M_{2n}(A), \quad t \mapsto w(t),$$

is a continuous path of unitaries such that $w(0) = w(1) = 1_{2n}$, and therefore w is a unitary in $\Omega_{2n}(A)$. Moreover,

$$q(t) = w(t)q(0)w^*(t) \qquad \text{for all } t, \tag{1}$$

since $v_1(t) \oplus v_2(t)$ commutes with $q(0) = 1_n \oplus 0_n$. There exists a unitary $w' \in M_{2n}(S(A)^\sim)$ such that $\psi_{2n}(w') = w$. Since

$$\psi_{2n}(p) = \psi_{2n}(w')\psi_{2n}(1_n \oplus 0_n)\psi_{2n}(w')^*$$

by Eq. (1) we have $p = w'(1_n \oplus 0_n)w'^*$. Hence, $[p] = [1_n]$, and $x = [p] - [1_n] = 0$. Thus, $\tilde{K}_0(S(A)) = 0$. $\square$

7.5.6. Theorem. *If A is a unital AF-algebra or a von Neumann algebra, then $\tilde{K}_1(A) = 0$.*

Proof. It suffices to show that $U_n(A) = U_n^0(A)$ for all n, by Lemma 7.5.5. If A is a unital AF-algebra, so is $M_n(A)$, and likewise if A is a von Neumann algebra, $M_n(A)$ is one also. Thus, the theorem is proved if we show that $U(A) = U^0(A)$. The von Neumann algebra case is given by Remark 4.2.2. That $U(A) = U^0(A)$ if A is finite-dimensional is given by Theorem 2.1.12. If A is a unital AF-algebra and $u \in U(A)$, then there is a finite-dimensional C*-subalgebra B of A containing the unit of A, and a unitary v in B such that $\|u - v\| < 1$. Hence, $\|1 - uv^*\| < 1$, so $uv^* \in U^0(A)$ by Theorem 2.1.12 again. Since $v \in U(B) = U^0(B) \subseteq U^0(A)$, this implies that $u \in U^0(A)$. $\square$

It is sometimes the case that the only effective means of showing that a C*-algebra is not an AF-algebra is to show that its $\tilde{K}_1$-group is non-zero.

It will be useful in a number of contexts to have an alternative way of looking at the suspension, and for this reason we introduce a new algebra: We denote by $\mathbf{S}$ the closed ideal of $C(\mathbf{T})$ consisting of all functions f such that $f(1) = 0$.

7.5.7. Theorem. *If A is a C*-algebra and γ is the function*

$$[0,1] \to \mathbf{T}, \quad t \mapsto e^{i2\pi t},$$

then there is a unique $*$-isomorphism γ_A from $A \otimes_* \mathbf{S}$ to $S(A)$ such that $\gamma_A(a \otimes f) = (f \circ \gamma)a$ for all $f \in \mathbf{S}$ and $a \in A$.

Proof. Uniqueness is obvious, so we show only existence. The map

$$A \times \mathbf{S} \to S(A), \quad (a, f) \mapsto (f \circ \gamma)a,$$

is bilinear, so it induces a unique linear map $\gamma_A \colon A \otimes \mathbf{S} \to S(A)$ such that $\gamma_A(a \otimes f) = (f \circ \gamma)a$ for all $f \in \mathbf{S}$ and $a \in A$. It is easy to check that γ_A is an injective $*$-homomorphism and therefore the function

$$p \colon A \otimes \mathbf{S} \to \mathbf{R}^+, \quad c \mapsto \|\gamma_A(c)\|,$$

is a C*-norm. Since $\mathbf{S}$ is abelian it is nuclear (Theorem 6.4.15), so p is the unique C*-norm on $A \otimes \mathbf{S}$. Hence, γ_A is isometric, and can therefore be extended to an isometric $*$-homomorphism from $A \otimes_* \mathbf{S}$ to $S(A)$ which we shall also denote by γ_A. To show that γ_A is surjective, it suffices to show that $S(A)$ is the closed linear span of the elements of the form fa where $a \in A$ and $f \in S(\mathbf{C})$. This follows from the easily verified fact that

$$S(A) \to A(0,1), \quad g \mapsto g_r,$$

is a $*$-isomorphism (where g_r denotes the restriction of g to $(0,1)$) and from an application of Lemma 6.4.16 to $A(0,1)$. $\square$

7.5.2. Remark. The *-isomorphism in Theorem 7.5.7 is natural in the sense that if $\varphi: A \to B$ is a *-homomorphism of C*-algebras, then the diagram

$$A \otimes_* S \xrightarrow{\varphi \otimes_* \mathrm{id}} B \otimes_* S$$

$$\downarrow \gamma_A \qquad\qquad \downarrow \gamma_B$$

$$S(A) \xrightarrow{S(\varphi)} S(B)$$

commutes.

7.5.8. Theorem. *If*

$$0 \to J \xrightarrow{j} A \xrightarrow{\varphi} B \to 0$$

is a short exact sequence of C-algebras, so is*

$$0 \to S(J) \xrightarrow{S(j)} S(A) \xrightarrow{S(\varphi)} S(B) \to 0.$$

Proof. By Theorem 6.5.2,

$$0 \to J \otimes_* S \xrightarrow{j \otimes_* \mathrm{id}} A \otimes_* S \xrightarrow{\varphi \otimes_* \mathrm{id}} B \otimes_* S \to 0$$

is a short exact sequence. A straightforward diagram chase using this and Remark 7.5.2 then shows that

$$0 \to S(J) \xrightarrow{S(j)} S(A) \xrightarrow{S(\varphi)} S(B) \to 0$$

is a short exact sequence.   □

As with $\tilde{K}_0$, the functor $\tilde{K}_1$ is weak exact:

7.5.9. Theorem. *Suppose that*

$$0 \to J \xrightarrow{j} A \xrightarrow{\varphi} B \to 0$$

is a short exact sequence of C-algebras. Then the sequence*

$$\tilde{K}_1(J) \xrightarrow{j_*} \tilde{K}_1(A) \xrightarrow{\varphi_*} \tilde{K}_1(B)$$

is exact.

Proof. This is immediate from Theorems 7.5.8 and 7.3.5.   □

We are going to derive a connection between $\tilde{K}_1$ to $\tilde{K}_0$. This requires some convoluted constructions.

If $\varphi_1: A_1 \to B$ and $\varphi_2: A_2 \to B$ are *-homomorphisms of C*-algebras, then

$$C = \{(a_1, a_2) \in A_1 \oplus A_2 \mid \varphi_1(a_1) = \varphi_2(a_2)\}$$

is a C*-subalgebra of $A_1 \oplus A_2$, called the *pullback* of A_1 and A_2 *along* φ_1 and φ_2.

If $\varphi: A \to B$ is a *-homomorphism of C*-algebras, we denote by Z_φ the pullback of A and $B[0,1]$ along the *-homomorphisms φ and

$$B[0,1] \to B, \quad f \mapsto f(0).$$

The surjective *-homomorphism

$$Z_\varphi \to A, \quad (a, f) \mapsto a,$$

is called the *projection* of Z_φ onto A, and the injective *-homomorphism

$$A \to Z_\varphi, \quad a \mapsto (a, \varphi(a))$$

is the *inclusion* of A in Z_φ. The *-homomorphism

$$\varepsilon: Z_\varphi \to B, \quad (a, f) \mapsto f(1),$$

is the *canonical* map from Z_φ to B. The kernel of ε is denoted by C_φ and is called the *mapping cone* of φ. Explicitly,

$$C_\varphi = \{(a, f) \in A \oplus B[0,1] \mid f(0) = \varphi(a), \ f(1) = 0\}.$$

The surjective *-homomorphism

$$C_\varphi \to A, \quad (a, f) \mapsto a,$$

is the *projection* from C_φ onto A.

7.5.10. Lemma. *Let* $\varphi: A \to B$ *be a *-homomorphism of C*-algebras and suppose that* $\tilde{\pi}: Z_\varphi \to A$ *is the projection, and* $i: A \to Z_\varphi$ *is the inclusion, *-homomorphism. Then* $\tilde{\pi} i = \mathrm{id}_A$, *and* $i\tilde{\pi}$ *is homotopic to* id_{Z_φ}.

Proof. It is obvious that $\tilde{\pi} i = \mathrm{id}$.

For $f \in B[0,1]$ and $t \in [0,1]$, define $f_t \in B[0,1]$ by $f_t(s) = f(ts)$. If (a, f) belongs to Z_φ, so does (a, f_t), and the map

$$\varphi_t: Z_\varphi \to Z_\varphi, \quad (a, f) \mapsto (a, f_t),$$

is a *-homomorphism. It is easily checked that $(\varphi_t)_t$ is a homotopy from $i\tilde{\pi}$ to id_{Z_φ}.   □

7.5.11. Lemma. *Let $\varphi: A \to B$ be a *-homomorphism of C*-algebras and let $\pi: C_\varphi \to A$ be the projection of C_φ onto A. Then the sequence*

$$\tilde{K}_0(C_\varphi) \xrightarrow{\pi_*} \tilde{K}_0(A) \xrightarrow{\varphi_*} \tilde{K}_0(B)$$

is exact.

Proof. Let $\tilde{\pi}: Z_\varphi \to A$ be the projection, $i: A \to Z_\varphi$ the inclusion, and $\varepsilon: Z_\varphi \to B$ the canonical map. If $j: C_\varphi \to Z_\varphi$ is the inclusion, then $\tilde{\pi}j = \pi$ and $\varepsilon i = \varphi$, so $\tilde{\pi}_* j_* = \pi_*$ and $\varepsilon_* i_* = \varphi_*$. By Lemma 7.5.10, $\tilde{\pi}i = \mathrm{id}_A$ and $i\tilde{\pi} \approx \mathrm{id}_{Z_\varphi}$, so $\tilde{\pi}_* i_* = \mathrm{id}$ and $i_* \tilde{\pi}_* = \mathrm{id}$ by Theorem 7.3.7. Hence, $\varepsilon_* = \varphi_* \tilde{\pi}_*$. Since

$$0 \to C_\varphi \xrightarrow{j} Z_\varphi \xrightarrow{\varepsilon} B \to 0$$

is a short exact sequence of C*-algebras, it follows from Theorem 7.3.5 that the sequence

$$\tilde{K}_0(C_\varphi) \xrightarrow{j_*} \tilde{K}_0(Z_\varphi) \xrightarrow{\varepsilon_*} \tilde{K}_0(B)$$

is exact. Hence, $\tilde{\pi}_*(\mathrm{im}(j_*)) = \tilde{\pi}_*(\ker(\varepsilon_*))$. Since $\tilde{\pi}_*(\mathrm{im}(j_*)) = \mathrm{im}(\pi_*)$ and $\tilde{\pi}_*(\ker(\varepsilon_*)) = \ker(\varphi_*)$, therefore $\mathrm{im}(\pi_*) = \ker(\varphi_*)$.   □

If $\varphi: A \to B$ is a *-homomorphism of C*-algebras, we call the injective *-homomorphism

$$k: S(B) \to C_\varphi, \quad f \mapsto (0, f),$$

the *inclusion* of $S(B)$ in C_φ. It is easily checked that

$$0 \to S(B) \xrightarrow{k} C_\varphi \xrightarrow{\pi} A \to 0$$

is a short exact sequence of C*-algebras, where $\pi: C_\varphi \to A$ is the projection. Given a short exact sequence of C*-algebras

$$0 \to J \xrightarrow{j} A \xrightarrow{\varphi} B \to 0,$$

we define an injective *-homomorphism

$$\hat{j}: J \to C_\varphi, \quad a \mapsto (j(a), 0),$$

which we call the *inclusion* of J in C_φ. Note that $j = \pi\hat{j}$. The surjective *-homomorphism

$$\hat{\varphi}: C_\varphi \to C(B), \quad (a, f) \mapsto f,$$

is the *projection* of C_φ onto $C(B)$. It is readily verified that

$$0 \to J \xrightarrow{\hat{j}} C_\varphi \xrightarrow{\hat{\varphi}} C(B) \to 0$$

is a short exact sequence.

7.5.12. Lemma. *Suppose that*

$$0 \to J \xrightarrow{j} A \xrightarrow{\varphi} B \to 0$$

is a short exact sequence of C*-algebras, and let $\hat{\jmath} \colon J \to C_\varphi$ be the inclusion. Then $\tilde{K}_0(C_{\hat{\jmath}}) = 0$, and the map $\hat{\jmath}_* \colon \tilde{K}_0(J) \to \tilde{K}_0(C_\varphi)$ is an isomorphism.

Proof. Set

$$D = \{ f \in C(C_\varphi) \mid f(0) \in \hat{\jmath}(J) \}.$$

Then D is a C*-subalgebra of $C(C_\varphi)$, and the map

$$\psi \colon C_{\hat{\jmath}} \to D, \quad (a, f) \mapsto f,$$

is a *-isomorphism. Hence, $\psi_* \colon \tilde{K}_0(C_{\hat{\jmath}}) \to \tilde{K}_0(D)$ is an isomorphism.

Let the map $\hat{\varphi} \colon C_\varphi \to C(B)$ be the projection, and denote by $\hat{\varphi}'$ the *-homomorphism

$$D \to S(C(B)), \quad f \mapsto \hat{\varphi} \circ f.$$

Suppose that $g \in S(C(B))$. Then the map

$$[0, 1] \to B, \quad t \mapsto g(t)(0),$$

belongs to $S(B)$, so by the surjectivity of $S(\varphi)$ (Theorem 7.5.8), there exists a map $f' \in S(A)$ such that $\varphi(f'(t)) = g(t)(0)$ for all $t \in [0, 1]$. The map

$$f \colon [0, 1] \to C_\varphi, \quad t \mapsto (f'(t), g(t)),$$

is continuous, and indeed $f \in D$ and $\hat{\varphi}'(f) = \hat{\varphi} \circ f = g$. Therefore, $\hat{\varphi}'$ is surjective.

The map

$$\hat{\jmath}' \colon C(J) \to D, \quad f \mapsto \hat{\jmath} \circ f,$$

is an injective *-homomorphism, and clearly $\hat{\varphi}' \hat{\jmath}' = 0$, so $\mathrm{im}(\hat{\jmath}') \subseteq \ker(\hat{\varphi}')$. Suppose f is an arbitrary element of $\ker(\hat{\varphi}')$. Then $\hat{\varphi} \circ f = 0$, so for all $t \in [0, 1]$ we have $f(t) = (f'(t), 0) \in A \oplus C(B)$ for some $f'(t) \in A$. But $\varphi(f'(t)) = 0$, so $f'(t) = j(h(t))$ for some element $h(t) \in J$. The map

$$h \colon [0, 1] \to J, \quad t \mapsto h(t),$$

is continuous, because f is continuous, and moreover, $h(1) = 0$, so $h \in C(J)$. Also, $\hat{\jmath}'(h)(t) = (\hat{\jmath} \circ h)(t) = (j(h(t)), 0) = f(t)$, so $\hat{\jmath}'(h) = f$. Hence, $\mathrm{im}(\hat{\jmath}') = \ker(\hat{\varphi}')$.

It follows from what we have just shown that

$$0 \to C(J) \xrightarrow{\hat{\jmath}'} D \xrightarrow{\hat{\varphi}'} S(C(B)) \to 0$$

is a short exact sequence. Therefore,

$$\tilde{K}_0(C(J)) \xrightarrow{\hat{j}'_*} \tilde{K}_0(D) \xrightarrow{\hat{\varphi}'_*} \tilde{K}_0(S(C(B)))$$

is exact by Theorem 7.3.5. But $\tilde{K}_0(C(J)) = 0$ and $\tilde{K}_0(S(C(B))) = 0$, since $C(J)$ and $S(C(B))$ are contractible. Hence, $0 = \text{im}(\hat{j}'_*) = \ker(\hat{\varphi}'_*) = \tilde{K}_0(D)$, so $\tilde{K}_0(C_{\hat{j}}) = 0$.

Using Theorem 7.3.5 again, since

$$0 \to J \xrightarrow{\hat{j}} C_\varphi \xrightarrow{\hat{\varphi}} C(B) \to 0$$

is a short exact sequence, it follows that the sequence

$$\tilde{K}_0(J) \xrightarrow{\hat{j}_*} \tilde{K}_0(C_\varphi) \xrightarrow{\hat{\varphi}_*} \tilde{K}_0(C(B)) = 0$$

is exact. Hence, $\text{im}(\hat{j}_*) = \ker(\hat{\varphi}_*) = \tilde{K}_0(C_\varphi)$, so $\hat{j}_*$ is surjective.

Finally, suppose that $\pi: C_{\hat{j}} \to J$ is the projection. Then

$$0 = \tilde{K}_0(C_{\hat{j}}) \xrightarrow{\pi_*} \tilde{K}_0(J) \xrightarrow{\hat{j}_*} \tilde{K}_0(C_\varphi)$$

is exact by Lemma 7.5.11, so $0 = \text{im}(\pi_*) = \ker(\hat{j}_*)$. Thus, $\hat{j}_*$ is injective, and therefore we have shown that it is an isomorphism. $\square$

If A is a C*-algebra and $f \in S(A)$, then the map

$$\kappa_A(f): [0,1] \to A, \quad t \mapsto f(1-t),$$

also belongs to $S(A)$. It is easily checked that

$$\kappa_A: S(A) \to S(A), \quad f \mapsto \kappa_A(f),$$

is a *-isomorphism such that $\kappa_A^2 = \text{id}$.

7.5.13. Lemma. *Suppose that*

$$0 \to J \xrightarrow{j} A \xrightarrow{\varphi} B \to 0$$

is a short exact sequence of C-algebras, so*

$$0 \to S(B) \xrightarrow{k} C_\varphi \xrightarrow{\pi} A \to 0$$

*is also a short exact sequence, where k and π are the inclusion and projection maps. Let $\hat{k}: S(B) \to C_\pi$ and $k': S(A) \to C_\pi$ be the inclusion maps associated to this second short exact sequence. Then $\hat{k}S(\varphi)$ and $k'\kappa_A$ are homotopic *-homomorphisms from $S(A)$ to C_π.*

Proof. For $f \in S(A)$ and $t \in [0,1]$, define $f_t \in C(A)$ by setting $f_t(s) = f(1 - t + st)$. The map

$$\varphi_t \colon S(A) \to C_\pi, \quad f \mapsto ((f_t(0), \varphi \circ (\kappa_A(f))_{1-t}), f_t),$$

is a $*$-homomorphism, $\varphi_0 = \hat{k}S(\varphi)\kappa_A$, and $\varphi_1 = k'$. It is straightforward to verify that $(\varphi_t)_t$ is a homotopy, and therefore $\hat{k}S(\varphi)\kappa_A \approx k'$, from which it follows that $\hat{k}S(\varphi) \approx k'\kappa_A$, using the fact that $\kappa_A^2 = \text{id}$. $\square$

Suppose that

$$0 \to J \xrightarrow{j} A \xrightarrow{\varphi} B \to 0 \tag{2}$$

is a short exact sequence of C*-algebras and $\hat{j} \colon J \to C_\varphi$ and $k \colon S(B) \to C_\varphi$ are the inclusion maps. We denote the composition $(\hat{j}_*)^{-1}k_*$ by ∂. Thus, ∂ is a homomorphism from $\tilde{K}_1(B)$ to $\tilde{K}_0(J)$, called the *connecting* homomorphism (relative to the short exact sequence (2)).

7.5.14. Theorem. *If*

$$0 \to J \xrightarrow{j} A \xrightarrow{\varphi} B \to 0$$

is a short exact sequence of C-algebras, then the sequence*

$$\tilde{K}_1(A) \xrightarrow{\varphi_*} \tilde{K}_1(B) \xrightarrow{\partial} \tilde{K}_0(J) \xrightarrow{j_*} \tilde{K}_0(A)$$

is exact.

Proof. Let k, π, k', and $\hat{k}$ be as in Lemma 7.5.13. Since $\pi\hat{j} = j$, we have $\pi_*\hat{j}_* = j_*$, and since

$$0 \to S(B) \xrightarrow{k} C_\varphi \xrightarrow{\pi} A \to 0$$

is a short exact sequence, the sequence

$$\tilde{K}_1(B) \xrightarrow{k_*} \tilde{K}_0(C_\varphi) \xrightarrow{\pi_*} \tilde{K}_0(A)$$

is exact by Theorem 7.3.5. Hence, $\text{im}(\partial) = \hat{j}_*^{-1}(\ker(\pi_*)) = \ker(j_*)$.

By Lemma 7.5.13 $\hat{k}S(\varphi) \approx k'\kappa_A$, so $\hat{k}_*S(\varphi)_* = k'_*(\kappa_A)_*$. If ∂' is the connecting homomorphism for the short exact sequence

$$0 \to S(B) \xrightarrow{k} C_\varphi \xrightarrow{\pi} A \to 0,$$

then

$$\tilde{K}_1(A) \xrightarrow{\partial'} \tilde{K}_1(B) \xrightarrow{k_*} \tilde{K}_0(C_\varphi)$$

is exact, by the first part of this proof. Since $\partial' = \hat{k}_*^{-1}k'_* = S(\varphi)_*(\kappa_A)_*$, we have $\ker(\partial) = \ker(k_*) = \text{im}(\partial') = \text{im}(S(\varphi)_*)$. $\square$

7.5.3. Remark. If

$$0 \to J \xrightarrow{j} A \xrightarrow{\varphi} B \to 0$$

is a short exact sequence of C*-algebras, we say that it *splits* if there exists a
*-homomorphism $\psi \colon B \to A$ such that $\varphi\psi = \mathrm{id}_B$. In this case the sequence

$$0 \to \tilde{K}_0(J) \xrightarrow{j_*} \tilde{K}_0(A) \xrightarrow{\varphi_*} \tilde{K}_0(B) \to 0$$

is a split short exact sequence also. That the equality $\mathrm{im}(j_*) = \ker(\varphi_*)$
holds is a consequence of Theorem 7.3.5, and that φ_* is surjective follows
from the fact that $\varphi_*\psi_* = \mathrm{id}$. Injectivity of j_* follows from the exactness
of the sequence

$$\tilde{K}_1(A) \xrightarrow{\varphi_*} \tilde{K}_1(B) \xrightarrow{\partial} \tilde{K}_0(J) \xrightarrow{j_*} \tilde{K}_0(A)$$

(Theorem 7.5.14) and the fact that $\tilde{K}_1(\varphi)$ is surjective ($\tilde{K}_1(\varphi)\tilde{K}_1(\psi) = \mathrm{id}$).

Recall that **A** denotes the Toeplitz algebra (see Section 3.5), and that
this algebra is generated by the unilateral shift u. We shall make frequent
use of the universal property of **A**: If v is an isometry in a unital C*-algebra
B, then there is a unique *-homomorphism $\varphi \colon \mathbf{A} \to B$ such that $\varphi(u) = v$
(Theorem 3.5.18). We denote $K(H^2)$ by **K** (H^2 is the Hardy space).

The unique *-homomorphism $\tau \colon \mathbf{A} \to \mathbf{C}$ such that $\tau(u) = 1$ is the
canonical map from **A** to **C**.

7.5.15. Theorem (Cuntz). *Let D be a unital C*-algebra and $\tau \colon \mathbf{A} \to \mathbf{C}$
be the canonical map. Then the map*

$$\tilde{K}_0(\mathbf{A} \otimes_* D) \xrightarrow{(\tau \otimes_* \mathrm{id})_*} \tilde{K}_0(\mathbf{C} \otimes_* D)$$

is an isomorphism.

Proof. If B is a C*-algebra, we shall simplify the notation throughout
this proof by writing B' for $B \otimes_* D$. If $\varphi \colon B \to C$ is a *-homomorphism
of C*-algebras, we shall write φ' for $\varphi \otimes_* \mathrm{id}_D$. Let $j \colon \mathbf{C} \to \mathbf{A}$ be the
unique unital *-homomorphism. Then $\tau j = \mathrm{id}$, so $\tau' j' = \mathrm{id}$, and therefore
$\tau'_* j'_* = \mathrm{id}$. Thus, we have only to show that $j'_* \tau'_* = \mathrm{id}$.

Let $e = 1 - uu^*$ where u is the unilateral shift on the standard orthonormal basis of H^2, and denote by ε the *-homomorphism

$$\mathbf{A} \to \mathbf{K} \otimes_* \mathbf{A}, \quad a \mapsto e \otimes a.$$

There is a *-isomorphism $\gamma \colon (\mathbf{K} \otimes_* \mathbf{A}) \otimes_* D \to \mathbf{K} \otimes_* (\mathbf{A} \otimes_* D)$ such that
$\gamma((b \otimes a) \otimes d) = b \otimes (a \otimes d)$ (*cf.* Exercise 6.9), and $(\gamma \varepsilon')_* \colon \tilde{K}_0(\mathbf{A} \otimes_* D) \to$
$\tilde{K}_0(\mathbf{K} \otimes_* (\mathbf{A} \otimes_* D))$ is clearly the canonical isomorphism. Hence, ε'_* is an

isomorphism. Since $\mathbf{A} \otimes 1 = \{a \otimes 1 \mid a \in \mathbf{A}\}$ is a C*-subalgebra, and $\mathbf{K} \otimes_* \mathbf{A}$ is a closed ideal, of $\mathbf{A} \otimes_* \mathbf{A}$, the set $B = \mathbf{A} \otimes 1 + \mathbf{K} \otimes_* \mathbf{A}$ is a C*-subalgebra of $\mathbf{A} \otimes_* \mathbf{A}$. Let $\pi: B \to B/(\mathbf{K} \otimes_* \mathbf{A})$ be the quotient *-homomorphism, and let θ denote the *-homomorphism

$$\mathbf{A} \to B, \quad a \mapsto a \otimes 1.$$

Denote by C the pullback of B and $\mathbf{A}$ along the *-homomorphisms π and $\pi\theta$. The maps

$$i: \mathbf{K} \otimes_* \mathbf{A} \to C, \quad b \mapsto (b, 0),$$

and

$$p: C \to \mathbf{A}, \quad (b, a) \mapsto a,$$

are, respectively, injective and surjective *-homomorphisms, and

$$0 \to \mathbf{K} \otimes_* \mathbf{A} \xrightarrow{i} C \xrightarrow{p} \mathbf{A} \to 0$$

is a short exact sequence which splits (the *-homomorphism

$$k: \mathbf{A} \to C, \quad a \mapsto (a \otimes 1, a),$$

is a right inverse for p). Since $\mathbf{A}$ is nuclear (*cf.* Example 6.5.1), it follows from Theorem 6.5.2 that

$$0 \to (\mathbf{K} \otimes_* \mathbf{A})' \xrightarrow{i'} C' \xrightarrow{p'} \mathbf{A}' \to 0$$

is a short exact sequence, and this also splits, since $p'k' = \text{id}$. Hence, by Remark 7.5.3, i'_* is injective. Set $\psi = i\varepsilon$. Then $\psi'_* = i'_* \varepsilon'_*$ is injective, since i'_* and ε'_* are. To show that $j'_* \tau'_* = \text{id}$, it therefore suffices to show that $\psi'_* j'_* \tau'_* = \psi'_*$.

Let

$$z_0 = u^2 u^{*2} \otimes 1 + eu^* \otimes u + ue \otimes u^* + e \otimes e$$

and

$$z_1 = u^2 u^{*2} \otimes 1 + eu^* \otimes 1 + ue \otimes 1.$$

Then z_0 and z_1 are symmetries in B, and if for each $t \in [0, 1]$ we set

$$u_t = -i \exp(i\pi(1 - t)z_0/2) \exp(i\pi t z_1/2),$$

then the map

$$[0, 1] \to B, \quad t \mapsto u_t,$$

is a continuous path of unitaries such that $u_0 = z_0$ and $u_1 = z_1$. Since $\pi(z_t) = 1$ for $t = 0, 1$ we have $\pi(u_t) = 1$ for all $t \in [0, 1]$.

Let $\varphi_t\colon \mathbf{A} \to C$ be the unique $*$-homomorphism such that $\varphi_t(u)$ is the isometry $(u_t(u \otimes 1), u)$ in C. Then $(\varphi_t)_t$ is a homotopy, and therefore $(\varphi_t')_t$ is a homotopy. If $v = u^2 u^*$, then v is an isometry in the C*-algebra $(1 - e)\mathbf{A}(1 - e)$, and $(v \otimes 1, u)$ is an isometry in the unital C*-algebra $\theta((1 - e)\mathbf{A}(1 - e)) \oplus \mathbf{A}$. Hence, there is a unique $*$-homomorphism φ from $\mathbf{A}$ into this C*-algebra such that $\varphi(u) = (v \otimes 1, u)$, and since $(v \otimes 1, u) \in C$, we may suppose that φ is a $*$-homomorphism from $\mathbf{A}$ to C.

Since $\psi(u) = (e \otimes u, 0)$, we have $\varphi(u)\psi(u) = \psi(u)\varphi(u) = \varphi(u)\psi(u^*) = \psi(u^*)\varphi(u) = 0$, from which it follows that φ and ψ are orthogonal. Therefore, φ' and ψ' are orthogonal, and φ and $\psi j\tau$ are orthogonal. Moreover, $\varphi_0 = \varphi + \psi$ and $\varphi_1 = \varphi + \psi j\tau$, so $\varphi_0' = \varphi' + \psi'$ and $\varphi_1' = \varphi' + \psi'j'\tau'$. Hence, by Theorem 7.3.7 and Remark 7.4.3, $\varphi_*' + \psi_*' = (\varphi' + \psi')_* = \varphi_{0*}' = \varphi_{1*}' = (\varphi' + \psi'j'\tau')_* = \varphi_*' + \psi_*'j_*'\tau_*'$, and therefore $\psi_*' = \psi_*'j_*'\tau_*'$. This proves the theorem. $\square$

Since B is $*$-isomorphic to $B \otimes_* C$ for any C*-algebra B, the preceding theorem implies in particular that $\tilde{K}_0(\mathbf{A}) = \tilde{K}_0(C)$, so $\tilde{K}_0(\mathbf{A}) = \mathbf{Z}$.

If $\tau\colon \mathbf{A} \to C$ is the canonical map, we denote its kernel by $\mathbf{A}_0$.

7.5.16. Corollary. *If D is an arbitrary C*-algebra, then $\tilde{K}_i(\mathbf{A}_0 \otimes_* D) = 0$ for $i = 0, 1$.*

Proof. First suppose that D is unital. If $j\colon \mathbf{A}_0 \to \mathbf{A}$ is the inclusion map, then

$$0 \to \mathbf{A}_0 \xrightarrow{j} \mathbf{A} \xrightarrow{\tau} C \to 0$$

is a short exact sequence that clearly splits, and therefore the corresponding sequence

$$0 \to \mathbf{A}_0' \xrightarrow{j'} \mathbf{A}' \xrightarrow{\tau'} C' \to 0$$

is a split short exact sequence (Theorem 6.5.2), where we retain the notation $B' = B \otimes_* D$, $\varphi' = \varphi \otimes_* \mathrm{id}$, used in the proof of Theorem 7.5.15. It follows from Remark 7.5.3 that

$$0 \to \tilde{K}_0(\mathbf{A}_0') \xrightarrow{j_*'} \tilde{K}_0(\mathbf{A}') \xrightarrow{\tau_*'} \tilde{K}_0(C') \to 0$$

is a short exact sequence, and since τ_*' is an isomorphism by Theorem 7.5.15 therefore $\tilde{K}_0(\mathbf{A}_0 \otimes_* D) = 0$.

Now suppose that D is not necessarily unital. The split canonical short exact sequence

$$0 \to D \to \tilde{D} \to C \to 0$$

gives rise to a split short exact sequence

$$0 \to D \otimes_* \mathbf{A}_0 \to \tilde{D} \otimes_* \mathbf{A}_0 \to C \otimes_* \mathbf{A}_0 \to 0,$$

and therefore

$$0 \to \tilde{K}_0(D \otimes_* A_0) \to \tilde{K}_0(\tilde{D} \otimes_* A_0) \to \tilde{K}_0(C \otimes_* A_0) \to 0$$

is a short exact sequence. Since (by Exercise 6.10) $B \otimes_* A_0$ is $*$-isomorphic to $A_0 \otimes_* B$ for any C*-algebra B, and $\tilde{K}_0(A_0 \otimes_* \tilde{D}) = 0$ by the first part of this proof, hence $\tilde{K}_0(A_0 \otimes_* D) = 0$.

Finally, by Theorem 7.5.7, $S(A_0 \otimes_* D)$ is $*$-isomorphic to $(A_0 \otimes_* D) \otimes_* S$ and this algebra in turn is $*$-isomorphic to $A_0 \otimes_* (D \otimes_* S)$ (by Exercise 6.9), so $\tilde{K}_1(A_0 \otimes_* D)$ is isomorphic to $\tilde{K}_0(A_0 \otimes_* (D \otimes_* S)) = 0$. $\qquad \square$

Let π be the unique $*$-homomorphism from A to $C(T)$ such that $\pi(u) = z$, where u is the unilateral shift and z is the inclusion function of T in C. Clearly, $\pi(A_0) \subseteq S$. We use the same symbol π to denote the $*$-homomorphism from A_0 to S got by restriction. It is easy to check that $K \subseteq A_0$. If $j \colon K \to A_0$ denotes the inclusion $*$-homomorphism, then

$$0 \to K \xrightarrow{j} A_0 \xrightarrow{\pi} S \to 0$$

is a short exact sequence, by Theorem 3.5.11. Hence, by Theorem 6.5.2, for each C*-algebra D the sequence

$$0 \to K \otimes_* D \xrightarrow{j \, \otimes_* \, \mathrm{id}} A_0 \otimes_* D \xrightarrow{\pi \, \otimes_* \, \mathrm{id}} S \otimes_* D \to 0$$

is a short exact sequence. By Exercise 6.10, there is a unique $*$-isomorphism $\theta \colon S \otimes_* D \to D \otimes_* S$ such that $\theta(f \otimes d) = d \otimes f$ for all $f \in S$ and $d \in D$. Denote by π_D the $*$-homomorphism from $A_0 \otimes_* D$ to $S(D)$ got by composing $\pi \otimes_* \mathrm{id}_D$, θ and γ_D (cf. Theorem 7.5.7), so $\pi_D = \gamma_D \theta(\pi \otimes_* \mathrm{id}_D)$. Then

$$0 \to K \otimes_* D \xrightarrow{j \, \otimes_* \, \mathrm{id}} A_0 \otimes_* D \xrightarrow{\pi_D} S(D) \to 0$$

is short exact, so we get a homomorphism $\partial \colon \tilde{K}_1(S(D)) \to \tilde{K}_0(K \otimes_* D)$ (the connecting homomorphism). We denote by β_D the homomorphism from $\tilde{K}_1(S(D))$ to $\tilde{K}_0(D)$ got by composing the maps

$$\tilde{K}_1(S(D)) \xrightarrow{\partial} \tilde{K}_0(K \otimes_* D) \xrightarrow{e_*^{-1}} \tilde{K}_0(D),$$

where $e_* \colon \tilde{K}_0(D) \to \tilde{K}_0(K \otimes_* D)$ is the canonical isomorphism.

7.5.17. Theorem (Bott Periodicity). *For each C*-algebra D, the map* $\beta_D \colon \tilde{K}_1(S(D)) \to \tilde{K}_0(D)$ *is an isomorphism.*

Proof. Applying Theorem 7.5.14 to the short exact sequence

$$0 \to \mathbf{K} \otimes_* D \xrightarrow{\; j \, \otimes_* \mathrm{id} \;} \mathbf{A}_0 \otimes_* D \xrightarrow{\; \pi_D \;} S(D) \to 0$$

gives exactness of the sequence

$$\tilde{K}_1(\mathbf{A}_0 \otimes_* D) \xrightarrow{\; \pi_{D*} \;} \tilde{K}_1(S(D)) \xrightarrow{\; \partial \;} \tilde{K}_0(\mathbf{K} \otimes_* D) \xrightarrow{\; (j \, \otimes_* \mathrm{id})_* \;} \tilde{K}_0(\mathbf{A}_0 \otimes_* D),$$

and $\tilde{K}_i(\mathbf{A}_0 \otimes_* D) = 0$ for $i = 0, 1$, by Corollary 7.5.16. Hence, ∂ and therefore β_D are isomorphisms. □

Let

$$
\begin{array}{ccccccccc}
0 & \to & J & \xrightarrow{\; j \;} & A & \xrightarrow{\; \varphi \;} & B & \to & 0 \\
 & & \downarrow \gamma & & \downarrow \alpha & & \downarrow \beta & & \\
0 & \to & J' & \xrightarrow{\; j' \;} & A' & \xrightarrow{\; \varphi' \;} & B' & \to & 0
\end{array}
$$

be a commutative diagram of *-homomorphisms of C*-algebras and suppose the top and bottom rows are short exact sequences with corresponding connecting homomorphisms ∂ and ∂'. Then the diagram

$$
\begin{array}{ccc}
\tilde{K}_1(B) & \xrightarrow{\; \beta_* \;} & \tilde{K}_1(B') \\
\downarrow \partial & & \downarrow \partial' \\
\tilde{K}_0(J) & \xrightarrow{\; \gamma_* \;} & \tilde{K}_0(J')
\end{array}
$$

commutes. The proof is a simple diagram chase.

If $\varphi : D \to D'$ is a *-homomorphism between C*-algebras, then the diagram

$$
\begin{array}{ccccccccc}
0 & \to & \mathbf{K} \otimes_* D & \xrightarrow{\; j \, \otimes_* \mathrm{id} \;} & \mathbf{A}_0 \otimes_* D & \xrightarrow{\; \pi_D \;} & S(D) & \to & 0 \\
 & & \downarrow \mathrm{id} \otimes_* \varphi & & \downarrow \mathrm{id} \otimes_* \varphi & & \downarrow S(\varphi) & & \\
0 & \to & \mathbf{K} \otimes_* D' & \xrightarrow{\; j \, \otimes_* \mathrm{id} \;} & \mathbf{A}_0 \otimes_* D' & \xrightarrow{\; \pi'_D \;} & S(D') & \to & 0
\end{array}
$$

commutes (this uses Remark 7.5.2), and the top and bottom rows are short exact sequences. If ∂ and ∂' are the corresponding connecting homomorphisms, then by the preceding observation the diagram

$$
\begin{array}{ccc}
\tilde{K}_1(S(D)) & \xrightarrow{\; \tilde{K}_1(S(\varphi)) \;} & \tilde{K}_1(S(D')) \\
\downarrow \partial & & \downarrow \partial' \\
\tilde{K}_0(\mathbf{K} \otimes_* D) & \xrightarrow{\; (\mathrm{id} \otimes_* \varphi)_* \;} & \tilde{K}_0(\mathbf{K} \otimes_* D')
\end{array}
$$

commutes. Also,

$$\begin{array}{ccc} \tilde{K}_0(D) & \xrightarrow{\varphi_*} & \tilde{K}_0(D') \\ \downarrow e & & \downarrow e' \\ \tilde{K}_0(\mathbf{K} \otimes_* D) & \xrightarrow{(\mathrm{id} \otimes_* \varphi)_*} & \tilde{K}_0(\mathbf{K} \otimes_* D') \end{array}$$

commutes, where e and e' are the canonical isomorphisms. Since $\beta_D = e^{-1}\partial$ and $\beta_{D'} = (e')^{-1}\partial'$, the diagram

$$\begin{array}{ccc} \tilde{K}_1(S(D)) & \xrightarrow{\tilde{K}_1(S(\varphi))} & \tilde{K}_1(S(D')) \\ \downarrow \beta_D & & \downarrow \beta_{D'} \\ \tilde{K}_0(D) & \xrightarrow{\varphi_*} & \tilde{K}_0(D') \end{array} \qquad (1)$$

commutes.

7.5.18. Theorem. *If*

$$0 \to J \xrightarrow{j} A \xrightarrow{\varphi} B \to 0$$

is a short exact sequence of C^*-algebras, the following sequence is exact:

$$\begin{array}{ccccc} \tilde{K}_0(J) & \xrightarrow{j_*} & \tilde{K}_0(A) & \xrightarrow{\varphi_*} & \tilde{K}_0(B) \\ \uparrow \partial & & & & \downarrow \partial \\ \tilde{K}_1(B) & \xleftarrow{\varphi_*} & \tilde{K}_1(A) & \xleftarrow{j_*} & \tilde{K}_1(J). \end{array}$$

The map $\partial \colon \tilde{K}_0(B) \to \tilde{K}_1(J)$ is defined to be the composition of the homomorphisms

$$\tilde{K}_0(B) \xrightarrow{\beta_B^{-1}} \tilde{K}_1(S(B)) \xrightarrow{\partial'} \tilde{K}_1(J),$$

where ∂' is the connecting homomorphism from the short exact sequence

$$0 \to S(J) \xrightarrow{S(j)} S(A) \xrightarrow{S(\varphi)} S(B) \to 0.$$

Proof. By Theorems 7.3.5, 7.5.9, and 7.5.14, we have only to show exactness at $\tilde{K}_0(B)$ and $\tilde{K}_1(J)$.

Now $\ker(\partial) = \beta_B(\ker(\partial')) = \beta_B(\mathrm{im}(\tilde{K}_1(S(\varphi))))$, by exactness of

$$\tilde{K}_1(S(A)) \xrightarrow{\tilde{K}_1(S(\varphi))} \tilde{K}_1(S(B)) \xrightarrow{\partial'} \tilde{K}_1(J).$$

But $\beta_B(\mathrm{im}(\tilde{K}_1(S(\varphi)))) = \mathrm{im}(\tilde{K}_0(\varphi))$, by commutativity of Diagram (1) preceding this theorem. Thus, we have exactness at $\tilde{K}_0(B)$.

Since $\mathrm{im}(\partial) = \mathrm{im}(\partial') = \ker(\tilde{K}_1(j))$, by exactness of

$$\tilde{K}_1(S(B)) \xrightarrow{\partial'} \tilde{K}_1(J) \xrightarrow{\tilde{K}_1(j)} \tilde{K}_1(A),$$

the theorem is proved.   □

7.5.4. Remark. If

$$0 \to J \xrightarrow{j} A \xrightarrow{\varphi} B \to 0$$

is a split short exact sequence of C*-algebras, then it follows easily from Theorem 7.5.18 that

$$0 \to \tilde{K}_1(J) \xrightarrow{j_*} \tilde{K}_1(A) \xrightarrow{\varphi_*} \tilde{K}_1(B) \to 0$$

is a split short exact sequence of groups. We have already seen the corresponding result for $\tilde{K}_0$ in Remark 7.5.3.

7.5.1. Example. If ε denotes the *-homomorphism

$$C(\mathbf{T}) \to \mathbf{C}, \quad f \mapsto f(1),$$

then

$$0 \to \mathbf{S} \xrightarrow{j} C(\mathbf{T}) \xrightarrow{\varepsilon} \mathbf{C} \to 0$$

is a split short exact sequence of C*-algebras, where j is the inclusion. Hence, if A is an arbitrary C*-algebra,

$$0 \to \mathbf{S} \otimes_* A \xrightarrow{j \otimes_* \mathrm{id}} C(\mathbf{T}) \otimes_* A \xrightarrow{\varepsilon \otimes_* \mathrm{id}} \mathbf{C} \otimes_* A \to 0$$

is a split short exact sequence, by Theorem 6.5.2. It follows from Remarks 7.5.3 and 7.5.4 that for $i = 0, 1$ the sequence

$$0 \to \tilde{K}_i(\mathbf{S} \otimes_* A) \xrightarrow{(j \otimes_* \mathrm{id})_*} \tilde{K}_i(C(\mathbf{T}) \otimes_* A) \xrightarrow{(\varepsilon \otimes_* \mathrm{id})_*} \tilde{K}_i(\mathbf{C} \otimes_* A) \to 0$$

is split short exact, and therefore, if $\cong$ denotes the relation "is isomorphic to," we have

$$\begin{aligned}
\tilde{K}_i(C(\mathbf{T}) \otimes_* A) &\cong \tilde{K}_i(\mathbf{S} \otimes_* A) \oplus \tilde{K}_i(\mathbf{C} \otimes_* A) \\
&\cong \tilde{K}_i(S(A)) \oplus \tilde{K}_i(A) \\
&\cong \tilde{K}_{1-i}(A) \oplus \tilde{K}_i(A) \\
&\cong \tilde{K}_0(A) \oplus \tilde{K}_1(A)
\end{aligned}$$

(the third $\cong$ is given by Theorem 7.5.17).

In particular, $\tilde{K}_i(C(\mathbf{T})) \cong \tilde{K}_0(\mathbf{C}) \oplus \tilde{K}_1(\mathbf{C}) \cong \mathbf{Z}$.

7. Exercises

1. Let A_1 and A_2 be C*-algebras. Show that
   (a) $\tilde{K}_1(A_1 \oplus A_2) = \tilde{K}_1(A_1) \oplus \tilde{K}_1(A_2)$;
   (b) $\tilde{K}_1(A_1) = \tilde{K}_1(\tilde{A}_1)$. Hence, extend Theorem 7.5.6 by showing that $\tilde{K}_1(A_1) = 0$ if A_1 is an AF-algebra (unital or not).

2. Let C be the Calkin algebra on an infinite-dimensional separable Hilbert space H. Calculate $\tilde{K}_i(K(H))$ and $\tilde{K}_i(C)$ for $i = 0, 1$.

3. Show that if $\varphi, \psi: A \to B$ are homotopic $*$-homomorphisms between C*-algebras, then $\tilde{K}_1(\varphi) = \tilde{K}_1(\psi)$.

4. Show that the functor $\tilde{K}_1$ is "continuous," that is, $\tilde{K}_1(\varinjlim A_n) = \varinjlim \tilde{K}_1(A_n)$. More precisely, suppose that A is the direct limit of a sequence of C*-algebras $(A_n, \varphi_n)_{n=1}^\infty$, and that G is the direct limit of the corresponding sequence $(\tilde{K}_1(A_n), \varphi_{n*})_{n=1}^\infty$. Suppose that $\varphi^n: A_n \to A$ and $\tau^n: \tilde{K}_1(A_n) \to G$ are the natural maps. Show that there is a unique isomorphism $\tau: G \to \tilde{K}_1(A)$ such that the diagram

$$\tilde{K}_1(A_n) \quad \xrightarrow{\ \tau^n\ } \quad G$$
$$\varphi_*^n \searrow \qquad \downarrow \tau$$
$$\tilde{K}_1(A)$$

commutes for all n.

5. Let H be a separable infinite-dimensional Hilbert space and e a rank-one projection in $K(H)$. If A is a C*-algebra, the $*$-homomorphism

$$\varphi: A \to K(H) \otimes_* A, \quad a \mapsto e \otimes a,$$

is independent of the choice of e up to homotopy. The map

$$e_* = \varphi_*: \tilde{K}_1(A) \to \tilde{K}_1(K(H) \otimes_* A)$$

is called the *canonical* map. Show that it is an isomorphism.

6. Let $\mathbf{A}$ be the Toeplitz algebra, and let $\tau: \mathbf{A} \to \mathbf{C}$ be the canonical $*$-homomorphism. Extend Theorem 7.5.15 by showing that for an arbitrary C*-algebra D, the map

$$\tilde{K}_i(\mathbf{A} \otimes_* D) \xrightarrow{\ (\tau \otimes_* \mathrm{id})_*\ } \tilde{K}_i(\mathbf{C} \otimes_* D)$$

is an isomorphism for $i = 0, 1$ (use Corollary 7.5.16).

7. Let τ be a tracial state on a C*-algebra A. Define $\tau_n: M_n(A) \to \mathbf{C}$ by setting $\tau_n(a) = \sum_{i=1}^n \tau(a_{ii})$ if $a = (a_{ij})$. Show that τ_n/n is a tracial state on the C*-algebra $M_n(A)$.

8. Let τ be a tracial state on a C*-algebra A, and denote by the same symbol the unique (tracial) state on $\tilde{A}$ extending A, and the tracial positive linear functionals τ_n on the C*-algebras $M_n(\tilde{A})$ obtained from τ as in Exercise 7.7. If $p, q \in P[\tilde{A}]$ and $p \approx q$, show that $\tau(p) = \tau(q)$. Show there is a well-defined homomorphism $\tau_*\colon \tilde{K}_0(A) \to \mathbf{C}$ such that $\tau_*([p] - [q]) = \tau(p) - \tau(q)$ for all $x = [p] - [q] \in \tilde{K}_0(A)$.

9. Let $s\colon \mathbf{N} \setminus \{0\} \to \mathbf{N} \setminus \{0\}$ and let M_s be the corresponding UHF algebra as in Sections 6.2 and 7.3. Denote by τ the unique tracial state of M_s. Calculate the range $\tau_*(\tilde{K}_0(M_s))$.

7. Addenda

Let A be a unital C*-algebra and let $\varphi_n\colon U_n(A) \to U_{n+1}(A)$ be the group homomorphism defined by setting

$$\varphi_n(a) = \begin{pmatrix} a & 0 \\ 0 & 1_1 \end{pmatrix}.$$

Since $\varphi_n(U_n^0(A))$ is contained in $U_{n+1}^0(A)$, we obtain an induced homomorphism $\psi_n\colon U_n(A)/U_n^0(A) \to U_{n+1}(A)/U_{n+1}^0(A)$. Denote by $K_1(A)$ the direct limit of the direct sequence of groups $(U_n(A)/U_n^0(A), \psi_n)_{n=1}^{\infty}$. There is an isomorphism from $K_1(A)$ onto $\tilde{K}_1(A)$.

An *order unit* for a partially ordered group G is an element $u \in G^+$ such that for each $x \in G^+$ there exists $n \geq 1$ such that $x \leq nu$.

A countable partially ordered group G is a *dimension* group if
(a) whenever $nx \in G^+$ and $n \geq 1$, then $x \in G^+$;
(b) whenever $x_1, x_2, y_1, y_2 \in G$ satisfy $x_i \leq y_j$ for all i and j, there exists $z \in G$ such that $x_i \leq z \leq y_j$ for all i and j.
Condition (b) is called the *Riesz interpolation* property.

If A is a unital AF-algebra, then $K_0(A)$ is a dimension group with an order unit, and in the reverse direction, if G is a dimension group with an order unit, there is a unital AF-algebra A such that $K_0(A)$ is order isomorphic to G (Effros–Handelman–Shen).

We say that C*-algebras A and B are *stably isomorphic* if there is a separable infinite-dimensional Hilbert space H such that $K(H) \otimes_* A$ and $K(H) \otimes_* B$ are *-isomorphic.

If A is an AF-algebra, one can make $\tilde{K}_0(A)$ into a partially ordered group in all cases, whether A is unital or not. AF-algebras A, B are stably isomorphic if and only if $\tilde{K}_0(A)$ and $\tilde{K}_0(B)$ are order isomorphic as partially ordered groups (Elliott). There is an analogue of Theorem 7.2.10 (also due to Elliott) for non-unital AF-algebras, where $\tilde{K}_0(A)$ has to be endowed with additional structure (a "scale").

References: [Eff], [Goo], [Bla].

Let $n > 1$ and let O_n be the C*-algebra generated by n isometries $v_1, \ldots, v_n$ such that $v_1 v_1^* + \cdots + v_n v_n^* = 1$. This algebra is called the *Cuntz* algebra. It is simple, and independent of the choice of $v_1, \ldots, v_n$ up to *-isomorphism. The K_0-group has torsion in the case of these algebras: $\tilde{K}_0(O_n) = \mathbf{Z}/(n-1)$. One also has $\tilde{K}_1(O_n) = 0$.

Let θ be an irrational number in $[0,1]$, and let A_θ denote the irrational rotation algebra (Exercise 3.8). The K-theory is given by $\tilde{K}_0(A_\theta) = \tilde{K}_1(A_\theta) = \mathbf{Z}^2$. There is a unique tracial state τ on A_θ, and $\tau_*(\tilde{K}_0(A_\theta)) = \mathbf{Z} + \mathbf{Z}\theta$.

References: [Bla], [Cun].

Appendix

In this appendix all vector spaces are relative to the field $\mathbf{K}$, which may be $\mathbf{R}$ or $\mathbf{C}$.

Let Γ be a non-empty family of seminorms on a vector space X. The smallest topology on X for which the operations of addition and scalar multiplication and all of the seminorms in Γ are continuous is called the topology *generated* by Γ. If $\varepsilon > 0$ and $p_1, \ldots, p_n \in \Gamma$, let

$$U(\varepsilon; p_1, \ldots, p_n) = \{x \in X \mid p_j(x) < \varepsilon \ (j = 1, \ldots, n)\}.$$

These sets form a basic system of neighbourhoods of 0 in X. If $(x_\lambda)_{\lambda \in \Lambda}$ is a net in X, then $(x_\lambda)_{\lambda \in \Lambda}$ converges to a point x of X if and only if $\lim_\lambda p(x_\lambda - x) = 0$ for all $p \in \Gamma$. A pair (X, τ) consisiting of a vector space X and a topology τ generated by a family of seminorms on X is called a *locally convex (topological vector) space*. There are other more geometric characterisations of these spaces but these will be irrelevant to us. It is easy to check that if Γ generates the topology on a locally convex space X, then X is Hausdorff if and only if Γ is *separating*, that is, for each non-zero element $x \in X$ there is a seminorm $p \in \Gamma$ such that $p(x) > 0$.

A.1. Theorem. *Let Γ be a family of seminorms generating the topology on a locally convex space X, and suppose that τ is a linear functional on X. The following conditions are equivalent:*

(1) *τ is continuous.*
(2) *There are a finite number of elements $p_1, \ldots, p_n$ of Γ and there is a positive number M such that*

$$|\tau(x)| \leq M \max_{1 \leq j \leq n} p_j(x) \qquad (x \in X).$$

Proof. The implication $(2) \Rightarrow (1)$ is clear. To show the reverse implication, suppose that Condition (1) holds; that is, τ is continuous. Then

267

the set $S = \{x \in X \mid |\tau(x)| < 1\}$ is a 0-neighbourhood in X, so there is a positive number ε and there are seminorms $p_1, \ldots, p_n \in \Gamma$ such that $U(\varepsilon; p_1, \ldots, p_n) \subseteq S$. If $p = \max_{1 \leq j \leq n} p_j$, then $p(x) < \varepsilon \Rightarrow |\tau(x)| < 1$. Set $M = 2/\varepsilon$. If $p(x) > 0$, we have $p(\frac{x}{Mp(x)}) = \varepsilon/2 < \varepsilon$, so $|\tau(\frac{x}{Mp(x)})| < 1$. Consequently, $|\tau(x)| \leq Mp(x)$ for all $x \in X$. $\square$

Let X be a normed vector space. The weak* topology on X^* is generated by the family of seminorms $p_x \colon \tau \mapsto |\tau(x)|$ $(x \in X)$. For $x \in X$, denote by $\hat{x}$ the linear functional on X^* defined by setting $\hat{x}(\tau) = \tau(x)$.

A.2. Theorem. *Let X be a normed vector space. Then a linear functional $\theta \colon X^* \to \mathbf{K}$ is weak* continuous if and only if $\theta = \hat{x}$ for some $x \in X$.*

Proof. We show the forward implication only, because the reverse implication is trivial. Suppose that θ is weak* continuous. By Theorem A.1, there exist vectors $x_1, \ldots, x_n \in X$ and there is a number $M \in \mathbf{R}^+$ such that

$$|\theta(\tau)| \leq M \max\{|\tau(x_1)|, \ldots, |\tau(x_n)|\} \qquad (\tau \in X^*).$$

Hence, θ equals 0 on $\ker(\hat{x}_1) \cap \ldots \cap \ker(\hat{x}_n)$. It follows by elementary linear algebra that $\theta = \lambda_1 \hat{x}_1 + \cdots + \lambda_n \hat{x}_n$ for some scalars $\lambda_1, \ldots, \lambda_n$. Thus, $\theta = \hat{x}$, where $x = \lambda_1 x_1 + \cdots + \lambda_n x_n$. $\square$

A.3. Theorem. *Let τ be a linear functional on a locally convex space X. Then τ is continuous if and only if $\ker(\tau)$ is closed in X.*

Proof. The forward implication is obvious. Suppose then that $Y = \ker(\tau)$ is closed. To show that τ is continuous, we may suppose it is non-zero. If $C = \tau^{-1}(1)$, there is a vector x_0 in C, so $C = x_0 + Y$. Hence, C is closed, and $X \setminus C$ is an open neighbourhood of 0. Let Γ be a generating family of seminorms for the topology of X. Then there is a positive number ε and there are elements $p_1, \ldots, p_n \in \Gamma$ such that $U = U(\varepsilon; p_1, \ldots, p_n) \subseteq X \setminus C$. For any point $x \in U$, choose $\gamma \in \mathbf{K}$ such that $|\gamma| = 1$ and $\gamma \tau(x) = |\tau(x)|$. Then $\gamma x \in U$ and $\mu = \tau(\gamma x) \in \mathbf{R}^+ \setminus \{1\}$. If $\mu > 1$, then $\gamma x/\mu \in U$, so $\tau(\gamma x/\mu) \neq 1$; that is, $1 \neq 1$, a contradiction. This argument proves that $\mu < 1$. Hence, $x \in U \Rightarrow |\tau(x)| < 1$, and therefore, for all $x \in X$,

$$|\tau(x)| \leq (2/\varepsilon) \max_{1 \leq j \leq n} p_j(x).$$

By Theorem A.1, τ is therefore continuous. $\square$

If Y is a vector subspace of a locally convex space X of codimension 1 in X, then there is a linear functional τ on X with kernel Y, so if Y is closed in X, then τ is continuous.

Let $x, y \in X$. The set $[x, y] = \{tx + (1-t)y \mid 0 \leq t \leq 1\}$ is connected in X, since it is the continuous image of the unit interval $[0, 1]$ in $\mathbf{R}$. If x, y are linearly independent, then $[x, y] \subseteq X \setminus \{0\}$.

If X is not one-dimensional, $X \setminus \{0\}$ is connected. For if $x, y \in X \setminus \{0\}$ are linearly independent, then they are connected by a continuous path in $X \setminus \{0\}$, as we have just observed, and if they are linearly dependent, there is a point $z \in X \setminus \{0\}$ such that z, x are linearly independent, and likewise for z, y, so we have a continuous path in $X \setminus \{0\}$ from x to z and then from z to y.

This observation is used in the following result.

A.4. Lemma. *Let X be a real locally convex space of dimension greater than one and suppose that C is a non-empty open convex set in X not containing 0. Then there is a non-zero element x of X such that $C \cap \mathbf{R}x = \emptyset$.*

Proof. The set $S = \cup_{t>0} tC$ is open in X. If x and $-x$ are both in S, then there are positive numbers t_1, t_2 and there are elements $x_1, x_2 \in C$ such that $x = t_1 x_1 = -t_2 x_2$. Convexity of C implies that $0 = (t_1 x_1 + t_2 x_2)/(t_1 + t_2) \in C$, which is impossible by hypothesis. Hence, $S \cap (-S) = \emptyset$. Since $X \setminus \{0\}$ is connected, it cannot be the union of the disjoint non-empty open sets S and $-S$, so there is a non-zero vector x of X such that neither x nor $-x$ belongs to S. Hence, $C \cap \mathbf{R}x = \emptyset$.   □

Let p be a seminorm on a vector space X, and let Y be a vector subspace of X. Define a seminorm p' on X/Y by setting

$$p'(x + Y) = \inf_{y \in Y} p(x + y).$$

If X is a locally convex space, so is X/Y when endowed with the largest topology making the quotient map $\pi \colon X \to X/Y$ continuous. For we may take a generating family Γ of seminorms for the topology of X such that $\max_{1 \leq j \leq n} p_j \in \Gamma$ if $p_1, \ldots, p_n \in \Gamma$, and then $\Gamma' = \{p' \mid p \in \Gamma\}$ is a generating family of seminorms for the topology of X/Y. To see this, it suffices to show that π is continuous and open when X/Y is endowed with the topology generated by Γ'. Continuity of π is clear, and to show that π is open it suffices to observe that the image of a basic set $U(\varepsilon; p)$ under π is the basic set $U(\varepsilon; p')$ in X/Y.

A.5. Theorem. *If τ is a non-zero continuous linear functional on a locally convex space X, then it is open.*

Proof. If $Y = \ker(\tau)$, then X/Y is a one-dimensional locally convex space, and the linear functional $\tau' \colon X/Y \to \mathbf{K}$, $x + Y \mapsto \tau(x)$, is a continuous linear isomorphism. Hence, X/Y is Hausdorff, as $\mathbf{K}$ is. Thus, if Γ is a generating family of seminorms of X/Y, then one of them, p say, must be non-zero. Using the fact that X/Y is one-dimensional, every element of Γ must therefore be of the form αp for some $\alpha \in \mathbf{R}^+$. Hence, X/Y is in fact a normed vector space with norm p inducing its topology. Since $p \circ \tau'^{-1}$

is a norm on $\mathbf{K}$, it is equivalent to the usual norm, and therefore τ' is a homeomorphism. Hence, τ is open, since it is the composition of τ' and the quotient map from X to X/Y, and both of these maps are open.   □

A.1. Remark. If X is a complex vector space, then for each real-linear functional $\rho\colon X \to \mathbf{R}$ there is a unique complex-linear functional $\tau\colon X \to \mathbf{C}$ such that $\operatorname{Re}(\tau) = \rho$ (set $\tau(x) = \rho(x) - i\rho(ix)$). We shall use this to reduce some arguments to the "real" case.

A.6. Theorem. *Let X be a locally convex space and C a non-empty open convex set of X not containing 0. Then there is a continuous linear functional τ on X such that $\operatorname{Re}(\tau(x)) > 0$ $(x \in C)$.*

Proof. By Remark A.1 we may suppose that the ground field is $\mathbf{R}$. Let Λ be the set of all closed vector subspaces Y of X disjoint from C, and make it into a poset using the partial ordering given by set inclusion. Observe that Λ is non-empty, since the closure of the zero space is an element. If S is any totally ordered set of elements of Λ, then $(\cup S)^-$ is an element of Λ majorising all elements of S. By Zorn's lemma, Λ admits a maximal element, Y say.

Suppose now that X/Y is not one-dimensional (and we shall deduce a contradiction). Let $\pi\colon X \to X/Y$ be the quotient map. Then $\pi(C)$ is a non-empty open convex subset of X/Y not containing 0, so by Lemma A.4 there is an element $x \in X$ such that $\pi(x)$ is non-zero and $\pi(C) \cap \mathbf{R}\pi(x) = \emptyset$. Put $Y_1 = Y + \mathbf{R}x$. Then $Y_1^- \in \Lambda$ and contains Y, so $Y_1^- = Y$ by maximality of Y. Hence, $x \in Y$, so $\pi(x) = 0$, a contradiction.

Therefore, X/Y must be one-dimensional; that is, Y is of codimension one in X. It follows that there is a linear functional τ on X with kernel Y, and τ is necessarily continuous by Theorem A.3. Since $\tau(C)$ is convex, it is an interval of $\mathbf{R}$, and because it does not contain 0, it is contained in $(0, +\infty)$ or $(-\infty, 0)$. By replacing τ with $-\tau$ if necessary, we can suppose that $\tau(C) \subseteq (0, +\infty)$, and this proves the result.   □

There are a number of closely related results in the literature that are called separation theorems. The following is one of them.

A.7. Theorem. *Let C be a non-empty closed convex set in a locally convex space X and $x \in X \setminus C$. Then there is a continuous linear functional τ on X and a real number t such that $\operatorname{Re}(\tau(y)) < t < \operatorname{Re}(\tau(x))$ for all $y \in C$.*

Proof. We may suppose that $\mathbf{K} = \mathbf{R}$. Let Γ be a generating family of seminorms for X. Since $-x + C$ is a closed set not containing 0, there is a positive number ε and there are elements $p_1, \ldots, p_n$ of Γ such that the set $U = U(\varepsilon; p_1, \ldots, p_n)$ is disjoint from $-x + C$. Hence, $W = x + U$ is open and disjoint from C. The set $W - C = \cup_{y \in C} W - y$ is open, and it is convex because W and C are convex. Moreover, it does not contain 0.

By Theorem A.6, there is a continuous linear functional τ on X such that $\tau(y) > 0$ for all $y \in W - C$. Now $U \not\subseteq \ker(\tau)$, since $X = \cup_{m=1}^{\infty} mU$, so there exists $z \in U$ such that $\tau(z) < 0$. Setting $t = \tau(x + z)$ and observing that $x + z \in W$, we get the inequality $\tau(x + z - y) > 0$ for all $y \in C$, so $\tau(x) > t > \tau(y)$. $\qquad\square$

A.8. Corollary. *Let C be a convex set in a locally convex space X. Then for any point $x \in X$, $x \in \bar{C}$ if and only if there is a net $(x_\lambda)_{\lambda \in \Lambda}$ in C such that $(\tau(x_\lambda))$ converges to $\tau(x)$ for all continuous linear functionals τ on X.*

Proof. Observe that $\bar{C}$ is convex, and apply Theorem A.7. $\qquad\square$

A.9. Corollary. *Let Y be a closed vector subspace of a locally convex space X and $x \in X \setminus Y$. Then there is a continuous linear functional τ on X such that $\tau(y) = 0$ $(y \in Y)$ and $\tau(x) = 1$.*

Proof. By Theorem A.7, there is a continuous linear functional ρ on X and a real number t such that $\text{Re}(\rho(y)) < t < \text{Re}(\rho(x))$ $(y \in Y)$. Since $0 \in Y$, therefore $t > 0$, and $\rho(x) \neq 0$. For each $n \in \mathbb{N}$ and $y \in Y$, we have $|n \, \text{Re}(\rho(y))| < t$. Hence, $\text{Re}(\rho(y)) = 0$ $(y \in Y)$. It follows $\rho = 0$ on Y. Set $\tau = \rho/\rho(x)$. $\qquad\square$

A.10. Corollary. *Let x, y be distinct points of a Hausdorff locally convex space X. Then there is a continuous linear functional τ on X such that $\tau(x) \neq \tau(y)$.*

Proof. Observe that $x - y \notin Y = \{0\}$, and apply Corollary A.9. $\qquad\square$

The usual form of the Hahn–Banach theorem asserts that if τ is a bounded linear functional on a vector subspace Y of a normed vector space X, then there is a bounded linear functional τ' on X extending τ and of the same norm. We prove an analogue of this for locally convex spaces.

If Y is a vector subspace of a locally convex space X, then Y is also a locally convex space when endowed with the relative topology. For if Γ is a generating family of seminorms for the topology of X, it is readily verified that $\Gamma' = \{p' \mid p \in \Gamma\}$ is a generating family of seminorms for the topology of Y, where p' is the restriction to Y of the seminorm p on X.

A.11. Theorem. *Let τ be a continuous linear functional on a vector subspace Y of a locally convex space X. Then there is a continuous linear functional τ' on X extending τ.*

Proof. Let Γ be a generating family of seminorms for the topology of X. Since τ is continuous, it follows from Theorem A.1 that there are seminorms $p_1, \ldots, p_n \in \Gamma$ and a positive number M such that $|\tau(y)| \leq p(y)$ for all $y \in Y$, where p is the continuous seminorm on X defined by $p(x) = M \max_{1 \le j \le n} p_j(x)$. If $Z = p^{-1}\{0\}$, then Z is a closed vector subspace of X, and X/Z is a normed vector space under the well-defined norm $\|x + Z\| = p(x)$. Let $\pi: X \to X/Z$ be the quotient map. Then

$$\rho: \pi(Y) \to \mathbf{K}, \quad \pi(y) \mapsto \tau(y),$$

is a well-defined norm-decreasing linear functional. By the Hahn–Banach theorem, there is a linear functional ρ' on X/Z extending ρ and also norm-decreasing. It follows that the function

$$\tau': X \to \mathbf{K}, \quad x \mapsto \rho'(x + Z),$$

is linear and $|\tau'(x)| \le p(x)$ ($x \in X$). Using Theorem A.1 again, τ' is continuous, and it clearly extends τ. $\qquad\square$

The intersection of a family of convex subsets of a locally convex space X is itself convex. Hence, if S is a subset of X, there is a smallest convex subset $\mathrm{co}(S)$ containing S. We call $\mathrm{co}(S)$ the *convex hull* of S. It is easily verified that

$$\mathrm{co}(S) = \{\sum_{i=1}^{n} t_i x_i \mid n \ge 1, \ t_1, \ldots, t_n \in \mathbf{R}^+, \ \sum_{i=1}^{n} t_i = 1, \ x_1, \ldots, x_n \in S\}.$$

We write $\overline{\mathrm{co}}(S)$ for the closure of $\mathrm{co}(S)$, and we observe that $\overline{\mathrm{co}}(S)$ is the smallest closed convex set of X containing S. We call $\overline{\mathrm{co}}(S)$ the *closed convex hull* of S.

A.2. Remark. If $C_1, \ldots, C_n$ are non-empty convex sets of X, then the convex hull $\mathrm{co}(C_1 \cup \ldots \cup C_n)$ is the set of all elements $t_1 x_1 + \cdots + t_n x_n$, where $t_1, \ldots, t_n$ are non-negative numbers such that $t_1 + \cdots + t_n = 1$ and $x_1, \ldots, x_n$ are in $C_1, \ldots, C_n$, respectively (the proof of this is an easy exercise).

A.12. Theorem. *Let $C_1, \ldots, C_n$ be convex compact sets in a locally convex space X. Then $\mathrm{co}(C_1 \cup \ldots \cup C_n)$ is compact.*

Proof. We may suppose that $C_1, \ldots, C_n$ are all non-empty. Let Δ denote the set of all non-negative numbers $t_1, \ldots, t_n$ such that $t_1 + \cdots + t_n = 1$. Clearly, Δ is compact in $\mathbf{R}^n$ and the map

$$\Delta \times C_1 \times \cdots \times C_n \to \mathrm{co}(C_1 \cup \ldots \cup C_n), \ (t_1, \ldots, t_n, x_1, \ldots, x_n) \mapsto \sum_{j=1}^{n} t_j x_j,$$

is a continuous surjection. Hence, $\mathrm{co}(C_1 \cup \ldots \cup C_n)$ is compact, since the product $\Delta \times C_1 \cdots \times C_n$ is compact. $\qquad\square$

A point x of a convex set C in a vector space X is an *extreme point* of C, if the condition $x = ty + (1 - t)z$, where $y, z \in C$ and $0 < t < 1$, implies that $x = y = z$. Equivalently, x is an extreme point of C if and only if, whenever $y, z \in C$ are such that $x = (y + z)/2$, we necessarily have $x = y = z$.

A non-empty convex subset F of C is a *face* of C if, whenever $x \in F$ and $y, z \in C$ and $x = ty + (1 - t)z$ for some $t \in (0, 1)$, we must have $y, z \in F$. If C is non-empty, it is a face of C, and a non-empty intersection of a family of faces of C is a face of C. A point $x \in C$ is an extreme point of C if and only if $\{x\}$ is a face of C. An extreme point of a face of C is an extreme point of C itself.

A.13. Lemma. *Let C be a non-empty convex compact set in a locally convex space X and suppose that τ is a continuous linear functional on X. Let M be the supremum of all $\mathrm{Re}(\tau(x))$ where x ranges over C. Then the set F of all $x \in C$ such that $\mathrm{Re}(\tau(x)) = M$ is a compact face of C.*

Proof. The set F is non-empty, since compactness of C implies that there is a point x_0 of C such that $M = \mathrm{Re}(\tau(x_0))$. It is clear that F is convex. It is also closed in C and therefore compact. Suppose that $x \in F$, and $y, z \in C$ and $x = ty + (1 - t)z$ for some $t \in (0, 1)$. Then

$$M = \mathrm{Re}(\tau(x)) = t\,\mathrm{Re}(\tau(y)) + (1 - t)\,\mathrm{Re}(\tau(z)).$$

If y or z is not in F, then $\mathrm{Re}(\tau(y))$ or $\mathrm{Re}(\tau(z))$ is less than M, so

$$t\,\mathrm{Re}(\tau(y)) + (1 - t)\,\mathrm{Re}(\tau(z)) < tM + (1 - t)M,$$

and therefore $M < M$, a contradiction. This argument shows that y, z are in F and consequently F is a face of C. □

The equation $C = \overline{\mathrm{co}}(E)$ in the following is the Krein–Milman theorem, one of the great results of functional analysis with a vast range of applications.

A.14. Theorem. *Let C be a non-empty convex compact set in a Hausdorff locally convex space X. Then the set E of extreme points of C is non-empty and*

$$C = \overline{\mathrm{co}}(E).$$

Moreover, if S is a closed set of C such that $\overline{\mathrm{co}}(S) = C$, then S contains E.

Proof. Let Λ denote the set of all compact faces of C. This is non-empty, because $C \in \Lambda$. We make Λ into a poset by setting $F \leq F'$ in Λ if $F' \subseteq F$. A totally ordered family of elements of Λ has the finite intersection property, so by compactness of C its intersection is non-empty, and is therefore a face of C. Hence, every totally ordered family of Λ is majorised by an element of Λ, so by Zorn's lemma there is a maximal element, F say, in Λ.

We claim that F is a singleton set. For suppose otherwise and let x, y be distinct elements of F. Then by Corollary A.10 there is a continuous linear functional τ on X such that $\mathrm{Re}(\tau(x)) \neq \mathrm{Re}(\tau(y))$. Set

$$M = \sup\{\mathrm{Re}(\tau(z)) \mid z \in F\}.$$

Then the set

$$F_0 = \{z \in F \mid \mathrm{Re}(\tau(z)) = M\}$$

is a compact face of F by Lemma A.13. Hence, F_0 is a face of C, so $F_0 \in \Lambda$, and therefore, by maximality of F, we have $F_0 = F$. Consequently, M is equal to $\mathrm{Re}(\tau(x))$ and $\mathrm{Re}(\tau(y))$, a contradiction, since these two numbers are distinct. Thus, F has to be a singleton, $\{x\}$ say. Hence, $x \in E$, so E is non-empty.

Suppose now that $\overline{\mathrm{co}}(E) \neq C$, so there is a point $z \in C \setminus \overline{\mathrm{co}}(E)$. By Theorem A.7, there is a continuous linear functional τ on X and a number $t \in \mathbf{R}$ such that $\mathrm{Re}(\tau(y)) < t < \mathrm{Re}(\tau(z))$ for all $y \in \overline{\mathrm{co}}(E)$. Set

$$M' = \sup\{\mathrm{Re}(\tau(y)) \mid y \in C\}$$

and

$$F' = \{y \in C \mid \mathrm{Re}(\tau(y)) = M'\}.$$

By Lemma A.13 again, F' is a compact face of C. It follows from the earlier part of this proof that F' has an extreme point, y say. Hence, y is an extreme point of C, and so $y \in E$. But $y \in F'$ implies that $M' = \mathrm{Re}(\tau(y))$, so $M' < t < \mathrm{Re}(\tau(z)) \leq M'$, a contradiction. This argument, therefore, proves that $\overline{\mathrm{co}}(E) = C$.

To show that an extreme point z of C lies in the set S, it suffices to show that for each neighbourhood U of 0 in X the intersection $(z + U) \cap S$ is non-empty. If Γ is a family of seminorms generating the topology of X, then it is clear that we may suppose that $U = U(\varepsilon; p_1, \ldots, p_m)$ for some positive number ε and for some seminorms $p_1, \ldots, p_m \in \Gamma$. Set $W = \frac{1}{2}U$, and observe that W is open. Now $S \subseteq \cup_{x \in S}(x + W)$, so by compactness of S there exist elements $y_1, \ldots, y_n \in S$ such that $S \subseteq (y_1 + W) \cup \ldots \cup (y_n + W)$. For $j = 1, \ldots, n$, set

$$S_j = (y_j + W^-) \cap C.$$

Since U is convex, so is $y_j + W^-$, and therefore, by convexity of C, the set S_j is also convex. It is clear that S_j is also compact and non-empty. Observe also that $S \subseteq S_1 \cup \ldots \cup S_n$. The set $K = \mathrm{co}(S_1 \cup \ldots \cup S_n)$ is compact by Theorem A.12, and since each set $S_j \subseteq K$, we have $S \subseteq K$. Hence, $C \subseteq K$, since $\overline{\mathrm{co}}(S) = C$. In particular, $z \in K$. By Remark A.2, we can write z in the form $z = t_1 x_1 + \cdots + t_n x_n$, where $t_1, \ldots, t_n \in \mathbf{R}^+$ and

$t_1 + \cdots + t_n = 1$, and $x_j \in S_j$ for $j = 1, \ldots, n$. Since $x_1, \ldots, x_n \in C$, and since z is an extreme point of C, it follows that $z = x_j$ for some index j (take any j such that $t_j > 0$). Hence, $z \in S_j$, and therefore, $z - y_j \in W^-$. Consequently, the set $(z - y_j + W) \cap W$ is non-empty. If x is an element, then $p_i(x) < \varepsilon/2$ and $p_i(z - y_j - x) < \varepsilon/2$, so $p_i(z - y_j) < \varepsilon$ for $i = 1, \ldots, m$. Hence, $y_j - z \in U$, and therefore, $(z + U) \cap S$ is non-empty. This proves the theorem.   □

Notes

These notes are very incomplete, and concerned only with those aspects of the theory most relevant to the material covered in this book.

Functional analysis began to evolve at the turn of the century out of the work of Fredholm, Fréchet, Hilbert, F. Riesz, and Volterra on integral equations, eigenvalue problems, and orthogonal expansions. The spectral theorem is due to Hilbert. The first abstract treatment of normed vector spaces is due to Banach in his 1920 thesis.

The theory of von Neumann algebras originated in a paper published in 1929 by von Neumann [vN], where he introduced these algebras under the name "rings of operators," and in which he proved his famous double commutant theorem. Applications to theoretical physics provided a motivation for von Neumann's interest. He treated the foundations of quantum mechanics from the point of view of operator algebras.

Gelfand and Naimark introduced the class of C*-algebras in 1943 [GN]. They proved one of the most fundamental results of the theory by showing that C*-algebras can be faithfully represented as closed self-adjoint algebras of operators on Hilbert spaces. The principal results of the early theory of C*-algebras are due to Fell, Glimm, Kadison, Kaplansky, Mackey, and Segal, among others.

The uniformly hyperfinite algebras (UHF algebras) form an important class of C*-algebras, and were first studied by Glimm in the early 1960s. The more general class of AF-algebras was introduced by Bratteli (early 1970s), who initiated their classification. This classification was completed by Elliott, whose formulation was not originally stated in K-theoretic terms (as it is in the present text).

The study of tensor products of C*-algebras was initiated by Turumaru in 1952, but major progress in the theory did not commence until the mid 1960s. Some important contributors to this area are Effros, Guichardet, Lance, and Takesaki.

277

Homological algebraic methods made a very significant impact on operator theory in the early 1970s with the work of Brown, Douglas, and Fillmore on the classification of essentially normal operators [BDF]. The K-theory of C*-algebras, and a powerful generalisation called KK-theory and due to Kasparov, have had profound applications to both C*-algebra theory, and to other areas, such as differential geometry. Two early and fundamental results are the Pimsner–Voiculescu six-term exact sequence for computing the K-theory of the crossed product of a C*-algebra with **Z** [PV], and Connes' theorem for computing the K-theory of crossed products of C*-algebras with **R** [Con 1].

Single operator theory and operator algebra theory are vastly more extensive then an introductory volume such as this can indicate. For a deeper understanding of the subject the reader is referred to [Bla], [Dix 1], [Dix 2], [KR 1], [KR 2], [Ped], [Sak], and [Tak].

References

[BDF] L. Brown, R. Douglas, and P. Fillmore, Unitary equivalence modulo the compact operators and extensions of C*-algebras, *Proc. Conf. on Operator Theory,* Springer Lecture Notes in Math. **345** (1973), 58–128.

[Bla] B. Blackadar, *K-Theory for Operator Algebras.* MSRI publications no. 5, Springer-Verlag, New York, 1986.

[BMSW] B. A. Barnes, G. J. Murphy, M. R. Smyth, and T. T. West, *Riesz and Fredholm Theory in Banach Algebras.* Research Notes in Mathematics 67, Pitman, London, 1982.

[Cnw 1] J. B. Conway, *Subnormal Operators.* Pitman, Boston, 1981.

[Cnw 2] J. B. Conway, *A Course in Functional Analysis.* Graduate Texts in Mathematics 96, Springer-Verlag, New York, 1985.

[Coh] D.L. Cohn, *Measure Theory.* Birkhauser, Boston, 1980.

[Con 1] A. Connes, An analogue of the Thom isomorphism for crossed products of a C*-algebra by an action of **R**, *Advances in Math.* **39** (1981), 31–55.

[Con 2] A. Connes, *Non Commutative Differential Geometry.* Chapter I: The Chern Character in K Homology. Chapter II: De Rham Homology and Non Commutative Algebra. *Publ. Math. I.H.E.S.* **62** (1986), 257–360.

[Cun] J. Cuntz, K-Theory for certain C*-algebras, *Ann of Math.* **113** (1981), 181–197.

[Dix 1] J. Dixmier, *Von Neumann Algebras.* North-Holland, Amsterdam, 1981.

[Dix 2] J. Dixmier, *C*-Algebras.* North-Holland, Amsterdam, 1982.

[Dou 1] R. G. Douglas, *Banach Algebra Techniques in Operator Theory.* Academic Press, New York, 1972.

[Dou 2] R. G. Douglas, On the C*-algebra of a one-parameter semigroup of isometries, *Acta Math.* **128** (1972), 143–152.

[Eff] E. Effros, *Dimensions and C*-Algebras.* CBMS Regional Conf. Ser. in Math., no. 46, Amer. Math. Soc., Providence, 1981.

[Enf] P. Enflo, A counterexample to the approximation problem in Banach spaces, *Acta Math.* **130** (1973), 309–317.

[GN] I. Gelfand and M. Naimark, On the embedding of normed rings into the ring of operators in Hilbert space, *Mat. Sb.* **12** (1943), 197–213.

[Goo] K. Goodearl, *Notes on Real and Complex C*-Algebras.* Shiva Publishing Ltd., Nantwich, 1982.

[Hal] P. R. Halmos, *A Hilbert Space Problem Book.* Springer-Verlag, New York, 1982.

[Kel] J. L. Kelley, *General Topology.* Springer-Verlag, New York, 1975.

[KR 1] R. V. Kadison and J. R. Ringrose, *Fundamentals of the Theory of Operator Algebras I.* Academic Press, New York, 1983.

[KR 2] R. V. Kadison and J. R. Ringrose, *Fundamentals of the Theory of Operator Algebras II.* Academic Press, New York, 1986.

[Lan] E. C. Lance, Tensor products and nuclear C*-algebras, *Operator Algebras and Applications* (ed. R. V. Kadison), Proc. Symp. Pure Math. Pt 1 **38** (1982), 379–399.

[Mur] G. J. Murphy, Ordered Groups and Toeplitz Algebras, *J. Operator Theory* **18** (1987), 303–326.

[Ped] G. K. Pedersen, *C*-Algebras and their Automorphism Groups.* Academic Press, London, 1979.

[PV] M. Pimsner and D. Voiculescu, Exact sequences for K-groups and Ext-groups of certain cross-products of C*-algebras, *J. Operator Theory* **4** (1980), 93–118.

[Rie] M. Rieffel, C*-algebras associated with irrational rotations, *Pacific J. Math.* **93** (1981), 415–429.

[Rud 1] W. Rudin, *Real and Complex Analysis.* McGraw-Hill, New York, 1966.

[Rud 2] W. Rudin, *Functional Analysis.* Tata McGraw-Hill, New Delhi, 1977.

[Sak] S. Sakai, *C*-Algebras and W*-Algebras.* Springer-Verlag, New York, 1971.

[Tak] M. Takesaki, *Theory of Operator Algebras I.* Springer-Verlag, New York, 1979.

[TL] A. E. Taylor and D. C. Lay, *Introduction to Functional Analysis.* Wiley, New York, 1980.

[Top] D. M. Topping, *Lectures on Von Neumann Algebras.* Van Nostrand, London, 1971.

[vN] J. von Neumann, Zur Algebra der Funktionaloperationen und Theorie der normalen Operatoren, *Math. Ann.* **102** (1929), 370–427.

[Wes] T. T. West, The decomposition of Riesz operators, *Proc. London Math. Soc. (3)* **16** (1966), 737–752.

Notation Index

Subject Index